ENGINEERING DESIGN FOR PROFIT

ELLIS HORWOOD SERIES IN MECHANICAL ENGINEERING

STRENGTH OF MATERIALS
Vol. 1: Fundamentals – Vol. 2: Applications
J. M. ALEXANDER, University College, Swansea
TECHNOLOGY OF ENGINEERING MANUFACTURE
J. M. ALEXANDER, University College, Swansea, G. W. ROWE, Birmingham University and R. C. BREWER
ELASTIC AND PLASTIC FRACTURE
A. G. ATKINS, University of Reading, and Y. W. MAI, University of Sydney
VIBRATION ANALYSIS AND CONTROL SYSTEM DYNAMICS
C. BEARDS, Imperial College of Science and Technology
STRUCTURAL VIBRATION ANALYSIS
C. BEARDS, Imperial College of Science and Technology
COMPUTER AIDED DESIGN AND MANUFACTURE 2nd Edition
C. B. BESANT, Imperial College of Science and Technology
BASIC LUBRICATION THEORY 3rd Edition
A. CAMERON, Imperial College of Science and Technology
SOUND AND SOURCES OF SOUND
A. P. DOWLING and J. E. FFOWCS-WILLIAMS, University of Cambridge
MECHANICAL FOUNDATIONS OF ENGINEERING SCIENCE
H. G. EDMUNDS, University of Exeter
ADVANCED MECHANICS OF MATERIALS 2nd Edition
Sir HUGH FORD, F.R.S., Imperial College of Science and Technology, and J. M. ALEXANDER, University College of Swansea.
MECHANICAL FOUNDATIONS OF ENGINEERING SCIENCE
H. G. EDMUNDS, Professor of Engineering Science, University of Exeter
ELASTICITY AND PLASTICITY IN ENGINEERING
Sir HUGH FORD, F.R.S. and R. T. FENNER, Imperial College of Science and Technology
DIESEL ENGINEERING PRINCIPLES AND PERFORMANCE
S. D. HADDAD, Associate Professor, Western Michigan University, USA, and Director of HTCS Co., UK, and N. WATSON, Reader in Mechanical Engineering, Imperial College of Science and Technology, University of London
DIESEL ENGINEERING DESIGN AND APPLICATIONS
S. D. HADDAD, Associate Professor, Western Michigan University, USA, and Director of HTCS Co., UK. and N. WATSON, Reader in Mechanical Engineering, Imperial College of Science and Technology, University of London
TECHNIQUES OF FINITE ELEMENTS
BRUCE M. IRONS, University of Calgary, and S. AHMAD, Bangladesh University, Dacca
FINITE ELEMENT PRIMER
BRUCE IRONS and N. SHRIVE, University of Calgary
CONTROL OF FLUID POWER: ANALYSIS AND DESIGN 2nd (Revised) Edition
D. McCLOY, Ulster Polytechnic, N. Ireland and H. R. MARTIN, University of Waterloo, Ontario, Canada
UNSTEADY FLUID FLOW
R. PARKER, University College, Swansea
DYNAMICS OF MECHANICAL SYSTEMS 2nd Edition
J. M. PRENTIS, University of Cambridge
ENERGY METHODS IN VIBRATION ANALYSIS
T. H. RICHARDS, University of Aston Birmingham
ENERGY METHODS IN STRESS ANALYSIS: With Intro. to Finite Element Techniques
T. H. RICHARDS, University of Aston in Birmingham
COMPUTATIONAL METHODS IN STRUCTURAL AND CONTINUUM MECHANICS
C. T. F. ROSS, Portsmouth Polytechnic
FINITE ELEMENT PROGRAMS FOR AXISYMMETRIC PROBLEMS IN ENGINEERING
C. T. F. ROSS, Portsmouth Polytechnic
ENGINEERING DESIGN FOR PERFORMANCE
K. SHERWIN, Liverpool University
ROBOTS AND TELECHIRS
M. W. THRING, Queen Mary College, University of London
MECHANICAL VIBRATIONS WITH APPLICATIONS
A. C. WALSHAW, Professor Emeritus, The University of Aston in Birmingham

ENGINEERING DESIGN FOR PROFIT

D. J. LEECH, M.Sc., F.I.Prod.E., M.I.Mech.E., C.Eng.
Department of Management Science and Statistics
University College of Swansea

and

B. T. TURNER, M.Sc., F.I.Mech.E., F.R.Ae.S., F.B.I.M., M.I.C.E., M.I.Prod.E., C.Eng.
Barry T. Turner Associates
Dunchurch, Warwickshire

ELLIS HORWOOD LIMITED
Publishers · Chichester

Halsted Press: a division of
JOHN WILEY & SONS
New York · Chichester · Brisbane · Toronto

First published in 1985 by
ELLIS HORWOOD LIMITED
Market Cross House, Cooper Street, Chichester, West Sussex, PO19 1EB, England

The publisher's colophon is reproduced from James Gillison's drawing of the ancient Market Cross, Chichester.

Distributors:
Australia New Zealand, South-east Asia:
Jacaranda-Wiley Ltd., Jacaranda Press,
JOHN WILEY & SONS INC.,
G.P.O. Box 859, Brisbane, Queensland 40001, Australia

Canada:
JOHN WILEY & SONS CANADA LIMITED
22 Worcester Road, Rexdale, Ontario, Canada.

Europe, Africa:
JOHN WILEY & SONS LIMITED
Baffins Lane, Chichester West Sussex, England.

North and South America and the rest of the world:
Halsted Press: a division of
JOHN WILEY & SONS
605 Third Avenue, New York, N.Y. 10016, U.S.A.

British Library Cataloguing in Publication Data
Leech, D.J.
Engineering design for profit. –
(Ellis Horwood series in engineering science)
1. Engineering design
I. Title II. Turner, Barry T.
620'.00425 TA174

Library of Congress Card No. 84-28990

ISBN 0-85312-847-2 (Ellis Horwood Limited – Library Edn.)
ISBN 0-85312-867-7 (Ellis Horwood Limited – Student Edn.)
ISBN 0-470-20190-8 (Halsted Press)

Typeset by Ellis Horwood Limited.
Printed in Great Britain by R.J. Acford, Chichester.

Table of Contents

SOME VIEWS ON DESIGN

infinite life
no cost
no mass
no size
Carnot efficiency
extreme beauty
no time to design or make.

The performing of a very complicated act of faith.

J.C. Jones

An ill-assorted collection of poorly-matching parts
forming a distressing whole

Datamation

Too big, too expensive, too late

AWRE Scientist

In recent years, management in this country has been greatly influenced by technology. But technical innovation is not an end in itself, and an innovative product does not *necessarily* sell. Our technological advances could have been exploited commercially much more effectively if senior industrial management had paid similar attention to design.

Margaret Thatcher

Introduction

The manufacturer knows that he is successful:

if he makes a profit on what he sells to his customers,

if his customers return to buy a second time and

if he does not offend a third party.

For the engineering designer, however, success is not so easily defined because he is not often in the position of selling his designs directly to a customer.

Most designers work in manufacturing companies, and their successes or failures will be seen, only indirectly, in the goods that the manufacturer sells. Unfortunately, it is not easy to put a monetary value on the design content of a manufactured article, and accountants seldom bother to do so. But the costs of design are obvious, and the costs of bad design even more obvious.

Profitable goods cannot be made without the designer, but when we talk of DESIGN FOR PROFIT we may be using terms which have not been defined and criteria which are largely negative.

The designer, himself, is at fault because he often seems to believe that merit lies in trivially ingenious solutions to problems that neither the customer nor the manufacturer has asked him to solve, and that he will become famous through his signature at the bottom of the scheme drawing.

In fact, a design scheme is only the first step in a design project, and it may be the easiest. It is certainly the cheapest stage of design in terms of direct cost, and its importance lies in the fact that it sets in train an increasingly expensive set of activities. A design is not finished until the product has been shown to work to the satisfaction of the manufacturer, the customer, and society. This should mean that the design is completed when the product meets the specification. Between the clever scheme and the proven product there may be an expenditure of a thousand million dollars or simply failure, because the money ran out. A few hours of invention may lead to a few hundred hours of scheming and drawing, a few thousand hours of development and proving, and, perhaps, several hundred million dollars of tooling and product support. A designer has done his job if each of these activities is no more expensive than it needs to be, and if the final pay-off is great enough to justify the effort.

The thesis of this book is that design costs money and is done for profit. More precisely, design costs money, but without it manufacturing cannot be profitable. These remarks seem so obvious that only in Britain would it be necessary to make them; but they amount to more than a simple homily because an understanding that design is for profit enables us to manage it.

Each activity in design costs money, but each activity is performed to achieve an objective that will be assessed in economic terms. The design specification will list the required performance, the environment, the number of products to be made, the reliability, the life of the product, codes of practice and legal constraints which must be observed, and many other factors. All of the objectives, even those concerned with safety or aesthetics, are quantifiable in terms of money. Managing design means making sure that those monetary objectives are met at an acceptable cost.

The first four chapters of the book show how the engineering designer is responsible for generating costs and benefits for the manufacturer, the customer, and society.

Chapter 5 sketches the process by which the finished design evolves from the first creative work.

Chapter 6 discusses the use of resources; the materials that the designer calls up and the man-hours that design consumes. Resource levelling and job turn-round time are shown to be concerns of design management.

Chapters 7, 8, and 9 discuss the place of design in the company's plan, the cash flow streams generated by the designer, and the monitoring of design costs.

Chpater 10 discusses the need for the designer to consider safety and reliability, and some ways of doing so.

Chapter 11 discusses the whole design process as a network of activities, the way in which each of these activities generates cost, and the ways in which permissible costs may be specified and monitored. Examples are given of the documentation which is used in industry to control the activities of design.

Chapter 12 discusses the management of design changes and how such changes are installed at minimum cost. In addition, configuration management and design review procedures are described.

Chapter 13 describes some of the benefits and systems of computer aided design.

The book concludes with a discussion of the education of designers, with particular reference to those disciplines which have not, traditionally, been taught within engineering curricula.

Where appropriate, a chapter opens with a summary of what is to be done, and concludes with a tabulation of the chapter's main findings. The busy manager will be able to determine the significance to him of a chapter and read the argument only when he knows of its relevance. The tabulated summary of findings and further references will, in some cases, provide the designer with a check list which will help to control the design process.

Some use has been made of computing in the book. Simulation has been used to demonstrate problems that arise in using resources to their full extent; a small CPM. program has been used to show the difficulty of resource levelling; a spreadsheet package has been used to calculate the present values of the cash flows that the designer generates, and a number of programs have been written and used in the calculation of life distribution, spares holding, scheduled replacement, and capital recovery. These programs have been designed to work on small computers such as the BBC Micro computer, and they can be made available to anyone who wishes to use them.

The book is addressed, firstly, to those designers in industry who wish to see their work as contributing to the performance of the firm which employs them or whose work requires that they manage the contributions of others in the technical organisation of the company.

The book is addressed, secondly, to those who teach design. Most university and polytechnic engineering degree schemes of study recognize that more is needed than engineering science, and, indeed, many undergraduate courses include some introductory lectures on such non-engineering subjects as economics, health and safety, industrial relations, or management methods. It is hoped that this book could provide a means of relating some non-engineering coursework to the technical work that the engineer will be required to do in industry. Where design is taught as a specific course within an engineering degree scheme the teacher may find that this book provides a useful framework for part of that course. Design management is now seen to be a subject to be taught in its own right in some colleges, and this book is offered to support such courses.

An expanding area of education is the training of those who will teach craft, design, and technology to sixth forms in schools. These training courses contain an element of design management, and this book could support such training.

The authors wish to thank Mr. Farhad Etemad, a research worker in the department of Management Science at the University College of Swansea. Mr. Etemad was responsible for almost all the computer programs which have been listed or refered to, in this book.

CHAPTER 1

What design costs

This chapter defines the design process and suggests that the technical requirements of a design specification are derived almost wholly from economic considerations. The management of design has two major objectives, the first of which is to ensure that the design department yields a good return on the money invested in it, and the second is to ensure that the products designed are profitable to the customer and the manufacturer.

1.1 MONEY AND DESIGN

- In an aircraft accessory firm a design team thought that they would solve a problem by spending £250,000 in eighteen months. In the event they took two years and spent £400,000. In anybody's book, £400,000 is worth looking at; in the aerospace business where the design cost could be 25% of the selling price of the product, an overspend of this size can mean the difference between profit and loss.

GOOD DESIGN COSTS MONEY – USUALLY MORE THAN WE EXPECTED

DESIGN COSTS ARE WORTH CONTROLLING

- Ford of America are said to have lost more than $50 million in recall costs for the Pinto, and to have paid $100 million damages in a case arising from the siting of the car's petrol tank.

DESIGN FAULTS COST MONEY

- In a company making storage racks, a design which cost about £3000 led to an expenditure of £250, 000 on tooling.

SPENDING MONEY ON DESIGN IS A COMMITMENT TO SPEND MUCH MORE MONEY LATER

- A company lost an order for a valve because the competitor's product was cheaper. The competitor's designer had drawn a valve which required one less operation on a lathe, and the difference in price was accounted for by this operation.

THE COST OF BAD DESIGN DOES NOT ALWAYS SHOW IN THE ACCOUNTS

These are all examples of the designer costing money. Examples of dramatic profits created by designers are less easy to find. We can all point to inventions which have been profitable, but invention is not design. A manufacturing company which employs bad designers will probably see its share of the market decline gradually until it is totally eclipsed.

Why, after all, does a company employ designers? The company is in business either to sell the designs or to sell the products that have been designed. In either case the objective of the company must be to make a profit. It is true that there may be other objectives such as to provide employment, to serve the community, to give the owner a job he likes, to capture a stated share of the market, to provide a satisfactory return on the assets employed, or just to survive; but all such objectives can only be achieved if the firm makes a profit (or at least require that it does not make a loss). Where an enterprise knowingly makes a loss but is supported by public money to provide a service (as for example where a manufacturing company is supported in order to provide employment), the profit criterion still applies if the grant is regarded as income.

When a company employs designers it is usually as a necessary preliminary to manufacture, and so design must be considered as an essential part of the overall profit-making enterprise.

1.2 WHAT IS DESIGN?

Design has been defined in a number of ways; the *Oxford English Dictionary* defines it as 'contrivance in accordance with a preconceived plan, adaptation of means to ends'. E. Matchett [1] has defined it as 'The optimum solution to the sum of the true needs of a particular set of circumstances'. The Feilden [2] Committee offered the following definition:

(i) Mechanical engineering design is the use of scientific principles, technical information and imagination in the definition of a mechanical structure, machine or system, to perform pre-specified functions with the maximum economy and efficiency.

(ii) The designer's responsibility covers the whole process from conception to the issue of detailed instructions for production and his interest continues throughout the designed life of the product in service.

P. J. Booker [3] described design as simulating what we want to make (or do) before we make (or do) it. This may be done as many times as necessary to feel confident in the final result. T. T. Wooderson [4] said 'Design is an iterative decision-making activity to produce plans by which resources are converted, preferably optimally, into systems or devices to meet human needs'. Bruce Archer [5] said 'The key element in the act of designing is the formation of a prescription or model for a finished work in advance of its embodiment.'

In a company which manufactures products with the intention of selling them for profit, the Feilden definition offers a basis for management and is the definition which will be assumed throughout this book.

A design project usually starts when someone identifies a market. Anyone may identify the market, although it is more likely to be a salesman, a customer, a service engineer, or a designer. Arbitrarily we may agree that the first formal work of design is the definition of the product to be designed, in terms of what it is expected to do, what environments it must perform in, and how much it will cost.

The final work of design is the supply of a set of instructions which define what is to be made. Usually, in mechanical engineering, this consists mainly of a set of drawings which define the geometry and material of each component, sub-assembly, and assembly of the product together with any instructions necessary to define processes that are to be used in manufacture. In fact, it is common for the drawings not to be used in full-scale production (although they may be used in the manufacture of prototypes), for usually the drawings are interpreted by production planners who produce manufacturing instructions which detail each operation required in manufacture. The man who cuts metal may not see the drawing of the component he is helping to make, but he will be given precise instructions about his contribution. Again, then, to define the completion of design as the production of a set of drawings is arbitrary.

Nevertheless, in many firms the design organization does encompass the work between (and including) the writing of the design specification and the production of a set of drawings. While the design organization may be the responsibility of the technical director of the company, the planning department may be part of the production organization which is responsible to the production director.

But in any design project that is not trivial, the drawings will not be right first time. Often what is drawn cannot be made, and when the drawings are corrected, what is made will not work. Even when laboratory development leads to a product which works, it is usually necessary to make more design changes before the specified life, reliability, operability, maintainability, price, and other qualities can be demonstrated as being met by the product. This need to correct drawings goes on even when the product has been sold and is in service with the customer, because it is inevitable that some design faults will remain undiscovered until the customer finds them. The full design process is therefore much more than drawing. It embraces many tasks from customer liaison and specification writing, through drawing, prototype manufacture, development, proving, product support, and production support to modification. Customer liaison and product support will involve the production of operating and maintenance manuals and may require the training of the customer operatives.

It is wrong to believe that the design process can exclude these functions, although it is tempting to argue that a good designer should get his drawings right first time. However, they never do, and even the greatest designers of history – such men as Watt, Whitney, Stephenson, Royce, and Mitchell – have all needed time and money to get their designs right. The rest of us are not likely to do better.

1.3 WHAT IS A DESIGNER?

If we acknowledge all the different activities which are part of design, we see the designer as a manager who welds those activities together. He is directly concerned with performance engineering, scheming, detailing, prototype manufacture, development, and the modification of drawings, and he must liaise with customers, salesmen, service engineers, production engineers, buyers, quality and reliability engineers, and with commercial departments. Nevertheless, confusion arises because, while we recognise that drawing is only a part of design, we customarily use the word 'designer' to describe the man who does creative work at the drawing board. Where confusion could arise, the term 'design draughtsman' will be used for such a man, and the word 'designer' will be used to describe someone who takes responsibility for the whole process. Sometimes the design draughtsman does take such responsibility, but it is not uncommon for the leader of a design team to come from the development laboratory or the performance office.

1.4 WHAT DOES DESIGN COST?

The cost of design varies considerably from company to company. Some, making simple products with little technical content, may spend on design only about 1% of the cost of running the company, while others, making advanced technical products with a good deal of innovation, may spend 30%. At the higher level there is clearly a case for the good management of design because, where profit is, say, 10% of running costs, bad management of the money allocated to design could turn profit into loss.

Good management, at the least, means that the design department must be treated as a cost and profit centre just as much as the production department. The average design worker will cost his employer something over £15 000 a year, and must clearly produce in a year, something which the customer is prepared to pay more than £15 000 for. Usually the cost of design is recovered by adding it to the price of the product, although there are circumstances in which the customer pays directly for the design.

Frequently, design work is abortive in that it does not lead to a saleable product, either because a competitor gets the job or because what first seemed a good idea turns out, on further analysis, to be not worth pursuing. Unsuccessful work cannot be eliminated, and the successful work must pay for it. All this requires careful budgeting and cost control if the design department is to justify its existence. Because some work will not be paid for, because the work varies in technical complexity, because different jobs are paid for in different ways, and because of the probabilistic nature of the income from design, the design budget formulation as part of the company's corporate plan and the control of design costs are usually more difficult problems than are encountered in other cost and profit centres such as the production department.

The formulation of the design budget and the control of its expenditure are clearly more significant where a high percentage of the company's turnover is spent on design, and there will be some level of technology at which over-spending in the design department has a direct effect on profitability. But the design department affects the economics of a company in many indirect ways which are, in the end, more significant than the direct costs of the draughtsmen and development engineers.

At a very early stage in the life of any project, the designer must do enough work to forecast the costs that will be incurred and contribute, at least, to the calculation of the revenue to be expected by the sale of the product. How much design effort should be spent before the company commits itself to a decision to go ahead with the project and to a price? So often the estimates of commercial success and the predictions of cost are optimistic – weapon systems, civil aeroplanes, atomic power stations, and many other complex systems provide well known examples of development costs which soared to levels which often eliminated profit [6]. Would more time (and money) spent at the early stages of design – before commitment – have either pointed to higher costs or predicted the problems which could have been solved early and cheaply on the drawing board instead of late and expensively in development and service? Downey [7] clearly came to the conclusion that, in the defence field at least, more design effort than is customary in the United Kingdom should be spent before the manufacturer commits himself to a price.

Part of the problem of managing design must, therefore, be to determine how much work should be done before the company commits itself to go ahead or stop work on a project. Usually this is done in several stages (Downey, for example, suggests a feasibility study followed by two stages of project definition, before embarking on the full design for production, and after any of these stages the decision may be taken to stop work on the project). This is not only a problem for day to day management but must be allowed for in setting the budget. If successful designs are to pay for unsuccessful designs, the amount to be recovered must be predicted when formulating the design budget.

1.5 WHAT DOES THE CUSTOMER WANT?

If the manufacturing company is to survive and prosper it must not merely sell its products to customers but continue to do so. Customers will buy new products if they have been satisfied in the past. New customers will buy if they see old customers satisfied. Assuming that the product that he has bought works, the customer judges it almost entirely by the cost of using and owning it†. It is not common to spend much effort rejecting designs which are technically infeasible, and provided the requirement is feasible there is no great difficulty in designing a product which will work. If a design team is asked to design an aeroplane which will fly at Mach 2, at 60 000 feet and carry a hundred passengers 3000 miles, the difficult problem is not to decide whether the aeroplane can be built. This question can be answered at comparitively little cost. The difficult questions are, 'How much will it cost to design the aeroplane?', 'How much will it cost to make the aeroplane?', 'How much will it cost to run the aeroplane?', and 'Will all these costs permit both the manufacturer and the customer to make a profit?'. The problem of carrying a passenger at Mach 2 is trivial compared with the problem of carrying a passenger at Mach 2 for 10p a mile.

A knowledgeable customer will not merely hope that the designer has considered the costs of use and ownership; he will have written maximum permissible values for these into the design specification which binds the designer to see their achievement. They will be written into the specification as the maximum permissible energy and manpower required to operate the product, the maximum cost of maintenance, the minimum acceptable reliability and life, and the required level of product support from the manufacture in terms of spares availability, servicing, and repair facilities.

The less sophisticated customers, too, are aware of the costs of use and ownership and buy where they see reliability. In the correspondence columns of *The Times* we see, from time to time, a letter from a man who has bought one car rather than another because of its reliability. Usually the correspondent is justifying his having bought a Japanese, French, or German car after problems with a British car, and usually the arguments offered are far from convincing; but the fact remains that the customer is making a judgement on reliability grounds to the embarrassment of the designer. This concern for reliability

---

†Brian Callick (Parkes (editor)) *Terotechnology Handbook,* HMSO, London, 1978) defines: *Cost of use:* costs of materials, energy and manpower incurred in using the product and the costs of loss of output due to malfunction or failure.
Cost of ownership: costs of acquisition, installation, commissioning, maintenance, modification and repair and of monitoring the condition and performance of the product.

on the part of the customer is reflected in the publication by the AA of the costs of owning different popular cars [8]. These costs, of course, involve more than reliability; they take account of the costs of spares, the costs of labour for servicing and repairs, the frequency of servicing and repairing, and the costs of petrol and oil. Some failures may be caused by bad workmanship in manufacture; but, even here, the designer is partly responsible, for good design considers the class of labour available, and makes good workmanship easier to achieve. Once again, the designer cannot win every battle, but he can improve the odds.

The main method of ensuring that the designer works to achieve acceptable costs of use and ownership is, of course, to use the discipline of the design specification. This is usually binding in its required levels of life, serviceability, reliability, spares availability, etc. More, however, is required than the mere statements of target values. Achievement of the specification requirements must be demonstrated, and this involves a whole network of tasks culminating in proof that the requirements are met. Endurance testing in adverse environments, with the probability of failure, requires access to specialized test equipment, the building of products which will be tested to destruction and which cannot subsequently be sold, the planning of development, the involvement of (and payment for) departments which collect and analyse field data, and much else. All these are design costs which must not only be accounted for but must be predicted to the extent that the budget to which the designer works is sufficiently accurate to permit cost control.

A major part of the cost of ownership is the cost of acquisition of the product. This is made up partly of the cost of manufacture plus profit for the manufacturer and partly of the cost of bringing the product to the quality at which it meets its specification. Typically, the purchase price of a product is about a fifth of the total cost of use and ownership. The RAF would argue that when they buy an aeroplane they spend only about 20% of what owning the aeroplane will ultimately cost them. At 20 pence per mile, a typical estimate of the cost of motoring, a little over 20% of the cost of using and owning a car is represented by its purchase price.

1.6 WHAT DOES THE MANUFACTURER WANT?

The cost of manufacture is important to the customer because it is reflected in the purchase price, and even if that is only 20% of the cost of using and owning a product, it is still a sizeable proportion of these costs and one on which savings can be real. The cost of manufacture is, however, likely to be of much more significance to the manufacturer because it represents a much higher percentage of his total costs. A 10% saving in manufacturing cost might represent a 2% saving to the customer or a 6% saving to the manufacturer. It is quite clear that many unnecessary costs in manufacturing a product derive directly from the design. The number of operations in a lathe, the difficulty of the operations, the difficulty of measuring, the difficulty of inspecting, whether a component is cast, forged, or machined from bar, and many similar features dictated by the designer's scheme, can make the difference between a competitive and a non-competitive product. It has often been argued that British designers are not as well versed in detailed manufacturing technology as those of some other countries, so that a basically good idea can be buried under poor manufacturing techniques.

It is certainly the case that in some companies, the planning office is distinct from the design office and under different direction, so that it is possible for the designer to be divorced completely from manufacturing problems until the production engineer finds himself unable to solve them without asking for a change in design.

In some fields, particularly that of the mass production of consumer durables, the designer's job is far less the design of the product than the design of the production line. The design of a car or washing machine may well derive from the design of a new production line rather than the reverse. The ergonomics of the line could site points of access and determine the shape of the product; the facilities available could determine methods of fastening – spot welds instead of screws, for example. Generally, of course, the problem is neither to design the manufacturing facilities, given the product, nor to design the product, given the manufacturing facilities, but to marry the two.

Disciplines which are sometimes imposed to ensure that manufacturing problems are taken into account, are the design review and the design audit†. Part of the review requires that drawings be checked by appropriate experts before they are issued for manufacture. Value engineering can also be used to impose the necessary discipline on the designer, and is often not much different from what will be done in a design review. Design review, value engineering, or just meetings round the drawing board with the methods engineer will cost time and money and require organizing. It is necessary to ensure that the pay-off is greater than the cost (hence the need to audit the review techniques).

The other major cost faced by the manufacturer – that of development – has been discussed as the cost of meeting the specification and of demonstrating that it is met. Clearly this cost varies with the level of the technology of the product and also with the production run. In the case of an aeroplane, such as the Concorde, with a development cost of the order of £10^9 and a production run of about 20, the development cost carried by each aeroplane is the major contribution to the cost of buying it.

---

†Later chapters (particularly Chapter 11) discuss design procedures, and Chapter 12 discusses design review methods.

1.7 SOME EXAMPLES OF DESIGN COSTS

The need to control the way money is spent on design is exemplified in Fig. 1.1. This shows the history of one design project, and we can see overrun in both cost and time. It is also clear that the estimates of the cost to completion were not only optimistic but increased in accuracy only slowly with the expenditure of design effort. We see that design cost 60% more than was estimated so that, if the price of the product had been based on the originally estimated design cost, the manufacturer would have been £140 000 out of pocket even before we consider the other costs that late delivery may generate – the loss of goodwill, the interruption of other programmes, the loss of operating income etc. – but no accountant will tell us what these are. The tendency to spend more on design than has been estimated is not confined to occasional projects, and Fig. 1.2 shows the distributions of estimated design costs and actual design costs for a sample of all projects for which the figures were available in a given year. We see that the tendency to underestimate design costs is, in fact, general. It should not be thought that the example is taken from a company which is bad at estimating or slow in developing its products. The fact that the figures were available shows that the company which supplied them has a sophisticated system for predicting and monitoring costs. As a result, although the estimates are frequently exceeded, estimating accuracy (together with an understanding of the technical work involved) has almost certainly improved since the introduction of the control system. The management procedure requires estimates from salesman and engineers, and, as Fig. 1.1 shows, some estimates are reviewed repeatedly during the course of the project. Computer back-up is necessary to give continuous monitoring of the expenditure on the project, week by week, by each of the design skills in each of the design tasks.

The computer programs, which owe something to PERT, assume a network of perhaps fifty activities in the whole process of design, from deciding to undertake the project to issuing production drawings. The decision to undertake a project is taken only after analysis has suggested that it will be expected to have a pay-off.

When a company is making products with comparatively low technology, the direct costs of design may seem of small importance, but design is often seen as a commitment to much larger costs, and poor design (or too little design effort) can generate the use of unnecessarily expensive manufacturing methods or generate errors in the prediction of manufacturing costs. This is exemplified by a company which makes components for storage systems. These components have a limited technical content, and the design costs amounted to less than 10% of the first year's cost in a project with a two year payback period. The market for the company's products demands large batch or mass production, and while the direct costs of design were about £3000, all spent in the first year, the costs of manufacture averaged over £50 000 a year for the five years of the project (five years is the estimated life of some of the moulds used in manufacture).

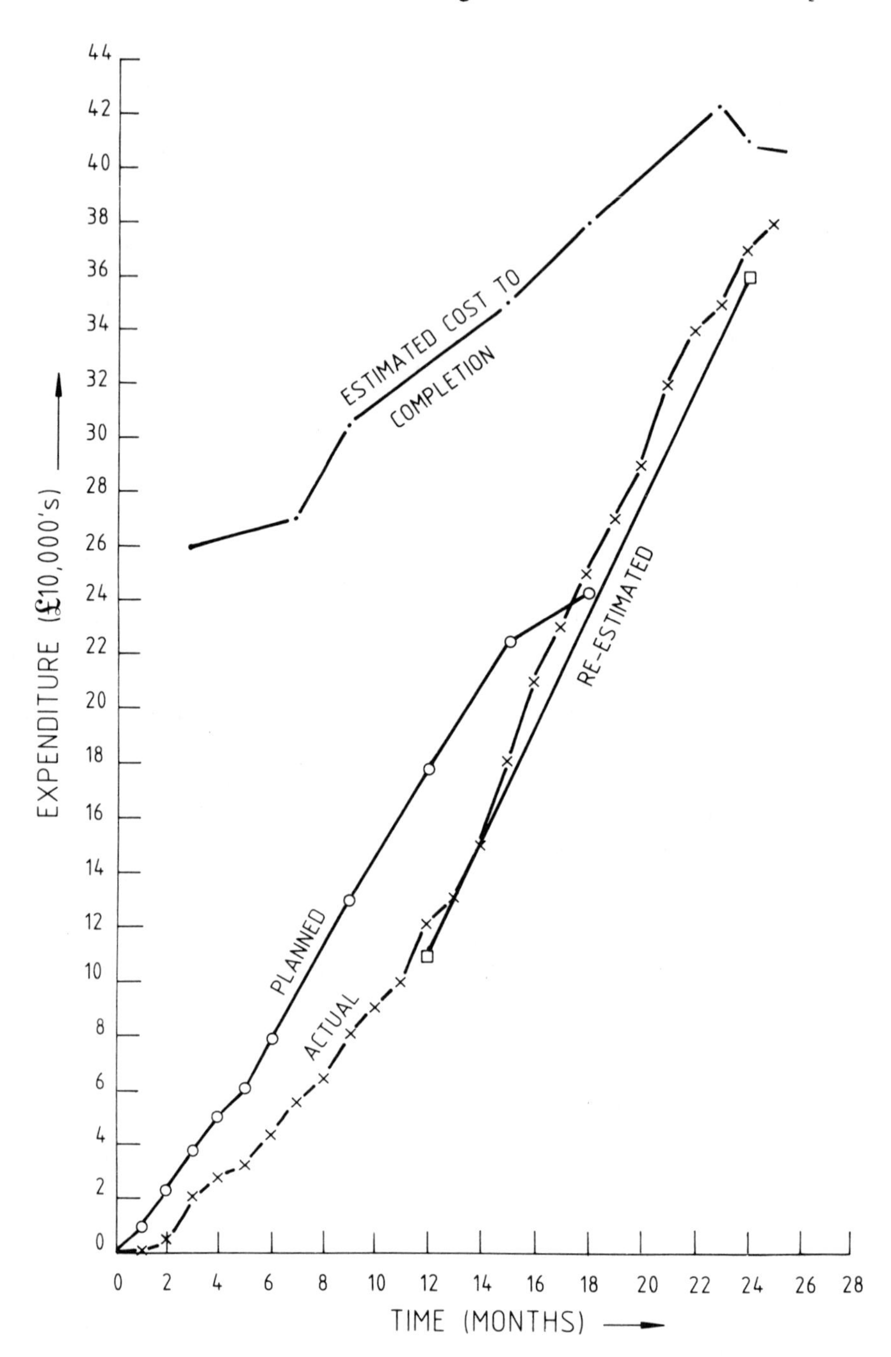
ESTIMATED COST TO COMPLETION
RE-ESTIMATED
PLANNED
ACTUAL
EXPENDITURE (£10,000's)
TIME (MONTHS)
0
2
4
6
8
10
12
14
16
18
20
22
24
26
28
30
32
34
36
38
40
42
44

Fig. 1.1.

Log 2 Cycles x Probability

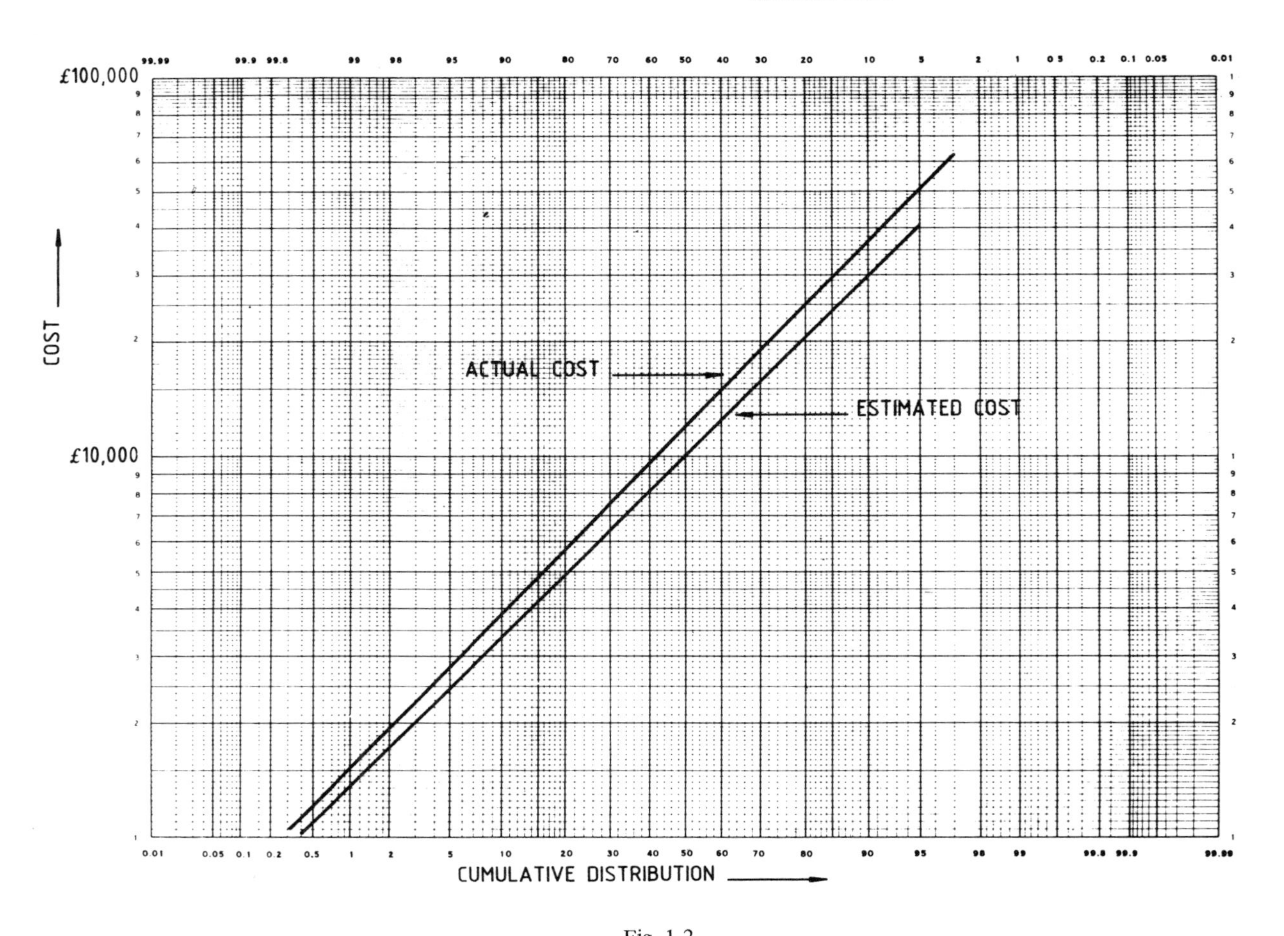
£100,000
£10,000
COST
ACTUAL COST
ESTIMATED COST
CUMULATIVE DISTRIBUTION

Fig. 1.2.

The great importance of design here, is to provide the information on which the manufacturing methods and cost estimates will be based. The same tendency to underestimate design costs is observed in this company, but this is less significant when design costs are so small a part of the company's expenses (repeated gross underestimating of design costs could, of course, lead to a department which is too small to generate enough work to keep the company going, but this is a problem which can be solved by *ad hoc* methods). The important point is that the design information, together with sales predictions, will enable the company to get the manufacturing method right. The design management system is therefore less complex than that in the high-technology company, although it aims at answering many similar problems. There is less need to monitor design department expenditure at frequent intervals and in many skill classifications. The re-estimates of costs are far more concerned with a prediction of manufacturing costs. Reviews are far more concerned with whether manufacturing costs and markets were accurately predicted.

Where a company is concerned with high-volume manufacture the economics of design are not concerned only with the high cost of tooling. One washing machine manufacturer takes great pride in the reliability of its products, and in its search for increasing reliability makes major model changes infrequently, and the current washing machine has not changed in a major way for fifteen years. Design projects have been generated by the desire to reduce the cost of the electric motor in the washing machine (much of the cost reduction was achieved by reducing the weight and material in the motor frame), by the need to increase the reliability of the switchgear, and by the need to redesign packing. The first two examples immediately generated design work. The last example arose from the opportunity to take advantage of cheap pick-a-back road/rail transport and became a design problem when the cost of redesigning the packing was compared with the cost of redesigning the washing machine to meet the new environment imposed by the new method of carriage. Every one of these design problems was generated by the need or opportunity to reduce the manufacturer's or the customer's costs. All the problems seem small, but the costs that they generated could not be ignored. Specifications are pragmatic, and requirements have been derived from an arbitrary decision to design a product which will last twenty years without a major overhaul. This requirement translates into 9000 operating cycles with some sub-assemblies or 7500 hours running with some others, and any modification is proved by test before being incorporated into machines for sale. Each apparently small change will therefore generate lengthy and expensive tests, and requires the previous investment in expensive test equipment. Testing new packaging (or the redesigned machine) in a new environment generates tests to determine the environment (in this case using ride recorders to determine the *g* levels encountered in transit) and then endurance testing of the machine in the appropriate acceleration and vibration conditions.

One lesson that can be learnt from this testing of washing machines is that a simple, pragmatic approach works. The testing procedure was generated, without recourse to advanced technology, by salesman and designers who simply thought of what the customer wants.

1.8 THE TOTALITY OF DESIGN

Any designer who hopes to be responsible for the creation of some saleable, useful artefact must be a generalist if his designs are more than trivial. The designer who is responsible for an aircraft engine will have to organize the design of the compressor, the combustion chamber, the turbine, the bearings, the controls, and many other sub-systems. It would not be possible for him to have specialist knowledge of every sub-system, and he could not afford to concentrate on only one problem of the engine. When he knows what he is about, he can buy the time of the specialist – the man who understands diffusers, the man who designs turbine blades, etc. Similarly, the designer who is responsible for designing a train, a boiler, a motor car, a power station, or any other, non-trivial system, is the manager of more specialist engineers, and to lead them he must have some knowledge of a broad span of technical subjects.

But a knowledge of technical subjects alone is not enough, for, as we have seen, it is important to be aware of the economic consequences of engineering decisions, and the senior designer must have an understanding of costs, markets and the management of time. To some extent it may be argued that to be a senior designer requires a broad knowledge of technical and managerial skills. The specialist must work for the generalist.

Even this spectrum of technical and managerial knowledge is inadequate because technical excellence, price, and delivery date are not the only factors which affect the saleability of a design. Fashion, aesthetic appeal, and ergonomics play their part and rarely find their way into the curriculum of an engineering course.

Ergonomics is now a well established discipline with texts (see, for example, [9] and [10]) that enable it to be taught as well as thermodynamics or strength of materials, and the reason that it is not a subject in many university or polytechnic engineering curricula may be lack of space rather than a lack of appreciation of its importance. The significance of good ergonomics in a design is clear when we consider both manufacturing costs and operating costs. If the ergonomics of a production line are good, assembly costs will be reduced, but the production line design will be constrained by the product being made. Again, ease of operation will reduce the customer's cost of use, but ease of operation has to be designed into the product almost from its conception.

Aesthetics is a subject quite unlike any of the other disciplines that the designer is expected to understand. It is not a science and cannot be developed logically. From the craft preached by William Morris, Art Nouveau, the ideas of the Bauhaus, and Art Deco, to the clean, utilitarian (Olivetti typewriter) designs of the fifties and the nostalgia (electric table lamps made to look like oil lamps) of the seventies, every movement has had its theory; a theory discarded and replaced by the next. Teaching industrial design seems possible only by the teaching of history, by the use of case studies, by practice in the studio. Usually the industrial designer is not an engineering designer, and this gives rise to the argument that the industrial designer puts a pretty case, and perhaps a pretty box, around what has already been designed by the engineer. There are many situations, however, in which the engineering designer must employ the services of the industrial designer. The motor car industry employs industrial designers, and clearly must do so, although there have been well publicized examples of the public's rejection of industrial designs which covered well engineered products.

1.9 CHAPTER SUMMARY

We may summarize the above in the following statements.

(i) Most design specification requirements, *including almost all the technical requirements,* are determined by economic considerations.

(ii) The design process involves a network of activities which include specification writing, scheming, detailing, prototype manufacture, testing, modifying, liaison with production management, liaison with product support staff, liaison with data collectors, and the production of manuals.

(iii) The design department is a cost and profit centre. Its budget must be formulated in advance and its cost monitored and controlled.

(iv) The budget must allow for design work which does not result in orders.

(v) Design management must attempt to optimize the work done before commitment to a project. This optimization consists of balancing the work which must be written off because it does not lead to an order with the work which must be done to make good forecasts of project success and project costs.

(vi) Design management must attempt to optimize the work done before drawings are issued. This optimization consists of balancing the cost of futher design work against the costs which that work may save in manufacture or use.

(vii) The design budget formulation and the cost control system must be such as to provide a management process which ensures that the customer's costs of use and ownership are properly considered. This management process must also provide the discipline which controls the costs of manufacture and development.

(viii) Predictions of costs and sales are probabilistic. Budget formulation and cost control systems must take the probabilistic nature of the predicted cash flow streams into account.

(ix) Apparently inexpensive design work may involve a commitment to expensive testing or expensive tooling.
(x) Design requires the prior investment in expensive test equipment.
(xi) A simple, pragmatic approach to proving a design will achieve many of the requirements without recourse to advanced technology.
(xii) The senior designer must be a generalist who employs specialists. The range of knowledge required by a designer extends beyond technical subjects.

SUBJECTS FOR DISCUSSION

(1) List several engineered items that you have bought in the last few years (a car, a television set, a lawn mower, . . ?). In each case, determine a technical aspect of the item which affected either the manufacturer's cost or your cost.

Discussion could centre around the reliability of your car, the replacement of steel by plastic in a washing machine, etc.

(2) List a number of activities which must be undertaken before a newly designed item should (a) be put on the market, and (b) be regarded as a successful design.

The list should include market research, laboratory testing, reliability analysis, methods engineering, . . . , product support, ...

(3) In the context of a product with which you are familiar, discuss how much of the design you would expect to be right when the drawings are first issued, and how many drawings will have to be modified before the product is saleable in a competitive market.

(4) Discuss the range of non-technical factors which could affect the saleability of a design. How far could the designer expect to involve himself in decisions which involve these factors?

The answer could include references to demography, methods of distribution, marketing, . . . , fashion, . . . , the nature of the product support available to the customer, ...

(5) List some non-technical subjects that must be understood by the designer, and justify them.

(6) How can a designer with general knowledge, control the work of specialists?

REFERENCES

[1] Matchett, E. & Briggs, A. H., 'Practical design based on method', *The design method,* S. A. Gregory (Ed.), Butterworths, London, 1966.

[2] Feilden, G. B. R. (Chairman), *Engineering design,* Report of a committee appointed by the Council for Scientific and Industrial Research, HMSO, London, 1963.

[3] Booker, P. J., *Conference on the Teaching of Engineering Design,* London, Institution of Engineering Designers, 1964.

[4] Wooderson, T. T., *Introduction to engineering design,* McGraw-Hill, New York, 1966.

[5] Archer, B., *Systematic method for designers,* London, Council of Industrial Design, 1968.

[6] Hartley, K. & Cubitt, J., *Cost escalation in the UK,* Eleventh Report from the Expenditure Committee, The Civil Service, London, HMSO, Vol. III, Appendices, p. 1054, July 1977.

[7] Downey, W. G., *Development cost estimating,* Report of the Steering Group for the Ministry of Aviation, HMSO, 1969.

[8] *Drive*; published monthly by the Automobile Association, Basingstoke.

[9] Oborne, D. J., *Ergonomics at work,* Wiley, Chichester, 1982.

[10] McCormick, E. J., *Human factors in engineering and design,* McGraw-Hill, New York, 1976.

CHAPTER 2

What the customer wants

This chapter discusses in comparatively general terms the costs of using and owning a product that are observed by the customer, and the way in which the designer may be held responsible for those costs. The analysis of these costs defines many of the tasks which the designer must perform, and also goes some way towards determining the cost of design and the likely pay-off. The costs of design are seen to be partly the costs of the men, machinery, and material used and partly the cost which results from bad design.

When a customer buys an engineering product, he wants something that will meet his technical requirements at the lowest cost [1]. This does not mean the product that is cheapest to buy, for as we have seen, even in the simple case of a motor car, buying the product may cost less than a quarter of the cost of owning it. When the customer writes his design specification he will of course, list the performance requirements and the environments in which this performance is to be achieved. The remainder of the specification will be concerned with ensuring that the performance requirements are met at the lowest cost to the customer. We may list those factors which affect the customer's cost of ownership and use.

2.1 THE COST OF ACQUISITION

2.1.1 The purchase price

Although the purchase price of the product is a fraction of the total cost of ownership, it is still significant. The customer will know if the design is such that the product costs more than it need for the purpose for which it is intended. If he does not, a competitor will soon tell him.

The selling price to which the designer must work is not usually mentioned in the specification. This is not unreasonable when the customer supplies the specification with an invitation to tender, because the price will not be determined until competing suppliers have submitted their proposals. What is more surprising is that, even after a price has been determined and made the subject of a contract, the designer is sometimes not told the manufacturing cost to which he is designing. This has happened in cases where those who negotiate the product price with the customer feel that they would be constrained in such negotiations if the designer were aware of the difference between the selling price and the expected manufacturing cost. However, if the designer does not work within the constraint of a maximum permissible manufacturing cost, he is contributing less than he should.

In many companies, serious attempts are made to ensure that a design scheme is such that economic manufacture has been considered. Ways of doing this vary in formality. At one extreme the designer may call for the advice of methods engineers, buyers, planners, etc. while the scheme is on the drawing board; there may be a formal meeting before the scheme (or the set of detail drawings) is released at which representatives of the manufacturing departments will be required to approve the design; the value engineer may study the design, or there may be a full design review system in the factory. The design review procedure will be discussed in Chapter 12.

2.1.2 Distribution costs

The cost of transporting the product from the manufacturer to the customer will add to the customer's bill. In some cases, the designer will be required to consider the method of transport (by, for example, incorporating slinging points in the design), but these cases are generally well defined.

Transport may provide a hazard, however, which must be recognized by the designer who may have to design the product or its packaging to meet environmental conditions during transport that are more severe than those that will be encountered during service (dropping from the tail board of a lorry, for example).

Typical of such problems was that encountered by a washing machine manufacturer and mentioned in Chapter 1. The cost of the packaging itself adds measurably to the selling price of the washing machine, and yet the cost of trying to recover the packaging is too great to be justified. Perhaps the machine itself can be designed in such a way that it will tolerate the transporting environment with cheaper packaging. This is a recurring problem because new methods of transport (for example the advent of pick-a-back systems) save money but generate new environmental problems. As we have seen, the solution to the problem can involve the use of ride recorders to measure the environment, subjecting the product to the environment (which means building the necessary vibrating and acceleration rigs) and testing the product after subjecting it to the hostile environment. In the United States, the results of such testing and of experience in using the new transport method are likely to result in approval of the packaging system used, and Fig. 2.1 is an extract from the *Uniform Freight Classification* [2]. A year of using the packaging and demonstrating its usefulness by keeping track of shipments and testing products after shipment may be required before the packaging method is approved for entry.

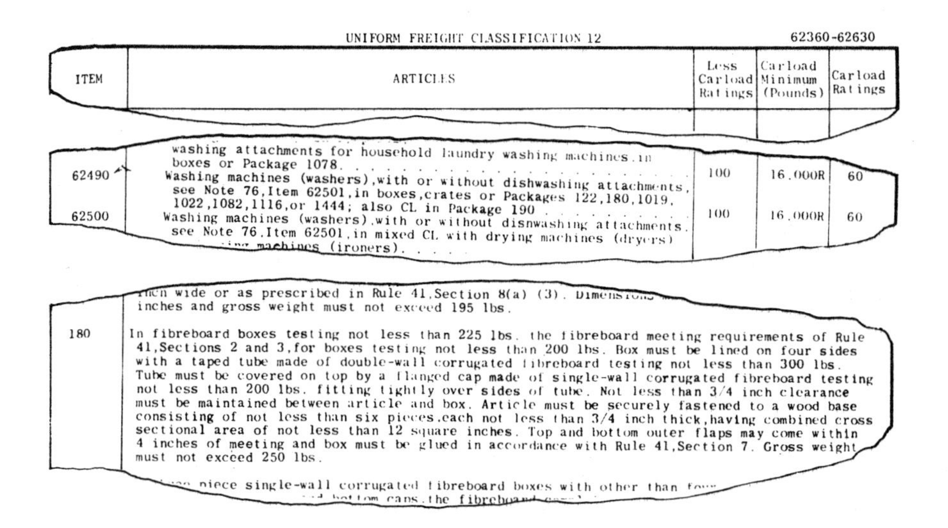

UNIFORM FREIGHT CLASSIFICATION 12 62360-62630

ITEM	ARTICLES	Less Carload Ratings	Carload Minimum (Pounds)	Carload Ratings
	washing attachments for household laundry washing machines, in boxes or Package 1078.			
62490	Washing machines (washers), with or without dishwashing attachments, see Note 76, Item 62501, in boxes, crates or Packages 122, 180, 1019, 1022, 1082, 1116, or 1444; also CL in Package 190	100	16,000R	60
62500	Washing machines (washers), with or without disnwashing attachments, see Note 76, Item 62501, in mixed CL with drying machines (dryers) ... machines (ironers). . . .	100	16,000R	60

	...inch wide or as prescribed in Rule 41, Section 8(a) (3). Dimensions ... inches and gross weight must not exceed 195 lbs.
180	In fibreboard boxes testing not less than 225 lbs. the fibreboard meeting requirements of Rule 41, Sections 2 and 3, for boxes testing not less than 200 lbs. Box must be lined on four sides with a taped tube made of double-wall corrugated fibreboard testing not less than 300 lbs. Tube must be covered on top by a flanged cap made of single-wall corrugated fibreboard testing not less than 200 lbs. fitting tightly over sides of tube. Not less than 3/4 inch clearance must be maintained between article and box. Article must be securely fastened to a wood base consisting of not less than six pieces, each not less than 3/4 inch thick, having combined cross sectional area of not less than 12 square inches. Top and bottom outer flaps may come within 4 inches of meeting and box must be glued in accordance with Rule 41, Section 7. Gross weight must not exceed 250 lbs.
	... piece single-wall corrugated fibreboard boxes with other than four ... bottom caps, the fibreboard ...

Fig. 2.1 – Extract from uniform freight classification.

The possibility of damage during transport may present the customer with another expense. Will the customer have to set up inspection and test procedures to check the quality of the products that he is receiving?

2.1.3 The cost of accepting the product

The customer will require proof that the product he has bought meets the specification. In particular, does the product perform within the required tolerances in all the specified environments? He can answer this question by setting up test facilities and testing the product to prove that it meets his requirements. Usually, however, the design organization will be better equipped to conduct such tests (many of which will, in any case, have to be done as a necessary part of the design process), and it will be cheaper to ask the manufacturer to supply proof that the requirements of the specification have been met. In this case, the customer will expect to be consulted on the test schedule before testing starts, and he will later require written declarations that the test requirements have been met. In all probability, the design specification will describe the form of the documentation in which such declarations are made. Some examples of such documentation are given in Chapter 11.

Some tests will have to be carried out by the customer to ensure that the product he has bought is compatible with the parent system. An aeroplane manufacturer who has bought a navigational aid will need to test the aeroplane with the aid installed, but this is not a repetition of the tests which prove that the navigation aid meets the specification. Nevertheless, the customer will expect (and will eventually have to pay for) technical support from the designer during such testing in the parent system.

In addition to a once and for all test to prove that the product, as designed, meets the requirements of the system, the customer may believe it to be necessary to set up a goods inwards inspection system to prove that all versions of the product that he receives have been manufactured to adequate quality requirements. This depends, to some extent, on the confidence that the customer has in the supplier.

2.1.4 The delivery programme

The customer will have his own programme, and the product that he is buying must be delivered to meet that programme or extra expense will be incurred. A customer building a new storage system will suffer a loss, or at least a reduced profit, from the system if such necessary parts as bins are not delivered on time. This will be for several reasons. A system which has been selected because it shows a profit will, because of the time value of money†, show less profit if it comes into service later than predicted. In certain circumstances, the coming into operation of the new system will be timed to coincide with the phasing out of an old system, so that failure to deliver components for the new system can result in the costs incurred by an idle factory. This would be the case if it were intended to install a new storage system, in place of the old, during the works holiday. The customer will, therefore, need to specify the delivery date of the product he is buying, and he may introduce into the contract, penalty clauses which require the manufacturer to compensate him for late delivery.

The customer may, in fact, need to specify more complicated delivery requirements than are implied by a single date. Consider the purchase of a new navigation aid for a new aeroplane. The customer will require space models for his mock up, prototype models for his laboratory test of the parent system, models to an agreed standard (which may not be the ultimately required standard) for his aeroplane test programme, and a regular delivery of products to match his aeroplane building programme. In such a case the specification will include a statement of the delivery requirements, and Fig. 2.2 is a quotation of such a statement from an aeroplane manufacturer's invitation to tender.

---

†A pound now is worth more than a pound in a year's time. This time value of money is discussed in Chapter 8.

Item	Quantity	Description	Delivery required by	Delivery to
1	1	Flying control system Test rig set Main air control valve	1st December 1981	Customer's factory at W –
2.1.	1	Wooden mock up representative of flight units	1st May 1981	Customer's factory at W –
2.2.	1	Metal space models of units for prototype aircraft.	1st December 1981	W –
2.3.	1	Metal space models of units for production aircraft.	1st July 1983	W –
3.1.	5	Functional units cleared for 100 hours' operation for prototype aircraft 1–5	1st February 1982 1st April 1982 1st June 1982 1st August 1982 1st October 1982	W – W – W – W – W –
4.1.	10	Functional units cleared for 100 hours' operation for use as spares during	1st March 1982 1st May 1982	W –
5.1.	300	Units to full production standard	Delivery programme a) September 1985 to March 1986. 5 sets per month. b) April 1986 to	Customers' factory at T –

Fig. 2.2.

The customer wants the product on time; he may not want it too early. Consider the supply of components to the car manufacturer. Because of the quantities of products involved, a delivery may be worth hundreds of thousands of pounds. If the customer has to pay for these earlier than necessary, he will tie up capital which could be earning money. Similarly, the manufacturer will not want to hold stock which could tie up capital and it is possible for quite complicated negotiations to go on between the supplier and the customer, each trying to avoid tying up his own capital and hoping that the other will carry a buffer stock of the products.

In general, however, in the case of a new design, the customer is more likely to be embarrassed by late delivery than by delivery that is too soon. Lateness may be caused by performance problems which occur and to which development does not provide a rapid cure, or it may be caused by unforeseen difficulties of manufacture or the long lead times of bought out parts. One of the difficulties of making small numbers of prototypes or small numbers of the product in preproduction runs is that suppliers of bought out components are not always willing to offer short lead times. The consequence can be, for example, that the designer will be forced to call up machined details where castings would be cheaper, in order to meet the customer's required delivery date. Such an enforced departure from what would normally be good engineering practice generates extra costs, not only because the direct manufacturing costs are greater but also because development may have to be repeated when design changes are incorporated to cheapen manufacture. An example of this problem arises from time to time in a company making aircraft electronic equipment. The equipment supplied often contains metal parts which, ideally, would be light alloy die castings, but it takes 28 weeks to get a casting after the drawing has been issued. This lead time is, itself, sufficient to prevent delivery of the finished product by the customer's required delivery date, but the situation becomes even worse if, as often happens during development, changes have to be made to the casting. There will be several hundred jobs alive in the factory at any one time, and it requires only that one or two be stuck in the system, waiting for castings, for the scheduling of all the jobs to become impossibly difficult. As a solution to this problem, the company frequently uses numerically controlled machine tools to machine parts from bar for prototype products and early production runs. From the receipt of a drawing to the supply of the first machined parts can be 16 hours, and if the part has to be modified during development, this can be done easily and quickly. But the cost of the N.C. mill is of the order of a quarter of a million dollars, and a programming team must be recruited and maintained. The designer is clearly faced with having to decide how to trade off the cost of numerically controlled machining against the cost of delays.

2.2 THE COST OF OPERATION

2.2.1 The cost of failing to meet the required performance

If the performance of the product is not as specified, the customer may find that his profit from its use is less than he hoped. A navigation system that does not work in the extremes of the aeroplane's environment may restrict the use of the aeroplane; a temperature control system which requires more droop than was expected to keep it stable may use more power than had been intended. If the required performance was incorrectly specified, then modification to the design specification may be required, but frequently the customer will be required to decide between extending the development programme, with the expense that it implies, and accepting a performance which is less than he had requested.

2.2.2 The cost of labour

The labour cost in operating a product will be:

the direct labour costs, which can be reduced if the designer considers the ergonomics of operation,

the cost of training the operative, which may be reduced by good design,

the cost of supervizing the operative, which may be reduced by good design, and

the cost of supporting the direct labour.

In the design of a factory storage system there are clear savings for good ergonomics which will reduce the effort required to operate the stores; a design which makes operation simple will reduce the time taken by operatives to operate the system and the supervision they will need, while good layout will reduce effort required by support staff with fork lift trucks etc.

Almost always, the designer will be required to supply operating manuals, and the customer will probably call for these in the specification. The provision of operating manuals is not an expense which the supplier can avoid to increase his profit. An understanding of the need for such manuals by the supplier and the customer will ensure that the one includes the cost of the manuals in his tender and the other sees the profit in buying the manuals.

2.2.3 The cost of power and materials

If there is a limit on the amount of fuel available or on the type of fuel, the customer will specify it. Obviously the less fuel used, the cheaper the product will be to operate, and newspaper advertisements demonstrate that manufacturers of cars believe that the customer wants the car that uses less petrol than others. But there can be more serious problems than this if, for example, the product requires a source of frequency controlled electrical power when only a frequency wild source is available, or a source of pneumatic power in freezing conditions when only a wet supply of air is available. In both cases the customer would have to spend money (possibly on further new designs) to create or modify the power sources.

2.3 THE COST OF MAINTENANCE

2.3.1 The cost of failure

When a product fails, the customer is involved in two major costs: the cost of repairing the product, and the cost of down time during, and while awaiting, repair. The more often failure occurs, the greater will be the cost to the customer.

A recent example of the cost of failure is the offer of the makers and the operators of the DC10 to award $60 000 000 to the dependents of those killed in the Chicago air disaster of 25 May 1979. (report in the *Financial Times* of 16 August 1979).

One of the more serious costs of failure can be the cost of ensuring that the product is safe. If the use or the failure of the product threatens life or property then the customer may be required to compensate those who suffer. He may be able to insure against such accidents, but the insurance company will require high standards of design and proof of reliability, and even then there will be high premium costs.

The cost is not always easy to measure. How, for example, does one calculate the value of keeping military aeroplanes flying? Nevertheless, if one reads (*The Guardian* 17 August 1979) that 'Last year, in its efforts to keep a reasonable number of the F-15s in the air, the U.S. Air Force cannibalized 15,474 parts and used a total of 47,898 maintenance man hours installing them', it becomes obvious that the cost of keeping aeroplanes flying can be very high. The same news item tells us that some F-111 aeroplanes '—— had had only about 270 hours' flying time. The rest of their life had been spent undergoing repairs'.

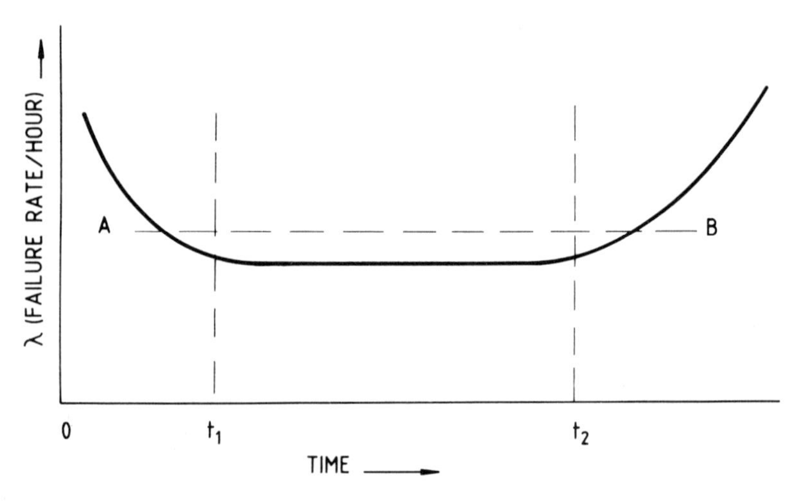

The period from time 0 until time t_1 is the period of infant mortality; the period after t_2 is the period of wear out.
Infant mortality is the early period of high failure rate caused by undetected faults in the raw materials and manufacture.
The wear out period starts when the failure rate increases because components are wearing out.
AB is the maximum permitted failure rate.

Fig. 2.3 – 'Bath tub' curve.

2.3.1.1 Frequency of failure

The frequency of failure is measured by the reliability of the product that is being designed, and the customer will specify the minimum reliability that he will accept. The method of specifying reliability will either be in terms of rate of failure (for example, the mean number of failures per hour) or the inverse of this, the mean time between failures (usually abreviated to MTBF). In the simplest cases (which will be discussed at greater length in Chapter 10) the variation of the failure rate during the life of a product is as shown in the 'bath tub' curve of Fig. 2.3. Where the failure rate is unacceptably high during the infant mortality period, it may be necessary to 'burn in' the products before sending them to the customer. An example might be lamp bulbs which could be burnt for a time in excess of the infant mortality period. Only those surviving this period would be delivered, and would be expected to have the low failure rate required.

Acceptable rates of failure are very low, and one typical quotation from a customer's specification reads, *'The product shall have a failure rate not exceeding 0.015 failures per 1000 hours of operation ---.'*

Not only will the customer require a specified reliability, but he will require that the achievement of this reliability is demonstrated. The designer must, therefore, have a source of information about the reliability that is demonstrated by products in service, and this implies that the design organization either contains a system for collecting and analysing data or has such a system associated with it. Normally such a system is run by a reliability engineer who may or may not be a member of the design organization, and the system may vary in complexity from a file containing information on all the failures of the company's products to a computer system with the information on file and built-in programs for analysing the data. The information is supplied by the service engineers, and a formal method of collecting and sending the information is essential.

The designer must also have test facilities available with which he can test and demonstrate the reliability of components for which data from the field are not available.

Where the cost of failure is extremely high (as for example where danger to life is concerned), even very high reliability may not be good enough. Sometimes redundant components can be used to increase already high reliabilities to acceptable levels of risk, but often it will be necessary to design the product so that failures are safe. This is not enough in itself, because it will be necessary to demonstrate that failures are safe. For example, it may be a requirement that a burst turbine wheel or compressor disc be contained within the machine casing. Theoretical demonstrations that this will happen are difficult and not very convincing. Practical demonstrations are more convincing, but are difficult and expensive.

2.3.1.2 The cost of repair

The designer can keep down the cost of repair if he makes sure that:

it is easy to diagnose faults,

details or sub-assemblies that are most likely to fail are easy to remove and replace,

special tools are not required, and

it is easy to check that a satisfactory repair has been effected.

Typical quotations from specifications are:

'. . . *the manufacturer shall state the fault diagnosis technique and the related test equipment that is required (for replacement in the field). Built in test equipment is to be preferred.'*

'. . . *sub-assemblies subject to wear or deterioration in service must be capable of being replaced easily. Such replacement parts will be held as spares and should not require pre-installation checks or calibration. The design should be such as to provide easily changed modular sub-assemblies.'*

These quoted statements leave much to be desired as specification requirements since, although they suggest trains of thought, they provide no criterion by which design achievements may be judged (they are rather like specifications which call for a component to be as light as possible). Nevertheless, they indicate a justifiable preoccupation by the customer with the costs of maintaining the product he has bought.

The designer is committed, then, to plan the methods of fault diagnosis and repair. In addition to considering these factors in the design of the product, he is usually committed to the writing and supply of maintenance manuals, and a quotation from a typical specification asks the manufacturer to supply '. . . , *Maintenance Manuals; Trouble Shooting Instructions; Instructions for In-situ Testing;* . . .'

In many cases, the customer will also require the manufacturer to provide training for the customer's maintenance engineers.

A repair is not completed until it is shown that the repaired product is working within the requirements of the specification, and this may mean that special test equipment has to be designed. If an altimeter were taken apart and repaired, it would be necessary to reset it after reassembly, and it would be difficult to check that the altimeter had been correctly repaired without some means of providing known levels of vacuum. The purchase of test equipment is an obvious cost to the customer which he must balance against the reduced cost of repairs. It will, of course, be profitable for the manufacturer to sell the supporting test equipment, but only if he can convince the customer that it is to his profit also. Sometimes it is desirable to build the testing system into the product itself.

2.3.1.3 The cost of waiting for repair

Often the cost of down time is more significant than the cost of the actual repair. An aeroplane may be grounded on a foreign aerodrome, with the cost of idle capital, the cost of hotel bills for passengers, airport dues, and many more expenses that far exceed the cost of the accessory that has failed. A production line in a factory may be brought to a standstill by the failure of a machine, and the cost of wages for idle operatives, the lost production, and the cost of idle capital, will far exceed the cost of the failed component.

A production line could cost £200m to build. If the money were borrowed at an interest of 10% and if the line worked three eight hour shifts a day, one hour's lost operation would cost over £3000 in loan charges alone.

The designer of the product must, therefore, work closely with those who design the system which ensures that there is a good back-up of spares. This, of course, goes with design for easy replacement of failed details and sub-assemblies, and is related to the reliability analysis.

In many cases it will be necessary to design a product support which will be hierarchical in structure. Consider an infantry weapon system. It is likely that the soldier operating weapon will be able to perform some repairs and adjustments, and he must be trained to diagnose the faults that he can remedy and must be given the necessary spares and tools to effect those repairs. Where the faults cannot be remedied by the operator of the weapon, it will be sent to a forward maintenance unit where the diagnostic equipment, labour skills, spares, tools, and test facilities permit maintenance at a greater depth. It is likely that there will be a further tier of maintenance facilities above this for repairs which can be effected only by the manufacturer. Each of these tiers in the maintenance system will have to carry spares, tools, etc., appropriate to the faults it will be required to remedy, and the spares will have to be held in numbers appropriate to the frequency with which each fault occurs. The maintenance system clearly has to be designed so that the cost of failure is minimized (or at least reduced to an acceptable, stated maximum cost), but once it has been designed the customer will have an incentive to continue to buy the product or close developments of it. An airline that is set up to maintain a particular engine, at many places in the world, would find it expensive to change to using a completely different engine, not least because of the cost of setting up a new maintenance system. Modifications to the original engine or improvements to it would permit the gradual development of the maintenance system, and this would be much less costly than step changes.

2.4 THE COST OF REPLACEMENT

The cost of complete replacement is really determined by the life of the product. Strictly speaking, at the end of the product's advertised life, it may not be necessary to replace it with a completely new one, for a major overhaul may bring the product to an 'as new' condition or at least to a state at which it may be given a new life. The effect of life on cost is obvious, for a car costing £4000 which lasts ten years is clearly cheaper for the customer than a car which costs £4000 and, other things being equal, lasts five years. There are situations in which the end of a product's life is not clearly defined, and there are many cars in use by second (or later) owners which the first owners considered to be at the end of their useful lives.

The customer will not object to buying a product which lasts longer than necessary†, but he will certainly specify the minimum life that he is prepared to accept. This minimum life will be one at which the customer judges that his cost of ownership will be acceptable. To some extent the life of the component is arbitrary because the advantages of a longer working life can usually be traded off against a greater cost of repairs or the cost of an increased failure rate. If we consider the bath tub curve of Fig. 2.3., the point B shows the maximum life that can be achieved with the specified failure rate, but clearly the life could be extended slightly at the cost of a short period of higher failure rate.

The relationships between maintenance cost and life and between reliability and life may be expressed in the specification by such requirements as:

'. . . the reliability requirement (of 0.015 failures per 1000 hours of operation) shall be achieved by the product throughout its declared serviceable life . . .'.
or
'The product shall achieve a minimum engineering cost per operating hour consistent with a parent system life of 2000 operating hours'.

This last quotation is really too vague to be a working requirement of a specification since it leaves the designer to determine the minimum engineering cost (whatever that is) and implies that a virtually impossible optimizing problem must be solved before designing can start. It is pointless to demand a minimum cost if the minimum cannot be determined. What the customer needs to specify is a maximum cost that he will accept and which can be determined externally as one which still leaves the product profitable to him.

Another specification is for a component which can be brought back to an 'as new' condition by a major overhaul. This specificaton, having listed the reliability requirements, says,

---

†There may be exceptions. An example is the case of a building which has exceeded its economic life but which can only be knocked down at great expense.

'. . . the product must have a safe fatigue life in excess of 30,000 hours.
'Evidence must be submitted by the manufacturer to justify the 30,000 hours operating life'.
'If it is not practical for the product to be on condition, the overhaul period after 2 years' service shall be 10,000 hours, established by sampling checks.'

The implication of this quotation is that it would not be practical to repair the product at the end of its fatigue life, but it would be permissible to overhaul it at intervals of 10 000 hours. There is the hope expressed that the product will eventually be 'on condition', which means that the product has an acceptable reliability over so long a period that replacement at a fixed period is not necessary. This, in turn, requires that there exists a method of inspection by which the condition of the product can be monitored and impending faults diagnosed. It is, for example, possible to diagnose wear in bearings before deterioration is sufficient to cause failure.

All these qualities that are specified by the customer require the designer to have available the data with which life and reliability can be predicted and the test equipment necessary to determine and prove the lives of components for which there are no service data. The designer is also required to ensure that *in-situ* inspection or testing is feasible in cases where 'on condition' products are offered.

Operating life is one phase in the total life of some products. There are systems which are only required to work at infrequent intervals (burglar alarms, for example), and products which are sold as spare parts must not deteriorate during periods in store. Sampled specifications require variously:

a shelf life of five years; a shelf life of ten years with checks at five year intervals; a shelf life of two years.

Although there are problems in specifying life, caused by difficulties of trading off life against maintenance costs and against reliability costs, the customer must establish acceptable combinations and state them. A further difficulty arises, however, because it is not possible to state the minimum acceptable life in deterministic terms. The failure of a product, like the death of a man, can occur at any time after birth, and the most that can be said about it is the probability with which it will occur at any given age. Fig. 2.4 shows the cumulative probabilities of belt lives collected from a number of failures in a continuously running process plant. The most we can say about the life of the belt from this information is the probability that it will last more than a given period. For example, we can see that we would expect 80% of belts to last more than 190 days, 50% to last more than 320 days, etc. Even this information is not wholly true, because the number of observations that we have, from which to determine the probabilities, will determine the confidence we have in any inferences we may draw from the the curve we have plotted. This difficulty is discussed

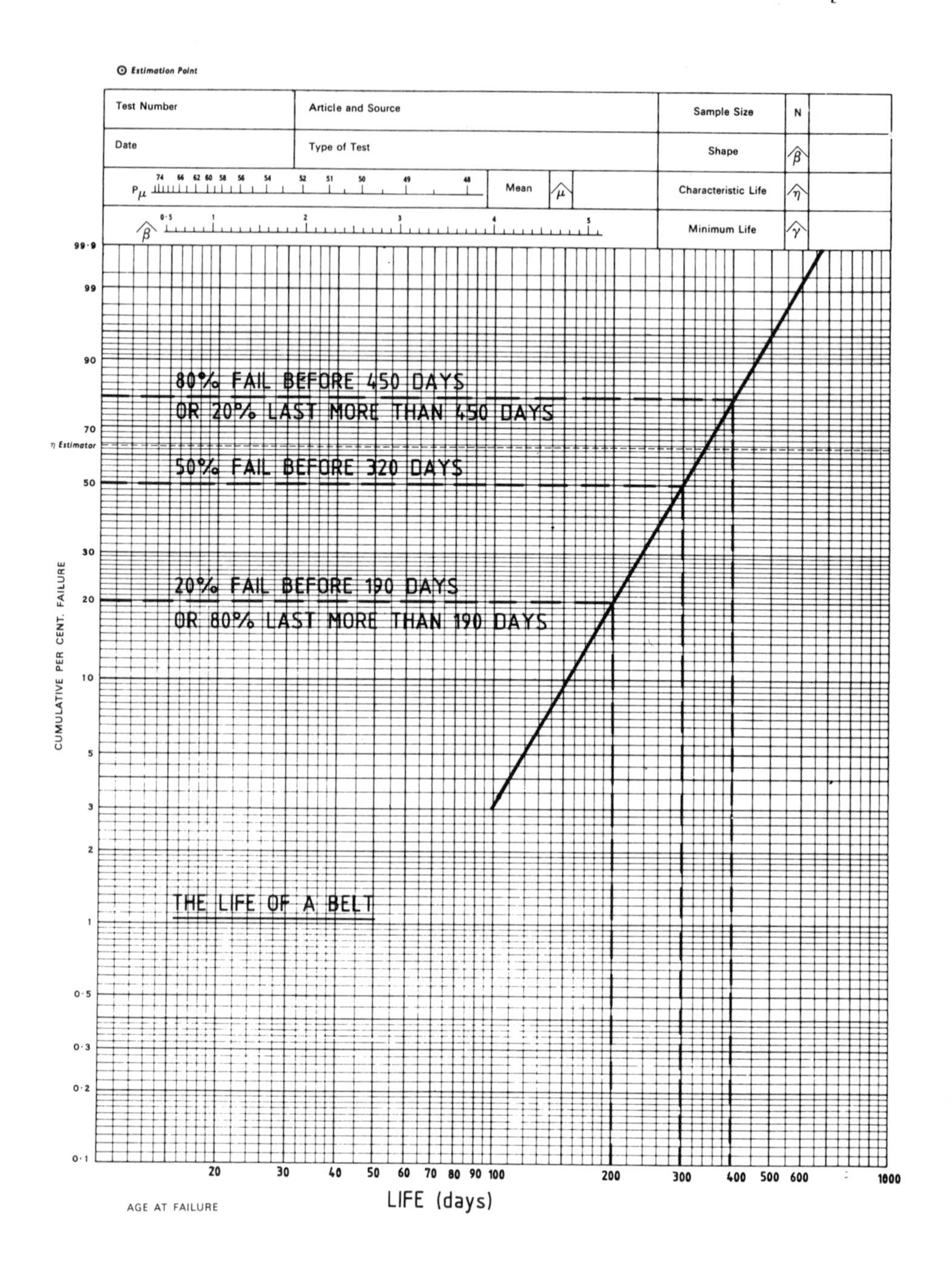

Fig. 2.4 – The life of a belt.

further in Chapter 10. Nevertheless, the fact that the only statements the designer can make about the life of a product are probabilistic does not absolve him from supplying the information. It means that the information must be given in terms of probabilities and confidence levels.

There are, of course, many customers who would not understand statements of probable failure rates, but even unsophisticated customers learn, in time, whether a product is reliable and set a value on reliability. The washing machine that has been discussed before sells at about 25% more than a competitor's product, and enough housewives are prepared to pay this higher price for the extra reliability they believe that they are getting (of course, the selling price may have to be higher to pay for the higher quality that goes into design and manufacture). The trade-off between the higher cost of quality and the lower cost of repair and down time was actually calculated for customers who intended to install the washing machine in laundrettes where the costs of repair and down time are calculable. Nevertheless, with washing machines as with cars or television sets, the customer will eventually find out which products are reliable and stop buying those which are not.

2.5 THE COST OF MODIFICATIONS

Sooner or later the product will have to be modified. This may be because service shows faults that were not apparent during development, and in this case it is probable that the manufacturer will have to pay for the cost of design, manufacture, and service involved. The customer will also face costs if the need for modificaton arose from failure to meet the specification, and he will want the manufacturer to compensate him for this.

The customer may wish to improve the performance, however, or change the specification, and in such a case he must expect to pay for the work involved. The design organization must be such that modifications can be controlled. This control is necessary to answer such questions as, 'When will the modification be done?', 'Is the modification necessary to make the product safe?', 'Can the modified product be supplied when a replacement is required anyway?', 'Must the modified component be supplied immediately?', 'Must the product be withdrawn from service until the modifications have been effected?', 'Is the modification desirable or essential?', 'Who will pay for the modification?'.

One form of modification implies neither faults of design nor faults of specification. This is the eventual stretching of the product to have a greater capacity than was originally required. Engines are developed to later marks with greater power and lower fuel consumption, aeroplanes are developed to later marks with greater carrying capacity and lower running costs. Even motor cars are developed to higher standards. The customer will clearly find more profit in using stretched designs in the future than in changing to products of novel design. A novel design will cost more to prove than a modification to an

A summary of the customer's costs of use and ownership and their effects on the designer's work

Cost observed by customer	Action taken by customer	Action required of designer	Support required by designer	Remarks
The cost of acquisition Purchase price	Negotiation of price and contractual commitment of manufacturer to the price	Design for economic manufacture Consultation with production engineers Consultation with buyers Value engineering Design review Good liaison with the production department	The production department must accept the need for supporting the design department	There is no real substitute for a designer who appreciates the problem. Design review may be too expensive a way of finding design errors
Distribution costs: transport costs, damage in transit, goods inwards inspection costs	Specification of quality of product on arrival Rejection of products failing goods inwards inspection Negotiation of F.O.B. prices	Design for transport Design to withstand transport environment (shocks, cold, etc.) or design of packing	Environment measuring equipment Environmental testing and test equipment	Test facilities are of course a part of the design department resources. They are specified in the capital budget after discussion between designers and the board
Acceptance tests: test facilities, test personnel, test of product in parent system	Specification that the designer supplies proof of performance Support expected during tests by customer	Agree test schedules with customer Supply evidence of compliance with test requirements Support customer testing	Test facilities	

Cost of late delivery	Specification of delivery programme Negotiating penalties for late delivery	Meeting the delivery programme may force the design of different standards of the product This could be the supply of early products which do not meet the life requirements or the use of dearer manufacturing methods to reduce lead times on early products or risking the early issue of drawings for long lead time components		It may not be possible for the designer to avoid risk and uneconomic manufacturing methods if the delivery programme is to be met. In this case allowance must be made in the price quoted to the customer before the expense is incurred
Operation The cost of operation, including the cost of training operators	Specification of power constraints Specification that the manufacturer supplies operating manuals and training facilities	Ergonomic design Provision of operating manuals	Technical authors Training facilities	Training operators and the provision of operating manuals can reasonably be paid for by the customer
Maintenance The cost of repair The cost of down time The cost of spares and spares holding The cost of special tools **The cost of special** test equipment	Specification of maximum acceptable failure rate Specification that reliability be demonstrated	Design for easy repair, *in situ* repair, easy replacement of failed sub-assemblies, fault diagnosis,	Data on failure of existing products Reliability engineering Service engineering reports Technical authors	

A summary of the customer's costs of use and ownership – *continued*

Cost observed by customer	Action taken by customer	Action required of designer	Support required by designer	Remarks
The cost of training maintenance men	Specification that the manufacturer supplies: maintenance manuals, test equipment Specification that the design is for easy replacement of failed sub-assemblies Specification that the manufacturer lists spares requirements Specification of a maximum permissible maintenance cost over the product life	easy test after repair, test by built-in instrumentation Provision of maintenance manuals Tests to demonstrate reliability Assistance with spares provisioning programme	Test facilities Stores	Maintenance manuals can be sold to the customer

Life Replacement cost or Cost of major overhaul	Specification of life Specificaton that life is demonstrated Specification that monitoring methods be provided	Life testing of product Determination of life properties from data collected in service Provision of condition monitoring methods Provision of condition monitoring tools	Test facilities Reliability engineering Service engineering Technical authors	
Changes The cost of modifications including out of service costs	Charge the manufacturer for the cost involved	Monitor the reason for the modification Design and prove the modification	Test facilities Service engineering Modification monitoring system	

The phrases shown underlined in the above table are examples of formal documentation which the designer is required to supply and which will be discussed in later chapters.

old one; it will require a complete change to the maintenance procedures and to the spares holding policy; operators and maintenance engineers will have to be retrained.

It is difficult to see how such continuity and evolution can be requirements of the design specification, for if the designer could foresee the better product, he would probably design it in the first place. Nevertheless, if a product is successful, it is usually stretched; and a designer, when faced with more than one solution to a particular problem, can choose the one which is more likely to permit enhanced performance in the future.

SUBJECTS FOR DISCUSSION

(1) Are there advantages to be gained from designing a structure or mechanism for a limited life or is it better to construct and manufacture for future generations?

(2) When an engineer is designing large plant for a developing country should he:

  (a) produce a simple design which makes the best use of local materials and abundant cheap labour or:

  (b) produce the most sophisticated possible design on the grounds that only the best is good enough for a developing country?

(3) Write down on paper a product you are familiar with.
Now list three ways in which you could enhance its sales appeal while maintaining its selling price.

(4) Name some new products in the appliance area that have had a slow evolutionary change but have gradually increased sales appeal through innovations and packaging.

(5) The community accepts the philosophy of 'built-in obsolence' for most of its major purchases but apparently does not expect to do more than minimum maintenance. Is this a good policy?

REFERENCES

[1] Leech, D. J., *Management of engineering design,* Wiley, Chichester, 1972, pp. 185, 200.

[2] *Uniform Freight Classificaion* No. 12 published by the Interstate Commerce Commission (a US Federal Agency) from the Tariff Publishing Officer, Room 1106, 222 South Riverside Plaza, Chicago, ILL.60606, USA.

CHAPTER 3

What the manufacturer wants

This chapter discusses in comparatively general terms the costs of selling, making, and supporting a product that are observed by the manufacturer and for which the designer will be held responsible. The analysis of these costs defines many of the designer's tasks. Each of these tasks will cost money for the men, machinery, and materials involved. In addition, where the design tasks are inadequately performed, the manufacturer will be involved in excessive costs for development of, manufacture of, and support for the project.

When a manufacturer initiates a project which involves design, he wants the result to be a product which he can sell at a profit. The selling price of the product is determined by the manufacturer only to a limited extent. To a much greater extent, the price of the product is determined by what the customer is prepared to pay for it, and this, in turn, is determined by what the product is worth to the customer. An airline knows how much it can charge passengers and will predict how much it costs to run an aeroplane, and it can then translate this informaton into how much it is worth paying for the aeroplane. A company which buys a new stillage system will have calculated how much it can afford to pay for the system if the investment is to show a profit[†]. A simplified version of the manufacturer's problem is, therefore, how to supply the product at a sufficiently low cost for there to be an adequate profit at the selling price that the customer is prepared to pay.

We may list those factors and activities which create cost for the manufacturer, and determine the disciplines in design which affect those costs.

3.1 THE COST OF MARKETING

Marketing is not merely the process of selling a product once it is built; it involves determining what the customer is likely to want to buy and in what

---

[†]These statements are an oversimplification of the calculation. The problem is discussed at greater length in Chapter 8.

quantities. The designer is frequently involved because, in any discussions with prospective customers, his expertise may be required to suggest possible solutions to customers' problems, to discuss costs of possible solutions or enhancements of existing designs, and to give opinions on the problems involved. Salesmen, in discussions with possible customers, may think it desirable to use designers to solve some of the customers' technical problems. At worst, this will gain the goodwill of the customer; at best, it will direct the solution of a customer's problem towards the purchase of equipment. Nevertheless, this use of design time costs money. Typically, salesmen will have a budget for the use of design hours which they will control.

Strictly speaking, therefore, such work is a cost borne by the sales department rather than the design department. Nevertheless, it is a cost which has to be met by the company, and one which requires the allocation of design resources. Typically a customer designing a new aeroplane may seek information from the manufacturers of navigational aids about recent developments, or the manufacturer of washing machines may seek advice from the manufacturers of storage systems. In such cases, questions may require the expertise of the designer to answer, but there is unlikely to be any commitment by the customer to an order at this stage. Maintenance of goodwill with possible customers, however, may well require that the service be offered.

Defining the requirements of a product may also be work with which the designers will assist the marketing staff, and this may be at such an early stage of the project that there is no forecast of payment for the work done.

Many projects are initiated by designers. This may be because the manufacturer has allocated a sum of money to 'open sky' research, which means that one or two members of the design team are given free rein to explore possible future products. More likely there will be some direction to the research – a team in a shoe-making factory may be required to study the possible application of new materials or adhesives, a team in an aircraft company may be required to study possible future methods of propulsion, a team in a company making storage systems may be studying new methods of stressing the racking.

Most of the above aids to marketing are required at a very early stage of any possible project, and obviously a high percentage of the work done will be, at best, only indirectly useful in that it may increase the customer's goodwill or improve designers' store of knowledge. The implication is, however, that a percentage of the design effort will be occupied permanently by work for which the expected pay-off cannot be defined.

3.2 SELLING COSTS

There are two different ways of selling a design. With many products, for example, cars, washing machines, racking, most machine tools, the product is designed, developed, and proved; the manufacturing facility is designed and built; the products are built; then the products are sold. There is no possibility of paying for the design until the products are made and sold, and the cost of design is simply one of the early investments in the project – much as buying a new machine tool might be. The designer would also be involved in the production of information which will help to sell the product. One would hope that the design costs would be recovered after a predicted number of the products had been sold, but it is common for a product to be designed, developed, and built but find no profitable market. In such a case the cost of the design must be written off. Frequently, in such cases, the cost of design is very small compared with the cost of tooling, and so perhaps is less remarked, although it should be asked whether more expenditure on design would have turned failure into success or would have predicted failure before money was spent on tooling.

At the other extreme the product is sold when only a small amount of design has been done. In this case the manufacturer has not sold the product so much as his ability to design and make it. Power stations, ships, military aeroplanes, and some aeroplane accessories are examples of products which are sold in this way. The customer makes known his requirements and the manufacturer authorizes sufficient design work to be able to make a technical proposal and state the price for which he will build the required product. Because only a fraction of the proposals will be successful (there will be competition), a fraction of the design effort will have no obvious pay-off and will have to be written off. This can be quite a large proportion of the design effort available. The fraction F which must be written off is given by

$$F = 1 - \frac{p}{p\,[1-f] + f} \quad \dagger$$

where p is the fraction of proposals which lead to orders and f is the fraction of the total design work that is done in an effort to get the order.

Typical values of p and f are 0.2 and 0.05 respectively, so that 17% of the design effort may be apparently wasted. This effort is not, in fact, wasted, because any attempt to reduce it would simply result in fewer orders. It is simply an expenditure that must be budgeted, although it may be possible to find an optimum amount of effort to spend on selling – an amount which reasonably provides the work needed to meet the company's objectives.

3.3 THE COST OF DESIGN

When a product has to be designed, clearly the cost of designing it must be faced. The problem is how much to spend. If too little is spent on design, there will be design faults which will result in high costs of manufacture and high costs of

† This formula and extensions of it are discussed in Chapter 6.

ownership and use. Spending more than some ill-defined optimum effort on design will not succeed in either ensuring that the best solution to the problem has been found or that the drawings will be without mistakes. Attempts to find all the errors or undesirable features in a set of drawing will not succeed, and will delay the issue of the drawings to the shops.

Each stage of design is more expensive than the previous one. The good idea which apparently solves the problem may take an hour to two to formulate; the scheme which is developed from this idea will take longer (perhaps a few hours, perhaps a few weeks); preparing the detail drawings and the assembly drawings will cost more than the scheme; manufacturing the prototype will cost still more; and developing the product will cost more again[†]. The cost of each stage of design does depend on the excellence of the work of the earlier stages.

A well-executed scheme will make it easy for the draughtsmen to produce the detail drawings. The use of standard components will reduce the draughtsmen's time, the manufacturing time, and the development time. Probably, more significantly, a good scheme will result in lower production costs than a bad one. The cost of the design is therefore made up of two parts: on the one hand we have the obvious direct costs of the man-hours involved, of the materials used, and of any equipment used[‡]. On the other hand we have the indirect costs added to later activities by the inadequate performance of an earlier activity. Most important of these is the unnecessary cost added to production by inadequate consideration of the design. This could perhaps be shown graphically as in Fig. 3.1.

This does not mean that the more we spend on design, the better. There will be diminishing returns, and a point will be reached where an extra pound's worth of design will save less than a pound in production costs. There will also be cut-off points dictated by the contracted delivery date, for if design continues beyond some determinable point there will be insufficient time to manufacture the product even if the drawings are without flaws. Chapter 12 discusses, in greater detail, ways of reviewing designs and ways of auditing the review systems which attempt to determine the point at which drawings should be released.

It is tempting to believe that in companies which manufacture large quantities of the products that they design, it would almost always pay them to spend more on the design than they do. The cost of recalling a large number of cars for free remedial work on brakes or steering would pay for a great deal of the

---

[†] The stages of design are discussed at greater length in Chapter 11.

[‡] The capital cost of equipment cannot be compared with the cost of labour and materials. We can, however, determine the annual equivalent cost of the equipment and compare this with the annual cost of labour and the annual cost of material. The annual equivalent cost of a capital investment is discussed in Chapter 8.

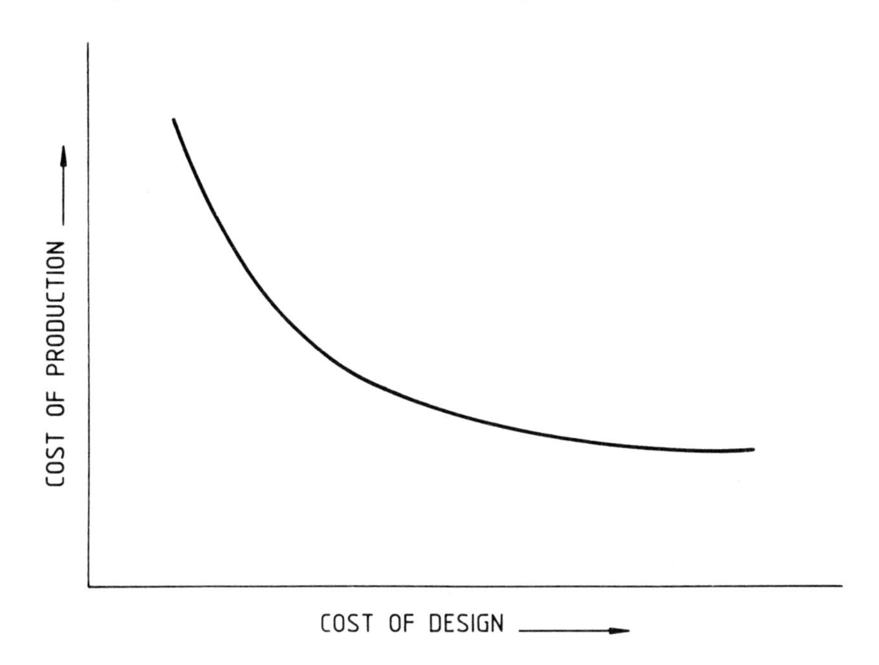

Fig. 3.1.

design effort needed to prevent the trouble in the first place. Unfortunately, a greater quantity of design will not necessarily avoid later embarassments unless the design effort is also of sufficient quality.

3.3.1 Some direct costs of design

We may list the main activities of design and agree that each will have its cost.

(i) Writing the design specification.
(ii) Preparing the scheme – the cost is always wholly in the cost of the skilled labour used.
(iii) Preparing the detail drawings and assembly drawings – the cost is almost wholly in the skilled labour used.
(iv) Making the prototype products – the cost is partly in the skilled, direct labour used, partly in the cost of equipment used, partly in the indirect labour (for example, in planning, scheduling, buying) used.
(v) Preparing the prototype testing and proving programmes.
(vi) Testing the prototype products – the cost is partly in the skilled labour used and partly in the cost of the equipment used.
(vii) Proving the prototype products – the cost is partly in skilled labour and partly in the cost of the equipment used.

(viii) Data collection and analysis – this is an important primary cost which can have significant secondary effects on the cost of development. The cost of data collection and analysis is mainly the cost of the skilled labour involved, but may include the cost of computer time.
(ix) Preparing service, maintenance, and operating manuals.
(x) Preparing the product support programme.
(xi) Devising the product support organization.

3.3.2 Some indirect costs of design

(i) Reduced sales – we have already seen that a fraction of the design effort must be written off since it does not result in sales. Poor performance will result in a lower fraction of the proposals being accepted by customers or a lower fraction of the products being sold. In either case there will be a smaller return to the manufacturer or a larger fraction of wasted design effort.
(ii) Litigation or the acceptance of losses through inadequate preparation of the specificaton – if the specificaton does not properly reflect what the customer wants, he may refuse to pay in full for the product. A comparable situation can arise if the processing of the specification is inadequate, so that work is done when the customer has not formally agreed that it is required.
(iii) Poor detail design or an unnecessary time spent in detail design because the scheme is either inadequately considered or inadequately instructs the draftsmen.
(iv) Unpredicted costs of prototype manufacture – these can arise from two causes. The designer may have not adequately considered the manufacturing methods when the scheme and drawings were produced, or the time constraint may have forced the designer to propose components which are not the most economical to manufacture. To some extent the first problem can be reduced by good design review procedures (although it cannot be eliminated). The second problem is really one of inadequately forecasting the cost of design. An example could be that a designer calls up a detail to be made wholly by machining from bar when he knows that a casting would be cheaper but would not be delivered in time to enable the production programme to be met. Since the decision to call up the less economical part is deliberate it should be predictable and allowed for in the predicted cost.
(v) Unpredicted costs of development and proving – to some extent the cost of development and proving is unpredictable or tests would be unnecessary. In many cases, however, a scheme will propose components or assemblies which are doubtful in performance and which can be replaced by others of known history. The use of standard parts and design reviews can reduce the problem. Inadequate liaison between the designer

and the development engineer will result in redundant tests or tests in illogical sequence. Most of this can be avoided by the preparation of sound test instructions. Proving tests, in particular, require careful design to establish the number of products to be tested to provide the necessary information, the number of products to be tested to destruction, and the extent to which the test environments reflect operating environments. The number of experiments required to predict life is discussed at greater length in Chapter 10. Testing costs may well be reduced if data are available from earlier products (data from the field or previous laboratory tests) which will demonstrate reliability and integrity or which will indicate areas of particularly sensitivity.

Among the costs which the manufacturer must consider are those for which the ownership of test equipment is responsible. Such equipment will have an equivalent annual cost (see Chapter 8) which will contribute to the overhead that the project must bear. Inadequate test facilities will mean that the product cannot be designed. Sophisticated, expensive, test facilities may pay for themselves quite easily, for they can reduce testing costs, permit cheaper product programmes, convince a customer of the integrity of the product and the manufacturing company's expertise, and in some cases be rented to other companies with the need for such facilities.

(vi) Unpredicted costs of manufacture – these are among the major unnecessary costs for which designers are responsible. Design review can help to reduce these costs if they ensure that the designer, when scheming, or the draughtsman, when detailing, liaises with manufacturing engineers.

In extreme cases, the designer may be required to design for specific manufacturing facilities. He may, for example, be concerned in the translation of his drawings into tapes for N.C. machines or the classification of his drawings for use in a Group Technology system.

Companies are becoming increasingly aware of the need for good liaison between the designer and the manufacturing engineer. Some maintain an 'up front' man, that is a manufacturing engineer, who is attached to the design department where his expertise is permanently available. This does not always work, because the designers will not necessarily consult the 'up front' man who, in any case, becomes alienated from his ex-colleagues. Some companies create a team to manage a design project, and this team will have in its membership at least a designer and a manufacturing engineer, and an important function of the project team will be to 'walk the product through the shops'. The use of a multidisciplinary project team (in some cases with a marketing engineer and a quality control engineer, as well as the designer and manufacturing engineer) seems to work in companies at all levels of technology, from the manufacture of washing machines to the manufacture of products for the aerospace industry.

Usually it is the direct cost of the operatives' labour that is seen to be wasted by poor design. Indirect labour may also be wasted, however, and inspection time is an example. A badly designed inspection procedure can cost much more to operate than one which is well designed. Examples are that a performance test can sometimes be replaced by the cheaper checking of dimensions, or the design is such that adjustments to meet the specification may be easily made (for example by the use of adjusting screws rather than packing washers). The discipline of designing the inspection procedure will also provide a yardstick for monitoring production methods (and possibly design quality), for it will provide the opportunity for measuring acceptance rates after production operations.

Many companies operate a system of 'scrap meetings' which are held to examine the ways in which components which fail inspection tests may be salvaged. One of the important outcomes of such meetings is the observation of high rejection rates for certain components and the discovery that these rates are a function of poor design which has created manufacturing difficulties. Clearly it is better if such faults can be found before manufacture commences.

The cost of inventory will be significantly affected by the designer's choice of materials and bought out parts. It will also be affected by the sub-assemblies he creates, because these will determine the work in progress inventory. Constant liaison between designer and planners is essential to ensure that sub-assemblies are essential in the building (or maintenance) process. Unnecessary sub-assemblies create redundant paperwork and redundant stores activities.

3.4 DISTRIBUTION COSTS

Although the method of delivering the product and the responsibility for delivering may be the subjects for a contract between the manufacturer and the customer, it is more likely that the manufacturer will have to ensure that his products reach the customer in good condition. A housewife who buys a washing machine will expect it to be delivered to the house in good working order and perhaps installed. If the RAF buys an instrument for use in an aeroplane, they will expect it to be delivered to their stores, without the need for inspection on arrival (although they may accept the need for special packing or assemblies to cope with the conditions met during transportation). Typical of such considerations is that an American manufacturer of washing machines instrumented a packing case to record the shocks received by their products when transported in railway box cars. This information enabled them both to design the product and its packing and to argue with the carriers about the conditions prevailing during transport.

3.5 PRODUCT SUPPORT

The manufacturer will be committed by the design specification or by contract to support his product in the field. Preparing the maintenance and operating manuals does to some extent provide the discipline necessary to force consideration of design for operation and maintenance. Because no maintenance or operating system will be perfect, modification will be necessary. Unpredicted failures will occur; unpredicted operating difficulties will occur; difficulties in repair will occur; spares will not be available because the pattern of breakdowns will differ from the predicted pattern; the maintenance procedures will be modified in the field as experience points to better organizational systems.

All this requires that there be good liaison between the operator and the designer. Usually this means between the service engineer and the designer. It means also that data on the maintenance and operation must be collected and analysed, and the analysis brought to the attention of the designer.

Modifications which arise from problems in the field are themselves expensive. they are expensive because, if the faults that inspire them are the designer's, the manufacturer will have to pay. He will have to pay for the new drawing work, the proving of the modificaton, the change to the manuals, the manufacturing involved in changing products manufactured to earlier drawings, the logistics involved, and for the possible loss of the customer's goodwill. The customer will find change expensive and will not welcome it. A system must therefore be used to control the modification of products[†].

3.6 FAILURE TO MEET THE SPECIFICATION

If the design is inadequate to the point that the product, when delivered, does not meet the specification, then the customer may reject it and demand his money back, he may accept it but at a reduced price, or he may accept it but only on the understanding that the manufacturer will continue to pay for design work and will modify the product at his own expense when the design problems have been solved. A further cost, which is difficult to assess, is the loss of future sales to a disappointed customer.

One such cost – product liability – which may involve high payments to injured customers or third parties, is becoming more important as legislation is introduced. Chapter 10 includes a short discussion on product liability.

3.7 FAILURE TO MEET COMPETITION

If designs meet their specifications but show a tendency to be poorer by some criterion than those of a competitor, business will decline. Although specifications are met, customers become aware of competing designs which are more elegant because they are lighter, simpler, or, for some other reason, cheaper.

---

† Such procedures and the paperwork involved will be discussed in some detail in Chapter 11.

A summary of the manufacturer's costs of design

I. *Budgeted costs of design*

An *indirect* cost must be recovered as an overhead
A *direct* cost may be charged to the product

Cost observed by manufacturer	Cost is direct or indirect	Effect of too little expenditure	Effect of too much expenditure	Remarks
(1) Aid to sales; help in marketing	Indirect	Reduced sales	Effort could be better used on jobs for which orders are assured or expected	Cost of the work as a fraction of the total design effort is $1 - \frac{p}{p(1-f)-f}$ where p is probability of getting an order and f is the fraction of the design done before an order is placed
(2) Preparing the specification	Indirect if the product is not sold Direct if the product is sold	Dissatisfied customer Expensive modification Redevelopment	Only small costs added to the product	
(3) Drawing the design scheme	Indirect if the product is not sold Direct if the	Reduced sales Increased costs of development and	Only small costs added to the product although ultimately the	Excellence is more likely to result from the employment of good quality designers

	product is sold	manufacture	delivery programme will be delayed	than from the extended use of poor quality designers. Design review by appropriate staff (manufacturing engineers, etc.) may find obvious errors
(4) Preparing the detail drawings	Usually direct	Expensive development Expensive manufacture Expensive product support	Only small costs added to the product although ultimately the delivery programme will be delayed	Excellence is more likely to result from the employment of good quality draughtsmen than from the extended use of poor quality draughtsmen. Design review may find obvious errors.
(5) Making the prototype	Usually direct			Time is largely dictated by the quality of design in the drawings but excess time spent in manufacture could be the result of poor planning or poor buying
(6) Preparing test schedules	Usually direct	Unnecessary time spent in development and proving tests	Only small costs added to the product	
(7) Testing and developing the prototype	Usually direct			Time is usually dictated by the test schedules and the equipment available

A summary of the manufacturer's costs of design – *continued*

Cost observed by manufacturer	Cost is direct or indirect	Effect of too little expenditure	Effect of too much expenditure	Remarks
(8) Proving (i.e. demonstrating life & integrity)	Usually direct			Time is usually dictated by the test schedules and the equipment available An investment in special test equipment for environmental testing will be essential
(9) Data Collection and analysis	Indirect	Unnecessary testing		This activity may be essential to demonstrate that contracted obligations have been met
(10) Preparing manuals	Direct	Unnecessary or unrecoverable costs in product support		
(11) Preparing the product support organization and the spares programme	Direct	Unnecessary or unrecoverable costs in product support		

II. *Costs caused by inadequate design*

(All must be indirect)

Cost observed by Manufacturer	Possible causes	Possible remedies	Remarks
(12) Reduced sales	Inadequate design effort in selling or marketing Poor reputation caused by earlier, poorly designed products	Spend a higher proportion of design effort in marketing or assisting sales Improve the quality of designers	
(13) Excess time producing detail drawings and assembly drawings	Inadequate time or quality of effort given to the preparation of the scheme	Spend more time producting the scheme or allocate a higher quality of effort to the scheme or allocate a higher quality of supervision to the draughtsmen Review the scheme	'In service' training of designers can help Design reviews (of the scheme) can help
(14) Excess time spent manufacturing the prototype	Inadequate time or quality of effort given to the preparation of the scheme Inadequate time or quality of effort given to the preparation of detail drawings Poor buying	Increase drawing time or raise the quality of effort Review the scheme Review the detail and assembly drawings Use more standard parts	The review body will include manufacturing engineers and buyers

A summary of the manufacturer's costs of design – *continued*

Cost observed by Manufacturer	Possible causes	Possible remedies	Remarks
(15) Excess time spent developing & proving the prototype	Inadequate time or quality of effort given to the preparation of the scheme, detail drawings, or test schedule Inadequate test facilities	Increase drawing time or raise the quality of effort. Review the secheme Review the detail drawings Review the test schedule Use more standard parts Buy appropriate test facilities	The review body will include manufacturing engineers and development engineers
(16) Excess time spent in manufacture	Inadequate time or quality of effort given to the preparation of the scheme or detail drawings Inadequate manufacturing facilities	Increase drawing time or raise the quality of effort Review the scheme Review the detail drawings Use more standard parts Buy appropriate manufacturing facilities	The review body will include manufacturing engineers, planners, and buyers
(17) Excess time spent in inspection or production tests	Inadequate time or quality of effort given to consideration of inspection procedures by the designer	Increase drawing time or raise the quality of effort Review the scheme Review the detail drawings	The review body will include development engineers and inspectors

(18) Excessive rejection of components and assemblies	Poor quality of design	Raise the quality of effort Review the scheme Review the detail drawings Hold scrap meetings	The review body will include manufacturing engineers, inspectors, and development engineers
(19) Excess inventory or work in progress	Poor design of assemblies Poor buying Poor production planning and control	Review the detail and assembly drawings	The review body will include planners and, buyers
(20) Damage during distribution	Poor specificaton writing Inadequate development	Package design	It will be necessary to collect environmental data
(21) Expensive product support	Poor design of assemblies Poor specificaton of the environment Inadequate knowledge of component life Poor manual writing Poor design of spares system	Increase time or quality of effort allocated to the preparation of manuals and to the design of product support systems	Sometimes the writing of manuals and the compiling of spares supply systems is regarded as a specialist function within the design organization Good liaison with service engineers & customers is essential
(22) Expensive modifications	Poor design generally Customer changes requirements	Improve quality of designers Establish a control system – where the modification arises from a changed requirement the customer will pay	

A summary of the manufacturer's costs of design – *continued*

Cost observed by Manufacturer	Possible causes	Possible remedies	Remarks
(23) Falling orders	Poor design generally	Improve quality of designers	
(24) Idle time	Poor project selection	Introduce a project appraisal procedure related to the resources available	Idle time is rarely obvious. Usually it is demonstrated by the apparent high consumption of man-hours by comparitively simple tasks (i.e., men allocate their time to the jobs available).

3.8 THE COST OF IDLE TIME

A problem, which arises particularly where a company is concerned with large projects of high technology, is one of maintaining an even work load. Consider the design of an advanced passenger aircraft or an atomic power station. Parts of the work will be more difficult than others, or, if not more difficult, will nevertheless require more effort. Designing and detailing the structure of the aeroplane will require many more designers and draughtsmen than, say, designing the cabin decor. Similarly, manufacturing the aeroplane will require distributions of skills, times, and capital resources which will be far from constant. We are unlikely to be able to fire men when we do not need them and hire them when we do; or if we can, it will be expensive process. We are committed, then, to paying idle men (and to paying for idle machines because the use of capital resources is also unlikely to be uniform). This problem may be solved (to some extent) by introducing other projects as resources become available; but, of course, it will be nearly impossible to find further projects which exactly use the idle time on the first project. The problem has some features in common with that of job shop scheduling, and will be discussed in Chapters 6 and 9.

SUBJECTS FOR DISCUSSION

(1) How much should a designer be held responsible for high manufacturing costs?

Give an example of an unnecessarily high product price for which bad design has been partly responsible.

(2) How important is the designer's role in selling?

How should this selling work be (a) budgeted for, and (b) paid for?

(3) Is there a difference between the designer and the detail draughtsman?

(4) List some activities, other than scheme drawing, in which a designer must be involved.

(i) Preparing test schedules
(ii) Briefing technical authors
(iii) ?
(iv) ?

(5) A specification has been described as a complete detailed technical description, supplementing a set of drawings to permit purchase or manufacture or construction. What part, if any, do economic factors play in specification writing?

(6) Imaginative design frequently draws engineering designers near to the limit of their knowledge and experience. This results, from time to time, in failures which are costly and an embarrassment to the designers concerned. How far should designers go in jeopardizing dependability in order to achieve economy of form and material?

CHAPTER 4

What society wants

This chapter discusses the requirements imposed by society – the legal requirements imposed by safety considerations or by possible damage to third parties and the influence of consumer groups. Costs are involved in meeting such requirements; greater costs may be involved in failure to meet them.

The customer and the manufacturer are not the only people involved when a designed product is bought and sold. Malfunction of the product could cause damage to the life or property of a third party; an ugly, noisy, or smelly product may damage the environment for others; consumer protection organizations may put constraints on what the customer will buy.

A house designed by an architect must meet the requirements of the building regulations [1], whatever the customer is prepared to accept.

An aeroplane must have a certificate of airworthiness before its purchaser is allowed to put it into airline service.

A motor car may not be sold unless its design meets certain safety requirements [2] that have not generally been dictated by the immediate customer.

A bad report in *Which* [3] is likely to reduce the sales of a washing machine.

A determined group of people can prevent the siting of an airport in Cublington [4] and perhaps influence the further technological requirements of aeroplanes.

The designer, then, has constraints imposed upon him that are not generated directly by the customer. These constraints are not necessarily unwelcome, because they can be a help to the designer rather than a hindrance as, for example, where a British Standard helps him to design a safe pipe flange. Failure to meet these constraints, however, can cost the designer's company millions of pounds.

Consider the following sources of constraints that are imposed on the designer:

(i) the imposition by government departments of safety requirements and codes of practice,

(ii) the imposition of punitive damages by courts in product liability cases,

(iii) the resistance of consumers to badly designed products when they are prompted by consumer protection groups, and
(iv) the resistance of the public to damage to their environment.

As example of constraints imposed by government departments we can cite those which arose from enquiries into the collapses of the Cleddau and Yarra bridges [5] and the Flixborough disaster [6]. The recommendations of the Merrison report led to the considerable expense of checking and, in some cases, modifying the design of a number of bridges already completed. The report also recommended that design procedures be modified, drew attention to the contractual statement of engineering responsibilities, set out design rules (while placing responsibility on the designer not to assume that compliance with the design rules alone would necessarily achieve a satisfactory performance), influenced the preparation of the British Standard *Code of Practice* for bridge design, and required the designer to be experienced in the field of stiffened plate steel structure design.

If we consider, more particularly, the report on the most recent of these disasters, Flixborough, we see further evidence of the responsibilities imposed on the designer, and how failure to appreciate those responsibilities is very expensive. The Flixborough disaster was an explosion, of warlike dimensions, at the Flixborough Works of Nypro (UK) Ltd on 1 June 1974. That this disaster was costly is obvious when we read that the works were virtually destroyed, 28 people were killed, 36 people were injured on the site, many were injured outside the works, and 1821 houses and 167 shops suffered damage.

The Court of Enquiry was appointed under Section 84 of the Factories Act 1961, although it is probable that this function of the Factories Act is now that of the Health and Safety at Work Etc. Act, 1974. The designer, as much as anyone else, must meet the requirements of the Health and Safety at Work Act, and it is interesting to note that, in many cases, the Act does not spell out what is safe but places responsibility on the managers and men to determine what is a safe machine and what is safe behaviour. The same is true of Occupational Safety and Health Administration in the United States.

A major cause of the disaster was thought to be a badly designed modification to the plant. The modification was a 'dog leg' pipe which replaced a reactor that had been removed for inspection. The flanges at the ends of the dog leg were connected to flanges in the original system so that in the modified layout, a bellows expansion joint was connected to each end of the new dog leg pipe.

The design was critized from many points of view, and among the criticisms were that

- '. . . . . no reference was made to the relevant British Standard (BS 3351:1971) or any other accepted standard;
- no reference was made to the designer's guide issued by the manufacturers of the bellows; . . . . .'

Both the British Standard and the designer's guide could be regarded as constraints which were imposed on the design, although it might be argued that they are really aids, providing the designer with necessary knowledge. The British Standard is probably somewhere between a constraint and a guide because, while it provides the designer with useful information on how the job should be done, it also dictates how the job should be tested. Designing to a British Standard may not always be mandatory, but a designer who ignores such a publication would have to be able to demonstrate that what he had done was as acceptable as what was suggested by the British Standard. Similarly, the designer who ignores the recommendations of the supplier of a bought out part would be very vulnerable if he could not demonstrate that he had a good reason for doing so.

The report of the Court of Enquiry demonstrates other constraints which are imposed on the design. It is argued, for example, that while 'No plant can be made absolutely safe any more than a car, aeroplane, or home can be made absolutely safe', the risks must not exceed '. . . . what at a given time is regarded as socially tolerable, . . .', and identifying risks and reducing them to tolerable levels '. . requires a conscious and constant effort on the part of everyone including those on the shop floor and their Unions'.

We see here the responsibility of the designer to assess the risk, and the clear involvement of others (other than the customer and supplier) in defining the risk. Although the Court of Enquiry was largely concerned with the disaster, and the modification is regarded as a major cause of the disaster, there is a clear admonition to the first designer of any product to consider constraints that have not been specified by the customer. The report offers a clear suggestion (Para. 203) that

- 'In many cases, there might be a better return (with regard to safety) from expenditure on making the original plant safer than by providing elaborate safety systems to deal with potential inadequacies'.

There were other comments which do not refer to constraints imposed on the designer but might, if noted, save a manufacturer or customer money or the pain of giving evidence at some future court of inquiry. Among these were references to

- the rewards of paying greater attention to the ergonomics of plant design,
- the conflicts which could arise between safe operation and the need to make a profit, and
- the desirability of paying attention, during design, to the terotechnology† of the plant.

†'A brand of technology that deals with the efficient installation, operation, and maintenance of equipment' (*Longman dictionary of the English Language,* 1984).

This example shows that the designer must write into his specification many requirements and conditions which do not derive directly from the customer's needs. Many of these requirements are perhaps nearer to design aids than to constraints, but they cannot be ignored. It is not possible to list all such requirements, because they vary from country to country and from industry to industry. Some requirements are concerned with safety and some with protection of the environment. It is the duty of the designer to acquaint himself with those requirements which are relevant to his field, and failure to do so can result in great cost to the manufacturer, the customer, or the country (the costs of the Flixborough Enquiry were expenses of the Minister in the execution of the Factories Act). Under the Health and Safety at Work Act Etc., the designer himself is not immune from penalty if he does a bad job.

The costs imposed by meeting (or in some cases, by not meeting) constraints imposed by government bodies can be very large. The cost of such disasters as Flixborough may be reflected in the costs of insuring plants; millions of dollars are at stake in determining the cost and responsibility for the nuclear accident at Three Mile Island on 28 March 1979 (although it looks as though the local consumer of electricity will end up by paying much of the money cost), and the estimated $500 000 for replacement generation [7] will be a very small part of those costs as the owners of the plant expect to spend $400m to put the plant back into service; the costs to the owners of DC-10 aircraft generated by the Federal Aviation Authority's temporary ban on their use is measured in millions of pounds (although whether the DC-10 problems were generated by design faults is questioned); the makers and operators of the DC-10 which crashed on 25 May 1979 have been reported (BBC news item 15 August 1979) as offering $60 000 000 to the dependents of those killed.

Product liability cases are increasingly common and a recent *Guardian* (18 July 1979) reports the award of $1 183 322 to the family of a man who died when his Jaguar XJ-6 was involved in an accident five years ago. 'The jury returned a majority verdict and said: "We find for the plaintiff and against the defendant, British Leyland".' During the trial, an issue was the siting of the car's fuel tanks. The newspaper report also cites cases in which '. . . $2.7 million was awarded to a man disabled in a Honda car which the jury ruled was structurally unsound and payments which Ford faces in the Pinto cases which hinge on the rear siting of the petrol tank. Ford are already said to have lost more than $50 million in recall costs for the Pinto, they have paid $100 millions in one case and face another 75 cases'.

Vaughn [8] discusses some aspects of product liability as it obtains in America and gives some examples, including one of liability resulting from failure to forecast the effect of the environment on the material used.

The law on product liability in the United States is, perhaps, more punitive to the designer than in Britain, although that is of small comfort to a British manufacturer selling his goods in America. In Britain, the customer may have a claim against the seller if what he buys is not fit for the particular purpose for which it is bought, although it would have been necessary to have made known to the seller the particular purpose that the product was bought for and to have relied on the seller's skill. Again, when goods are sold by description the customer may have a claim against the seller if what he buys is not of merchantable quality. The seller is not always the manufacturer, however, although it is likely to be the case where a product is designed specially for a customer. Where the manufacturer does not sell his product directly to the customer he, nevertheless, has a duty to the customer and to third parties who may be injured, and these duties may arise through guarantees or advertisements which may lead the customer to expect something better than he gets. Perhaps more important in the context of design is that the manufacturer may be responsible for damage caused by negligence, and negligence can be in design.

It could be argued that product liability is an example of costs to the customer which are passed on to the manufacturer when the design (or some other action of the manufacturer) of the product is found to be at fault. Nevertheless, this is an area in which the law plays a part and in which the law is beginning to define some of the constraints which must be recognized by the designer. In fact, the manufacturer may be responsible for damage to a third party (a bystander injured in an accident to a car with a design fault, for example) if the design is found to be at fault.

The costs of design faults discussed above have involved the intervention of the law or of government bodies. Many bodies without any official standing exist, however, to protect the consumer and, incidentally, to penalize the bad designer or his employer.

During the past twenty or so years, consumers have become increasingly critical of badly designed products. One of the most famous champions of the consumer is Ralph Nader in the USA, who pioneered campaigns for safer motor cars; campaigns which have led to design changes [9]. Perhaps even more effective, although less dramatic in impact, are groups such as the Consumers' Association who publish the magazine *Which* monthly, in which products are reviewed and may be selected as bad buys, may be criticized as being unsafe or unreliable, or given poor marks for performance. In 1975, 650 000 members of the Consumers' Association received *Which* every month, and it is likely that many more read it. It is difficult to believe that the judgements of the Association do not influence the sales of a product. What designer would like to see the television set that he has designed given the lowest grade (out of 17) for picture quality, the dishwasher he has created given a poor rating for drying and poor for washing, the juicer that he has designed listed as potentially dangerous? Who would not want to see his iron listed as the best buy or his electric light bulb as having no failures in less than 700 hours out of fifteen bulbs tested? The consumer is becoming more knowledgeable and more demanding, and his special interest magazines guide him in the choice of purchase. An example is articles in *What*

Hi-Fi? [10] which not only review products but discuss the code of practice developed by the Radio, Electrical and Television Retailers Association and the Office of Fair Trading. Spares holding by the retailer (and spares provisioning starts with the designer), reliability, and cost of repairs of hi-fi systems are discussed with, not surprisingly, a demonstration of the possible trade-offs between reliability, maintainability, and spares availability.

There is an increasing awareness of the damage that is being done to the environment by the poor application of high technology. Pressure groups with real or imagined grievances can prevent the application of a design or force expensive redesign. The objection of groups of people to the noise they expected from Concorde was a factor which reduced the number of aeroplanes eventually sold. The resistance of people to the dumping of atomic waste may force a re-think or even an abandoning of the use of atomic power.

An example of the effect of environmentalists on designers is seen in heavy lorry design in Britain. Such societies as the Council for the Preservation of Rural England and the Society for the Protection of Ancient Buildings have pointed to the high social costs caused by heavy lorries [11]. These costs result from vibration, impact damage, accidents, road deaths, and visual intrusion. A campaign was waged essentially to prevent the appearance on British roads of commercial vehicles with gross operating weights above 44 tons. The lobbying of members of Parliament by interested societies, the work of the Road Research Laboratory analysing accidents and causes of road wear, and the research of working parties set up by government departments eventually led to the determination of the British Government to resist proposals (in the E.E.C) to increase maximum axle weight and the size of lorries. A similar campaign was fought by the Wing Airport Resistance Association successfully to prevent the implementation of the *Roskill Report's* [12] recommended siting of the third London airport. The result of the campaign of lobbying and efforts to sway public opinion was general acceptance of the belief that any inland site within reach of London is a political impossibility. This example was given to show the power of the general population when it campaigns against what it believes to be likely to damage the environment. The result of the campaign may well have far reaching effects for designers, however, because realization of the power of society could change the concept of air travel and, indeed, change the design of aircraft. One incidental result of the campaign against Roskill was the discrediting of much of the cost benefit analysis used.

A more obvious and direct instruction to the designer arises from legislation to protect the environment in the United States. One of the best known environmental problems in the United States was the prevalence of fog caused largely by emission from cars in the South Coastal Basin of California. California, and later the federal government, imposed progressively more severe regulations on the permissible emission from motor cars of unburnt hydrocarbons, carbon monoxide, oxides of nitrogen, and other pollutants which were causes of the photochemical 'smog'. This smog does not only create a fog but causes eye irritation, damages lungs, and destroys vegetation. In addition, pollutants from motor car exhausts cause cancer. There are many actions that the designer may take that will reduce the emission of pollutants. These range from reducing the size of motor cars to adding exhaust gas reactors, but few methods are available which will solve the problem by themselves, and even partial solutions create further design problems. A typical modification to engines is the catalytic reactor – which is poisoned by tetraethyl lead. This means that an engine using a catalytic reactor must use lead-free petrol and must therefore have a low compression ratio to use fuel of a low octane value or have access to high octane petrol that is lead-free and expensive. The solution has generally been to reduce the compression ratios of engines and to use lead-free petrol (through a small petrol tank entry to prevent the insertion of the larger, older, petrol pump nozzles carrying leaded fuel). Other obvious contributory solutions are to increase performance while reducing fuel consumption, and this may be done by improved engine design, by improved transmission, by reducing body weight, etc. But however he does it, the designer must solve the problem of reducing pollutant emission from his motor car engine to the level permitted by U.S. Congress (Public Law 91–604) if he wants to sell his car in the United States. Ten years of engineering have already gone into solving the problem which was, perhaps, largely caused by profligacy in the early generation.

The costs of satisfying society may seem high; they include

- the costs of meeting tougher specifications (including the cost of finding or writing those specifications),
- the cost of analysis and testing and, in some cases,
- the cost of abandoning a project after money has been spent and it has been shown that the difficulties cannot be overcome at an acceptable cost.

The costs of not satisfying society are probabilistic but likely to be much higher; these include

- crippling damages if the product fails dangerously,
- loss of market if the consumer organizations do not like your product, and
- the cost of a product that society rejects.

Summarizing the requirements of society we may ask:

is there a British Standard,
a manufacturers instruction,
a Code of Practice, or
a published general requirement that must be considered in your design?

Your professional institution (e.g. the Institution of Mechanical Engineers, Institution of Civil Engineers) will know.

Have you checked whether the Health and Safety at Work, Etc. Act affects your design?

The Health and Safety Executive will know.

Will your design damage the health or property of its user or a third party? Will you be told its faults as a defendant being sued for damages?

Have you done a failure analysis of your design?

How would a consumer protection organization test your product? Will your tests be as informative?

What effect will your design have on the environment? Will it be bad enough to prompt someone (or a pressure group) to fight its use?

SUBJECTS FOR DISCUSSION

(1) Discuss the merits of international and national standards – how do they contribute to economic, viable designs? When should a company set its own standards?

(2) Industry extends and improves the fabric and facilities of society as well as providing employment for many people. Should industry be used as an economic regulator or its skilled resources kept fully employed? Discuss the advantages and disadvantages of these alternative policies.

(3) To what extent is it reasonable to attempt to protect the customer from bad design, by legal constraints?

(4) Give an example of an enquiry, generated by public concern, which may influence designer's work.

(i) The Sizewell B enquiry
(ii) ?

(5) Could the activities of 'Green Peace' influence the work of designers?

(6) How readily should a designer accept codes of practice?

REFERENCES

[1] *Building Regulations,* 1976, HMSO, London.

[2] British Standards, particularly B.S.A.U. series, British Standards Institution, British Standards House, 2 Park St., London, W1.

[3] *Which* is the magazine of the Consumers' Association, 14 Buckingham St, London, WC2 N6DS.

[4] *Hansard,* Parliamentary Debates, House of Commons Official Report, Fifth Series, 26th April 1971.

[5] A. W. Merrison, *Report of the Committee of Inquiry into the Basis of Design and Erection of Steel Box Girder Bridges,* London, HMSO 1973.

[6] Roger Jocelyn Parker, Q. C. (Chairman), *The Flixborough Disaster,* Report of the Court of Enquiry set up by the Department of Employment, London, HMSO, 1975.

[7] *Nuclear Engineering International,* **24,** No. 284, London, April 1979.

[8] Richard C. Vaughn, *Quality control,* Iowa State University Press, Ames, Iowa, 1974.

[9] William Haddon, Jr., 'Our delicate costly cars', *The consumer and corporate accountability,* Ralph Nader (Editor), Harcourt Brace Jovanovich, New York, 1973.

[10] *What Hi-Fi?* **8,** No. 4, pp. 109–111, January 1984. Haymarket Pub. Co., London.

[11] Kimber, Richardson, & Brookes, The Juggernauts: public opposition to heavy lorries, In: Kimber & Richardson, *Campaigning for the Environment,* Routledge & Kegan Paul, London, 1974.

[12] Mr Justice Roskill (Chairman), Commission on the Third London Airport, *Report,* HMSO, London 1971.

CHAPTER 5

The design process

This chapter sketches the process of evolving a finished design from the initial creative solution to the problem posed. The innovative process is compared with the iterative process in which new design solutions are obtained by modifications to existing solutions to similar, simpler problems. The need for communication by the designer and the complex nature of the total design project are emphasized, and the design brief is discussed.

Some main general design objectives are discussed.

5.1 THE NATURE OF DESIGN

All design work is to some extent creative, requiring intuition as well as technical knowledge, but some needs much less invention than others, and many designs tend to evolve gradually as a series of modifications of the same basic plan. Most engineering products are designed to fit into a system. An electrical component may be designed to fit into the electrical supply of a house; a lorry may be designed to fit into Europe's road and legal systems; a hydraulic pit prop may be designed to fit into a mining system. As systems change, so the designs of the products must evolve to cope with the changes.

At the other end of the spectrum there are innovations. Here, mere modifications will not do, and the risks of failing to solve the technical problems or of spending too much money on their solution increase.

Between these two extremes there may be combinations of the new and the old. Indeed, it may be argued that the difference in creativity between one design and another is always one of degree; that the Watt engine would not have come about without the system changes brought about by the Newcomen engine, that nuclear power fitted into an existing and developing electricity generating and distributing system. Nevertheless, we have only to consider the decade of failure of the Watt engine, during which money had to be poured into development to realize how greatly the financial risk can vary from one design project to another, and how that risk is associated with technical problems.

To manage design adequately we must develop special management procedures so that we can predict the likely outcomes and economic consequences of possible technical decisions.

Engineering design covers a wide field, varies from product to product, and the process depends on the degree of innovation involved. The differences of procedure are mainly differences of emphasis, however, and all design follows a common sequence of activities. Design is almost always an iterative, problem-solving process which has some similarity with the scientific method (see Fig. 5.1).

In science, the observation of phenomena leads us to formulate a problem. In design, our observation leads us to the perception of a need. In science we attempt to generate possible solutions to the problem that we have formulated,

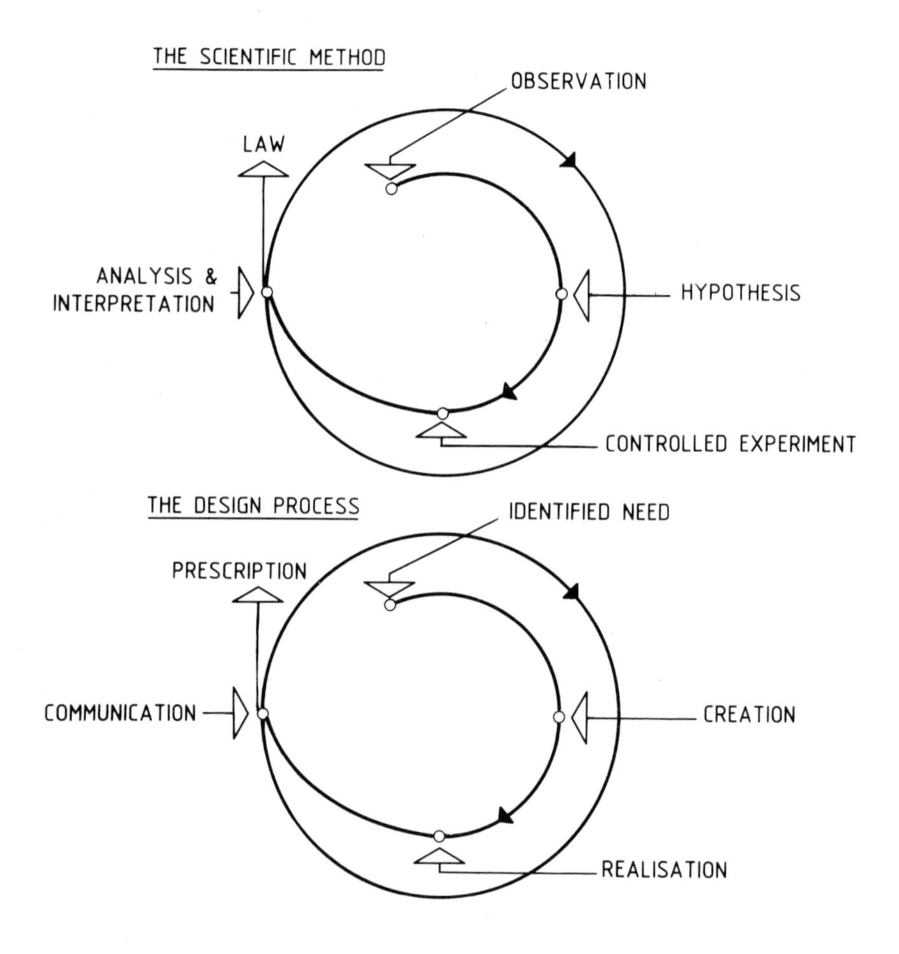

Fig. 5.1 – The scientific method.

while in design we attempt to generate possible ways of meeting the need that we have perceived. In science we attempt to judge our possible solutions by experiment (note that the experiment must be devised so that our contending solution will be proved wrong if it is false; we cannot prove that our solution is correct). In design we test our proposed ways of meeting the need, and our tests must be devised so that, if possible, they prove that our proposal will not work. In science, it is probable that experiment will show that our suggested solution to the problem is inadequate and will have to be modified. In design, it is probable that test will show the proposal to be inadequate and that modification or even redesign is necessary.

In one sense, it does not matter if the scientist is wrong, because he is merely seeking satisfaction and the admiration of his colleagues and, in any case, history shows us that the scientist never will be right and his quest is an unending one. The designer, however, lives by his ability to solve his problems sufficiently well to sell the results. If he is not successful, he will not be worth employing.

The final stage of design is communication, that is giving the instructions for making the product. The cycle will start with verbal statements, specifications and contract documents which will form the design brief. These will be the raw material from which designers produce layout drawings and sketches, etc. Finally, the manufacturing instructions will be conveyed, mostly by means of detail drawings in two dimensions, but also by written manufacturing specifications, three dimensional models and programmes for numerically controlled machine tools.

It will be seen that every design project follows a definite pattern, starting with abstract ideas such as identified needs and moving to ideas about things which are translated into an engineering prescription that transforms resources into usable products. As we proceed from ideas to the hardware, so the language changes.

5.2 THE LANGUAGE OF DESIGN

The types of language used are shown in Fig. 5.2. Creative work is often carried out by idea-sketching (talking to oneself in pictures). Generally, the higher the degree of abstraction in the language, the greater is its applicability to creation. That is, original thinking in its initial stages. The abstract languages are useful for problem solving and can reduce the time needed, but more concrete languages are needed for communication with other people. A new product (see Fig. 5.3(a)) will start with a statement of the need or the perception of an opportunity to exploit a new development, which will lead to a design requirement specification. After examination of this, possible ways of meeting it would be considered; one of them would finally be selected; and the chosen design detailed and produced. Feedback from the customer in the form of defect reports would provide further information for improved designs. The process may be very complex,

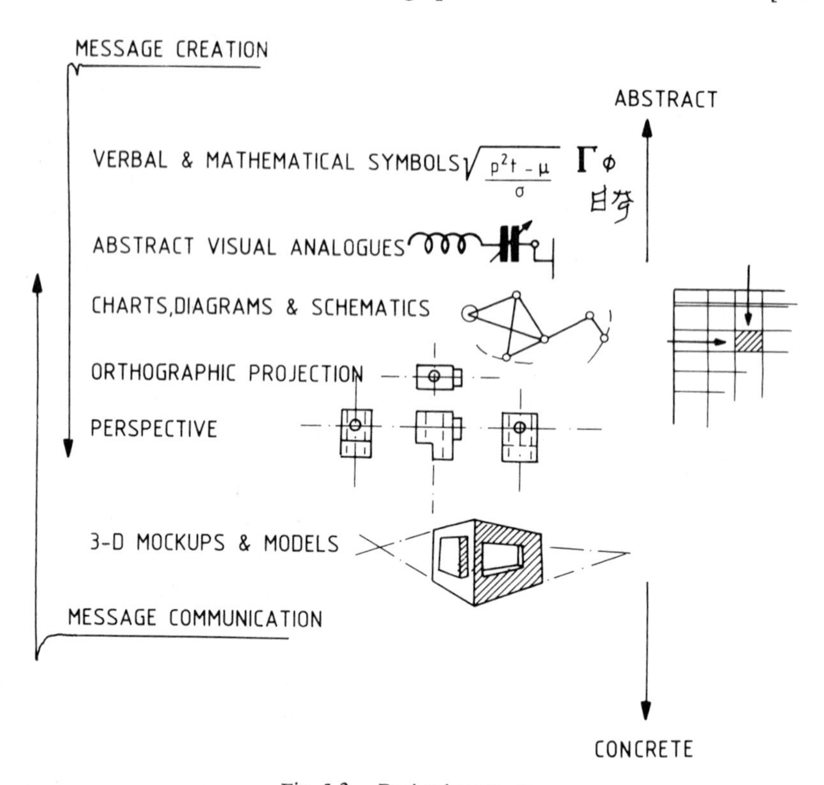

Fig. 5.2 – Design languages.

and a number of stages are passed through, some of which are suggested in Figs. 5.3(b), 5.3(c), and 5.3(d).

In these diagrams the main headings represent the main stages set out in Fig. 5.3(a). Hence in the box New Design in Fig. 5.3(a) we have I.F.E.S.R. standing for Investigate, Formulate, Examine, Select, Record as shown in Fig. 5.3(b).

Performing these tasks costs the money directly required to pay for the man-hours. Performing them badly leads to the production of goods which are not profitable to own or to manufacture. It is necessary, therefore, to formalize the design process into a system in which the outcome and cost of each stage may be monitored and controlled. This is discussed in Chapters 9 and 11.

In looking briefly at this process, one can see how easy it is for the communication of the design to be muddled up with its creation. It is important to recognize that these are essentially different phases of the process. Creative design requires accurate input information, from both inside and outside the design department, and all communication is best studied in terms of the recipient. Ideally, the engineering design message would be conveyed numerically, suitable for direct input to a N.C. manufacturing machine. But in most cases design intent has to be conveyed to human beings.

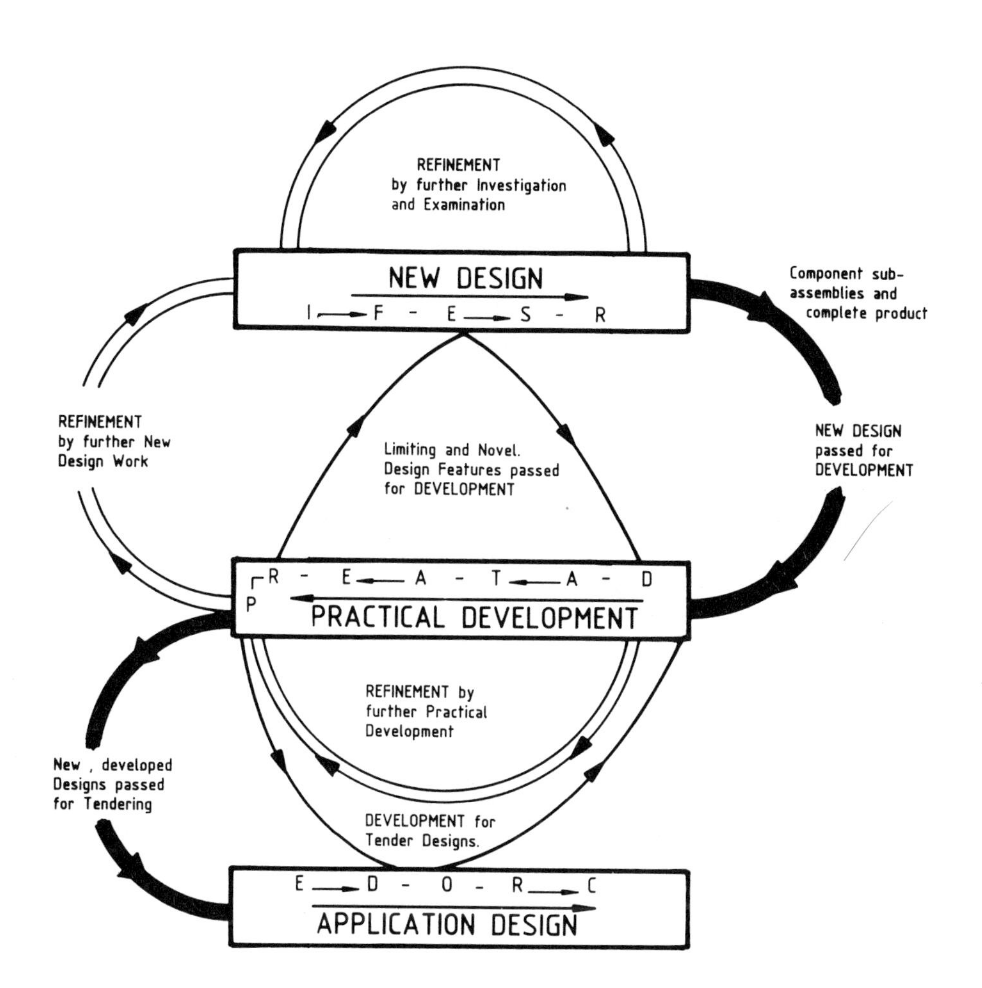

Fig. 5.3(a) – The design cycle.

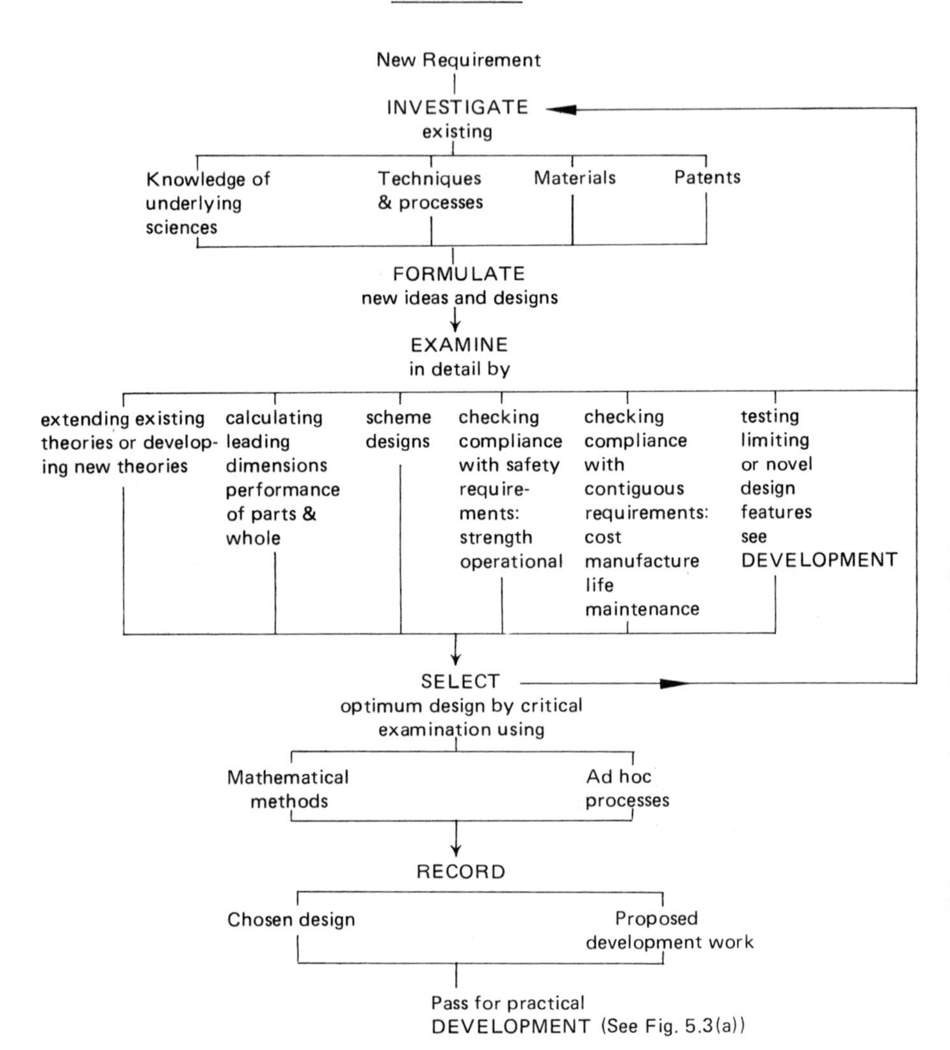
CREATIVE DESIGN WORK
NEW DESIGN
New Requirement
INVESTIGATE existing
Knowledge of underlying sciences
Techniques & processes
Materials
Patents
FORMULATE new ideas and designs
EXAMINE in detail by
extending existing theories or developing new theories
calculating leading dimensions performance of parts & whole
scheme designs
checking compliance with safety requirements: strength operational
checking compliance with contiguous requirements: cost manufacture life maintenance
testing limiting or novel design features see DEVELOPMENT
SELECT optimum design by critical examination using
Mathematical methods
Ad hoc processes
RECORD
Chosen design
Proposed development work
Pass for practical DEVELOPMENT (See Fig. 5.3(a))

Fig. 5.3(b).

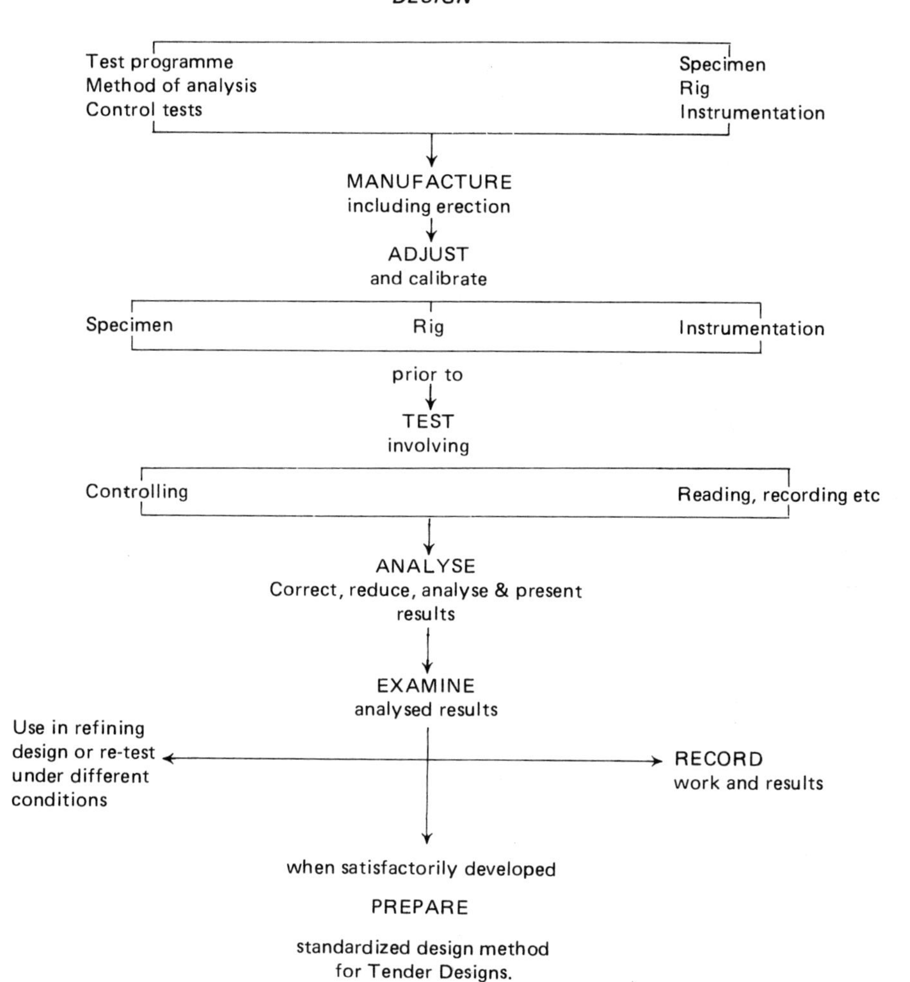
PRACTICAL DEVELOPMENT
Scheme for Prototype
DESIGN
Test programme
Method of analysis
Control tests
Specimen
Rig
Instrumentation
MANUFACTURE
including erection
ADJUST
and calibrate
Specimen
Rig
Instrumentation
prior to
TEST
involving
Controlling
Reading, recording etc
ANALYSE
Correct, reduce, analyse & present results
EXAMINE
analysed results
Use in refining design or re-test under different conditions
RECORD
work and results
when satisfactorily developed
PREPARE
standardized design method for Tender Designs.

Fig. 5.3(c).

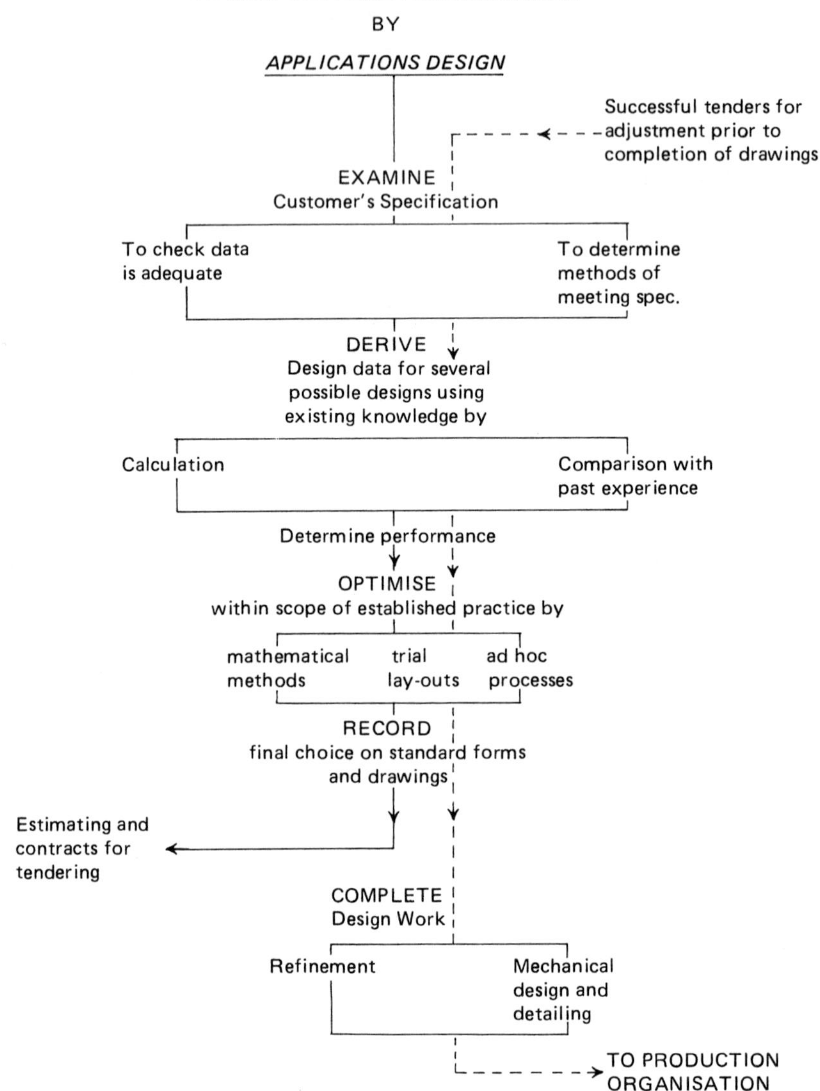
FULLY DEVELOPED NEW DESIGN IS ADAPTED
TO MEET CUSTOMER REQUIREMENTS
BY
APPLICATIONS DESIGN
Successful tenders for
adjustment prior to
completion of drawings
EXAMINE
Customer's Specification
To check data
is adequate
To determine
methods of
meeting spec.
DERIVE
Design data for several
possible designs using
existing knowledge by
Calculation
Comparison with
past experience
Determine performance
OPTIMISE
within scope of established practice by
mathematical
methods
trial
lay-outs
ad hoc
processes
RECORD
final choice on standard forms
and drawings
Estimating and
contracts for
tendering
COMPLETE
Design Work
Refinement
Mechanical
design and
detailing
TO PRODUCTION
ORGANISATION

Fig. 5.3(d).

During the last half century there have been tremendous developments in methods of communication, and these have produced changes which depend more on the medium than the use to which it is put. Two main trends are apparent; an increase in the speed of communication, due largely to advances in electronics, and an increase in the efficiency of communication brought about by O & M techniques and communication technology.

Nevertheless there is no doubt that communication skill is required from designers; they must be able to express themselves clearly and persuasively orally, graphically, and in writing.

5.3 THE TOTAL DESIGN CONCEPT

To achieve successful design engineering results it is necessary to apply the total or integrated design concept. Here the planning of every stage in the operation is carried out meticulously by a team. It is the team manager who must decide priorities and reconcile conflicting issues.

Civilization is based on specialization, and literally hundreds of collaborators were required to design the Severn bridge or Concorde. But it is too easy to split up the design work and let the sub-units work in isolation so that their part products fail to create a harmonious entity when they are put together.

In pre-war days a more leisurely pace allowed client, consultant, architect, structural and mechanical engineers, etc. to work largely independently. Today, the increased tempo and sophistication of engineering means that this is no longer possible. Any project has to be centrally planned and directed through teamwork which is more essential than ever. It is interesting to note that judges for the Design Council's award for capital goods had the remit to look for 'the total excellence of design', for example comprising innovation, function, economy, efficiency in use and manufacture, as well as good ergonomics and appearance.

However, the total design concept implies not only the coordinated effort of the whole design team but its integration into the departmental and information structure of the enterprise as a whole. Design, being perhaps the most crucial activity in an engineering company, must be of vital concern to management; particularly as regards the company's overall policy.

5.4 THE DESIGN BRIEF

If design is to be a purposeful activity the goal or goals must be identified. Any business is polarized round the customer. As has been discussed, the right product is one that meets the customer's need, that sells in a competitive market and wins a profit. There will be many other functions within the company, such as accounts, personnel, and management services, but all these are to support the main business of designing, building, and selling hardware for profit.

Methods of studying the market vary from market research to industrial espionage. In the military field, requirements are determined by the staff requirements which are set down in detail according to the strategic needs and knowledge of enemy development. In the civil field, where large sums of money are involved it is becoming customary to carry out some form of cost/benefit analysis. Increasing use is also being made of technological forecasting.

All these aids reduce the risk and uncertainty involved in designing new products, but any company will need to study its own product ranges and their life cycles in order to formulate their design policy. Many well designed products have, however, failed because they were poorly marketed or were launched at the wrong time. To combat the former there must be a publicity and promotion programme, while to avoid the latter there must be adequate planning throughout the design and manufacturing processes.

That progressive companies must have an overall design policy for innovation and progressive development goes without saying, but even minor policy statements on new products or systems often require courageous decisions to be taken, and the consequences of these decisions may very largely determine the survival and prosperity of the enterproses concerned.

Today, 85% of engineering business volume comes from products which were unknown about ten years ago. Furthermore, there is an ever increasing tempo of technological change; the lag between discovery and use is beginning to shorten. Products and systems are becoming more complex, and this leads to high research and development costs and the need for expensive test equipment. The result of this quickening pace of technology is that new techniques of surveillance, screening, and forecasting are essential if any degree of confidence is to be achieved in launching a new design.

Even when policy decisions have been taken and communicated to a design group there still remains a need for clear detail policies. Hence design policy can be considered under the two headings of company design policy and design group policy.

At athe company level in a large enterprise there is a need to coordinate the various divisions with regard to their future products. What should they be making in five or ten years' time? What should be the market – is it to be global, continental, or national?

Too often there is a mismatch between a firm and its environment; little is done early enough to decide what business the company is going to be in. If the company has to tender for 'turnkey' contracts it is essential that a policy of progressive product improvement, in line with what the overall system requires, should be pursued. But even for concerns selling domestic products, development direction must be specified at the company level to ensure a commanding sales position for the future.

There is good reason for believing that some firms would rather run to be second than take the risk of being first. In some cases this could be not so much a design policy as a way of death, although, if the second is close to the first, it could pay off (for example the Comet v. the Boeing 707). A design policy to bring prosperity to an enterprise demands something more than new ideas and courage alone. Restrictive practices such as cartelization that prevent newcomers from entering an industry and the high cost of new capital are factors which militate against innovation. If the present production methods and machinery yield a comfortable profit, then incentives to redesign are small. In some cases, if innovation is in the national interest, Government action will be required.

With the ever increasing need to export, many British firms will need to reformulate their design policies. For example, some large British customers demand products that are uncompetitive in foreign markets. It might pay to consider a dual or modular approach to satisfy both foreign and home customers, or to design a product so that extra finish or sophistication can be added to a less stringent basic design, thus meeting the needs of both markets. Alternatively, separate designs may be required.

British Standards may be fine for this country (when they are eventually issued) but too rigid for foreign needs. Such factors can lead to over-pernickety design which hampers exports.

It will be necessary to revive the industrial dynamic which this country possessed in the past by applying a scientific approach to the problem of design innovation. Every company in the manufacturing industry should create a mechanism which will constantly revitalize its designs by applying fresh ideas. The final recommendation of a report on technological innovation [1] says that every effort should be made to rationalize the innovative efforts of British firms and to build up international companies, which while not retarding decisions through undue consultation and committee work, could enjoy the advantages of scale and large markets. These large markets will only be exploited with superior designs and reduced lead times for development.

A forward-thinking unit or product policy committee should be formed in every organization to take into account past product performance and the economic, social, and other external influences which will affect the product's future. Designers should be included in the unit as well as representatives of the marketing, research, and production departments. Such a committee would be responsible for providing the design brief. This brief would embrace such factors as an estimate of the development work necessary to establish the technical feasibility of a new product, a survey of the market to determine probable sales and profits, and an estimate of the capital investment required. For this work, such data must be collected from inside and outside the company.

5.5 SOME DESIGN OBJECTIVES

The objectives of any design are to make profits for the manufacturer and the customer, within any constraints imposed by society. However, it is often convenient to establish objectives which are easier to assess than cost or profit but which are clearly translatable into money. Such objectives will usually be of local application in that they will apply more to one design than another or more to one part of a design than another. The following are some factors which are often seen to be major objectives for a designer.

5.5.1 Simplicity

In general, the simplest design that meets the specification should be used. The policy should always be to reduce the number of parts and make the product as small as is compatible with other requirements. The techniques of value engineering can be applied to this aspect with great effect. Sometimes, models are a useful aid to achieving simplicity.

5.5.2 Lightness

Unnecessary weight in an aeroplane increases the rate at which it will use fuel and will reduce its payload. This obviously reduces the owner's profit, but it also reduces the price that he is prepared to pay the manufacturer.

Even outside the aerospace industry, weight targets should be set and continuously monitored, if only because less weight means less material and less material means less money. Probably less material also means less processing (less machining, for example). In heavy electrical machines, the active material, copper and coreplate, cannot be reduced if performance guarantees are to be met, but it has been shown that on some stator frames material wastage can be as high as three times the minimum weight. All a company's past designs should be analysed so that realistic targets can be set. A typical example for a generator is given in Table 5.1.

5.5.3 Standardization

The use of standard parts, where possible, should be part of every company's design policy. Wherever possible, a tried and proven part should be used in a new system or as a sub-assembly in a new product. Non-standard parts require proving and tend to have higher production, inventory, and work-in-progress costs, as well as to create difficulties in planning and inspection.

A typical example of the relative costs of a standard machine and a special design for a small industrial motor is shown in Fig. 5.4.

Control of the use of non-standard items may be given to a committee to whom justificaion must be given.

Drawings should be coded, not by arbitrary piece numbers but by shape and size, so that it is easy to retrieve past designs of parts for use in new products (this also helps if the company wishes to use group technology). Similarly, preferred sizes of raw materials should be used in new designs since this encourages cost reduction by bulk buying.

Table 5.1

Material, labour and factory overhead as percentage of basic product cost.

		Material	Labour	Factory overhead	Total	
Field system	Frame	5.46	0.44	0.95	6.85	
	Main pole punchings and assembly	3.08	0.44	0.95	4.47	
	Main field coils	2.69	0.22	0.46	3.37	
	Compole punchings and assembly	0.90	0.27	0.60	1.77	
	Compole Coils	1.56	0.40	0.84	2.80	
	Compensating winding	3.30	0.64	1.40	5.34	
	Field connection	0.55	0.19	0.42	1.16	25.75
Armature	Shaft	6.95	0.51	1.11	8.57	
	Armature hub	0.81	0.40	0.86	2.07	
	Armature core and endplates	4.32	0.75	1.67	8.74	
	Armature winding equalizers and assembly	3.84	3.25	7.08	14.17	
	Half coupling	1.79	0.25	0.55	2.59	34.14
Commutator	Commutator hub and baffle	2.36	0.16	0.34	2.86	
	End rings	3.42	0.12	0.28	3.82	
	Commutator bars and risers	6.10	1.52	3.30	10.92	
	Commutator assembly	0.43	1.15	2.52	4.10	21.70
Brushgear	Brushgear	2.08	0.99	2.14	5.21	5.21
Covers	Endbells	0.92	0.91	1.99	3.82	3.82
Pedestals	Pedestals and bush	1.87	0.39	0.84	3.10	3.10
Miscellaneous	General erection, painting, etc.	0.29	1.85	4.13	6.27	6.27
Total		52.72	14.85	32.43	100.00	100.00

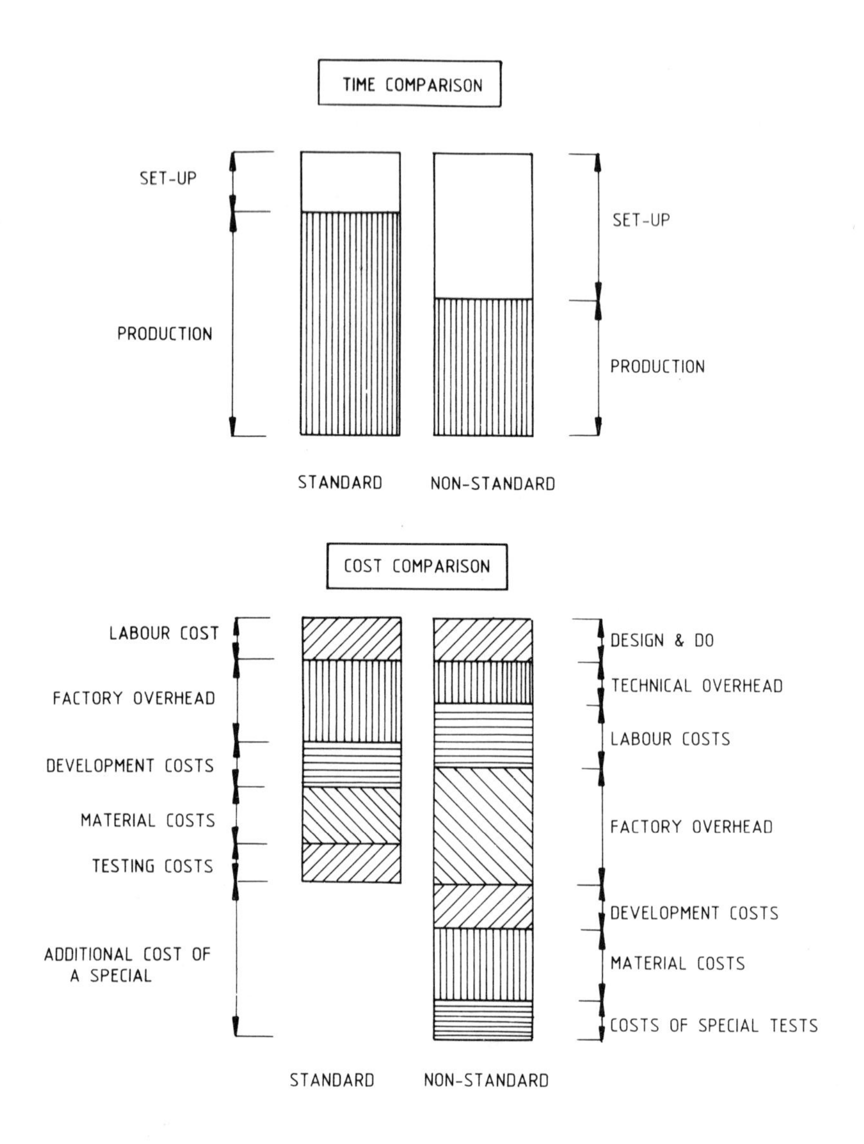

Fig. 5.4 – Standard v. one-off design.

Careful consideration should be given to the setting up of company standards so that overlapping is avoided. Preferred numbers and geometric series of sizes should be used. Where some variety is required for specials, the design should be such that these can be fitted into a standard framework on the final assembly.

5.5.4 Flexibility

With an entirely new design of a product such as a prime mover, design policy should lay down possible uprating requirements for growth of performance. If the initial design aims at too high an efficiency, it can lead to an inflexible product. For I.C. engines it may sometimes pay to provide additional strength for the scantlings in order to allow for higher compression ratios later on. In some cases this conflicts with cost and weight considerations, and compromises have to be made.

5.5.5 Appropriate tolerances

Designers tend to choose their materials carefully from the points of view of availability, strength, and durability, although they may need to review the alternatives by considering the cost for the quality. But equally, tolerances must be set carefully and appropriate finishes selected. Fig. 5.5 indicates graphically the relative machining costs, and can be used as a guide in selecting the most economic tolerance that will meet a design requirement.

What every designer must do as a matter of policy is to select the widest possible tolerances on the trivial items and the needed tolerances on the vital. Too often costs escalate because designer draughtsmen play safe by setting tight tolerances where they are not certain what they should be. Again, some draughtsmen feel that a drawing is not complete unless tolerances and finishes are specified for every part. The result is that the Product Department seeks relaxations which are given only after unnecessary, time-wasting examination. Some investigations have revealed that a large percentage of scrapped components is due to error in the interpretation of tolerances as well as the setting of tolerances that are too tight.

A further consequence of close tolerancing is the effect it has on competitive bidding. The inclusion of close tolerances on a drawing automatically reduces the number of sub-contractors, particularly in the sheet-metal and steel fabrication industries, who are able or willing to undertake the work.

The same applies to tolerances on plates and drilled holes, as can be seen from Table 5.2. Designers often give more attention to designing the shape and form of a component than they devote to the economics of manufacturing the part. The result is that costs are higher than they need be. Table 5.3 gives the cost of close-tolerance machining operations, and the large increase can immediately be seen where grinding is involved.

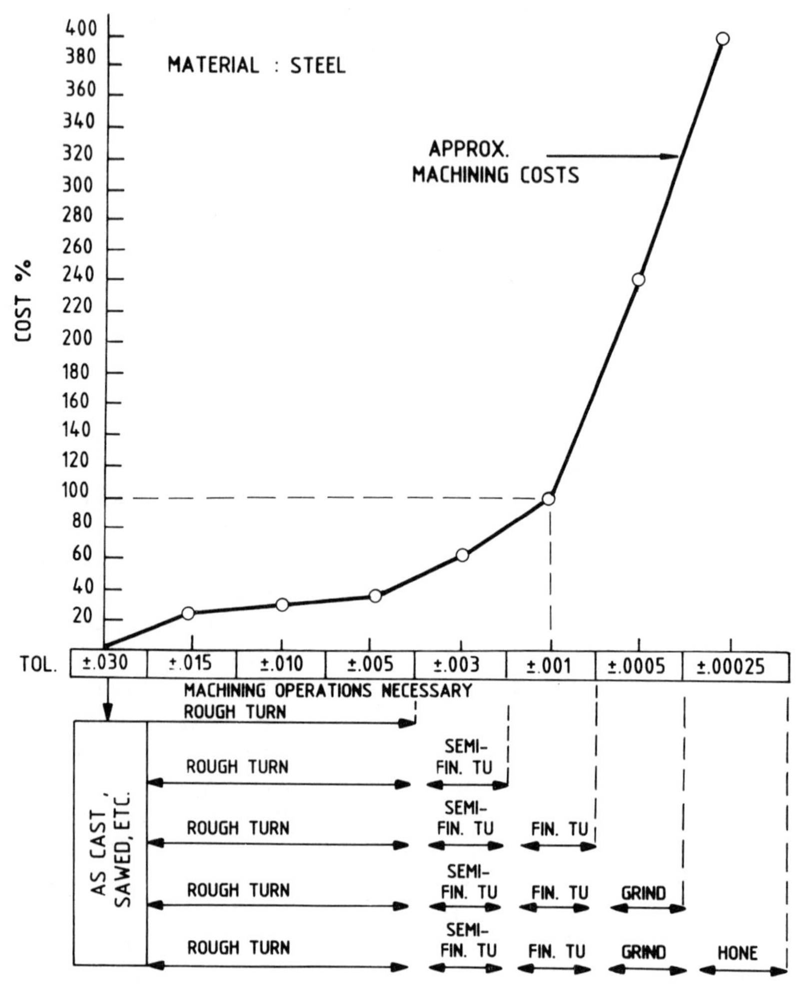

Fig. 5.5 – Relative machining costs.

The best policy is to ensure that drawings are prepared to good commercial practice with the minimum use of close tolerances. The choice of material should be determined by what is adequately strong for the function to be performed when the shape is of low weight and cheap to make.

It must be recognized that real cost reduction starts with the designer who must be kept informed of new methods of manufacture so that the best combination of production process, finish, and tolerances can be chosen.

Table 5.2
Tolerance on plates and drilled holes

Method	Lowest tolerance possible (in.)	Comparative cost	
		Tool cost	Labour cost
Punch (using template)	+.004	100%	100%
Drill (using jig)	+.001 −.002	175%	300%
Drill and ream (using jig and bushings)	+.0006 −.0000	225%	400%
Bore (using fixture) including punch, rough and finish to size on boring machine	+.0004 −.0000	540%	700%
Hone or lap (using fixture) including punch and bore	+.0002 −.0000	730%	1100%

NOTE: Total labour costs inclusive of material set-up costs based on 200 pieces 3.50 in × 5.00 in × 0.18 in thick brass plate.

Table 5.3

Cost data for a range of close-tolerance machining operations. Based on quantities of 1000 units

Typical examples Material: bright drawn mild steel bar	Tolerance (in.)	Operations	Approx. cost (labour & tools)
Shaft diameter	± .002	Cut off (c/o)	100%
	± .001	Turn, c/o	425%
	+.0000 −.0005	Turn, c/o, grind	460%
Shaft diameter (more than one dimension to be turned)	± .002	Turn, form, c/o	100%
	± .001	Turn, form, c/o	130%
	+.0000 −.0005	Turn, form, c/o grind	140%
Length of shaft	± .02	c/o	100%
	± .005	c/o	110%
	± .001	Face & c/o	150%
Total dia. runout	± .005	Turn, form & c/o	100%
	± .001	Turn, form & c/o	130%
	± .0002	Turn, form, c/o grind	165%
Straightness per foot round stock	± .005	Straighten	100%
	± .001	Straighten & grind	375%
	± .0002	Straighten & grind	1,250%
Face run-out	± .005	Turn & c/o	100%
	± .001	Turn, face & c/o	200%
	± .0005	Turn, form, c/o grind	400%

5.5.6 Proper use of manpower

Design policy must take account of the manpower available and whether the men have been trained for the work. More often than not, little provision is made for training or retraining. All designers concerned with the project must be instructed in the complete history and requirements of the project, and there should be an acknowledged policy of ensuring that designers are taught about new techniques and tools as they become available.

It is also necessary to have operator training under way well before product delivery. Initially, this will have to be done at the designer's works, perhaps making use of models or simulators. Certainly handbooks will have to be prepared, and it should always be a company's policy to issue them before the product itself is delivered. For many engineering products the handbooks are forgotten or inadequate or produced too late.

In summary, it may be said that design is a corporate creative activity requiring a holistic approach which demands a diverse number of skills in order to achieve sound communication of the design intent and to pay due attention to such matters as simplicity, standardization, flexibility, and requisite tolerancing. In addition there has to be adequate training (see Chapter 14) of both designers and customer operators if total success is to be obtained.

SUBJECTS FOR DISCUSSION

(1) Discuss the relative merits of innovation and gradual design development.
Give examples of useful innovation which has occurred within gradual product development.

(2) How and why does the designer talk to himself?
How and why does the designer talk to the customer, the manufacturer, the inspector, . . . , . . . , the service engineer?

(3) Sketch a methodology for designing a specified product to the point where manufacturing drawings are issued.

(4) List a number of properties by which a design is judged:

(i) lightness,
(ii) standardized,
(iii) . . . ,
(iv) . . . .

How may these parameters replace profit as an objective?

REFERENCE

[1] ACARD (Advisory Council for Applied Research and Development), *Industrial innovation*, HMSO, London, 1979.

CHAPTER 6

The use of resources

The first part of the chapter discusses the percentage of a designer's time that must be written off because of variable requirements, the impossibility of scheduling work without waiting time, the necessity or working on jobs which are ultimately unsuccessful, and the tendency to underestimate costs.

The second part deals with information resources. Designers tend to live off relevant timely information which they retrieve or generate for subsequent use. There are many sources of good design data which all designers must be aware of, and some of the main ones are listed.

The chapter ends with some aspects of material resources. Since the cost of any product or device is attributable to the cost of the material used in manufacture, and this material cost is generally over 50% of the total cost, the designer must select and use all materials demanded in his work with great care, paying due regard to price and availability.

6.1 THE USE OF HUMAN RESOURCES

6.1.1 Resource levelling

If we attempt to manage projects so that they are completed as soon as possible, they generally use resources very unevenly. Consider the simple network of a design job, taken from Leech [1] and shown in Fig. 6.1.

If we consider the men required to do each of the tasks we can ask the computer to print out the number of men required during every period of the project's duration. The specified manpower requirements and the profile of manpower usage are shown in Fig. 6.1 as the computer printed them. The level of resource that is required is seen to be very uneven, and this creates considerable expense. Designers, draughtsmen, development engineers, technicians, etc. cannot be hired and fired at a moment's notice; and, generally, it will be necessary to maintain, permanently, the maximum number of men required at any period of a project, knowing that for much of the time those men may be idle.

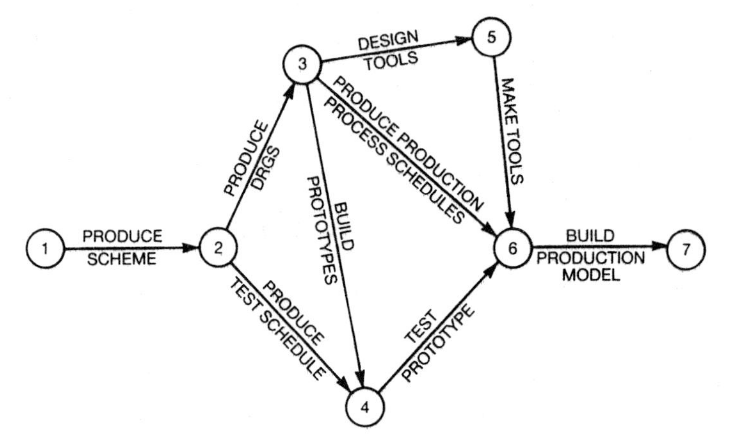

BN	EN	AT	ES	LS	EF	LF	TS	FS	R
1	2	3	0	0	3	3	0	0	2
2	3	4	3	3	7	7	0	0	6
2	4	1	3	12	4	13	9	9	1
3	4	6	7	7	13	13	0	0	5
3	5	2	7	12	9	14	5	0	2
3	6	3	7	14	10	17	7	7	2
4	6	4	13	13	17	17	0	0	2
5	6	3	9	14	12	17	5	5	2
6	7	4	17	17	21	21	0	0	5

CRITICAL PATH

1 – 2
2 – 3
3 – 4
4 – 6
6 – 7

BN = BEGINNING NODE
EN = END NODE
AT = ACTIVITY TIME
ES = EARLY START
LS = LATE START
EF = EARLY FINISH
LF = LATE FINISH
TS = TOTAL SLACK
FS = FREE SLACK
R = RESOURCE REQUIRED
NO. RES. NUMBER OF RESOURCES BEING USED
ACT. IN HAND = ACTIVITIES BEING WORKED ON AT THE TIME

TIME	NO. RES.	ACT. IN HAND
1	2	1 – 2
2	2	1 – 2
3	2	1 – 2
4	7	2 – 3 2 – 4
5	6	2 – 3
6	6	2 – 3
7	6	2 – 3
8	9	3 – 4 3 – 5 3 – 6
9	9	3 – 4 3 – 5 3 – 6
10	9	3 – 4 3 – 6 5 – 6

TIME	NO. RES.	ACT. IN HAND
11	7	3 – 4 5 – 6
12	7	3 – 4 5 – 6
13	5	3 – 4
14	2	4 – 6
15	2	4 – 6
16	2	4 – 6
17	2	4 – 6
18	5	6 – 7
19	5	6 – 7
20	5	6 – 7
21	5	6 – 7

Fig. 6.1.

There are several ways of tackling this problem. The first would be to attempt to reschedule the start of each activity, making use of the slack time as far as possible, in order to make the profile of resource use more level. Fig. 6.2 shows two results of rescheduling activities to take advantage of slack times. In one case, resources were levelled to some extent without delaying the project completion date, while in the other case, allowing the completion date to be delayed, levelled resources still further. These examples are much simpler than most design projects that are met in industry, and the computer was used merely as an aid to trial and error. Nevertheless, they demonstrate one of the difficulties of design management; a difficulty which becomes greater as projects become more complex.

One method that we use to avoid too irregular a use of manpower is to increase the number of projects in the hope that one project will use the manpower that is not required by another. This introduces much more complexity into design management because it is by no means obvious that projects can be so timed and meshed together that while one is using a resource heavily, others will have little need for that same resource. The difficulty of meeting delivery dates will be greatly increased as the number of projects is increased and as attempts are made to level resource needs. Where the company is concerned with one or two large projects at any one time (for example, designing and building an aeroplane, a power station, or a bridge), it may be possible to find a number of small projects which will use the resources that the major project does not require from time to time. This does, of course, mean that the smaller, minor projects will not be managed to meet tight due dates but to make use of resources as they become available. These minor projects may be the design and development of new products, but are more likely to be changes to old. They will be cost reduction exercises and product enhancement exercises, and are minor only in the sense that, individually, they are smaller users of resources than major projects. Since a company cannot survive unless it continues to reduce the cost of its products and enhance their quality, these small projects are not minor in the sense of being unimportant. Attempts to select projects which will make use of idle resources must be associated with the procedures which are used more generally to generate, select, and monitor the progress of design projects, and which are discussed elsewhere, but particularly in the early part of Chapter 11.

6.1.2 The Use of a man's time

Even a hard-working designer is unlikely to be able to offer a full hour's work for every hour that he is paid. In the simplest situation of one man giving his attention to a series of jobs as they arrive, any attempt to use the whole of his time profitably will result in unacceptable job turn-round times. A simple demonstration of this is offered by queueing theory.

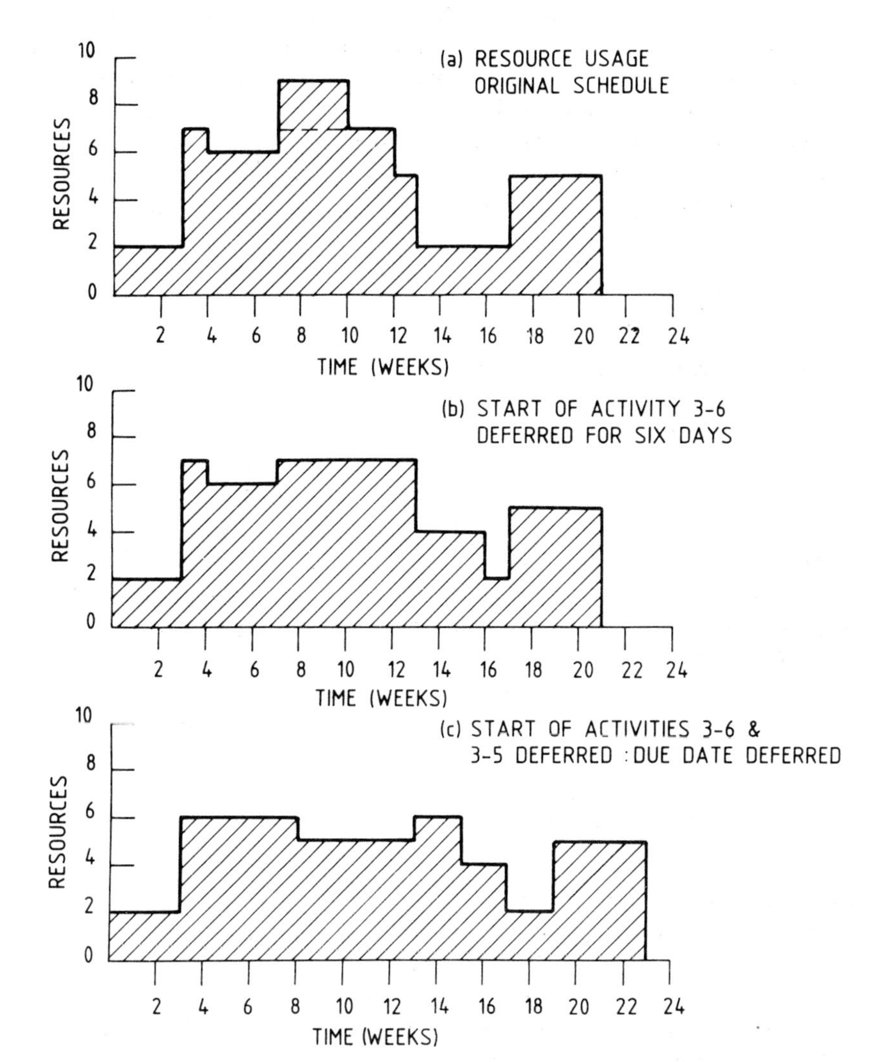

RESOURCE PROFILES OF A DESIGN PROJECT
(a) STARTING ALL ACTIVITIES ON EARLY START DAYS
(b) TAKING ADVANTAGE OF SLACK TIMES
(c) ALLOWING DUE DATE TO DRIFT
OUTPUTS OBTAINED BY TRIAL AND ERROR USING AUTHOR'S CPM PACKAGE

(a) REQUIRES 50% MORE STAFF THAN (c)

Fig. 6.2.

If we can accept the idea that design jobs arrive randomly and the time taken to do a job is probabilistic and taken from a negative exponential distribution, simple queueing theory provides us with formulae which yield some information about job turn-round time (that is, the time between the job arriving in the office and its completion). If, in these conditions, jobs arrive at the designer's workplace at an average rate of L per week, and if the mean time to do a job is $1/M$ weeks then, on average, a job will wait in the system for

$$\frac{L}{M(M-L)}$$

weeks before receiving attention. The turn-round time of the job will thus average

$$\frac{L}{M(M-L)}+\frac{1}{M}$$

weeks. These formulae are justified in any elementary book on operational research such as Fabrycky, Ghare, and Torgersen [2].

If we redefine $R = L/M$, then R is the fraction of a man's time that we can expect to use profitably. Clearly, R cannot be greater than one, or there will be more work arriving than the man can cope with. In fact, if we plot the average turn-round time for a job as a function of R we see that R must be very much less than one if reasonable turn-round times are to be achieved. Fig. 6.3 shows k as a function of R where:

$$k=\frac{\text{average job-round time}}{\text{average job time}},$$

and we see that if a man is loaded to work usefully for 80% of his available time, a job will be in the system for five times as long as it takes to do. That is, it will be waiting to be done for about four times as long as it takes to do. Perhaps this would be acceptable if the average time taken to do a job were, say, a week or a fortnight, but if the mean work content of a job were, say, six months, the average turn-round time would be two and a half years, and we would be well advised to increase staff to achieve a saleable delivery date.

One of the problems that is shown by this simple application of queuing theory is the apparently disproportionate effect of making changes. Suppose, for example, that the development of a product shows that design changes have to be made, and suppose that these changes increase, by 5%, the man-hours needed. Then, if the technical department is already heavily loaded, turn-round times could be doubled from ten times the work content of the job to twenty times. As changes are inevitable if the company is to stay in business, we cannot refuse to make them. What we must do is appreciate the

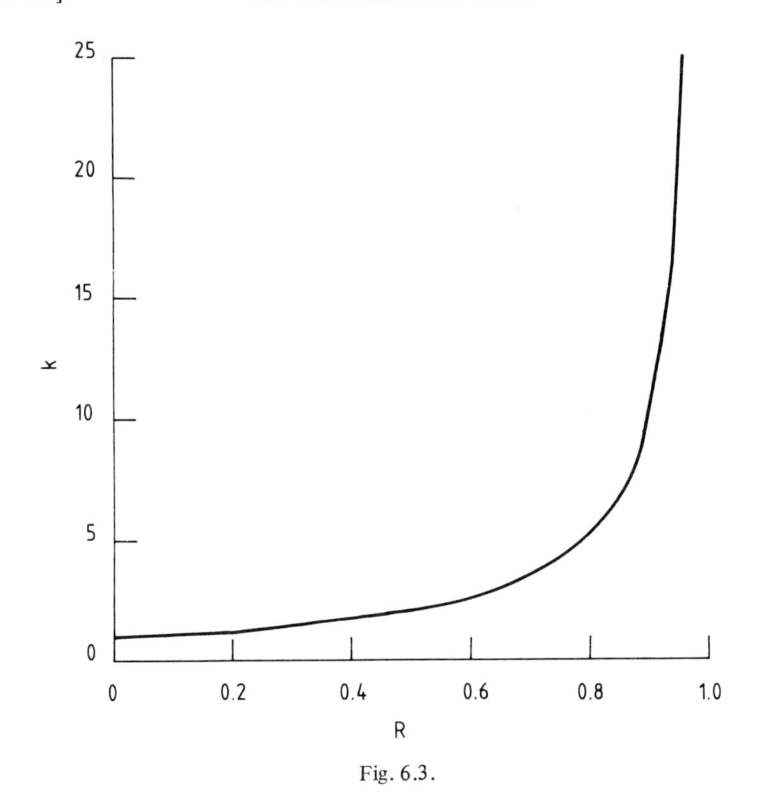

Fig. 6.3.

effect that changes will have on our manpower requirements, and budget for manpower use accordingly. The design manager must be aware of the number of men (or any other resources) who will be concerned with changes, as well as knowing the effort that the first design of a new, major product is likely to absorb.

There can be few firms in which this model of jobs queueing for attention by one designer is realistic. Generally, any design job will consist of a number of tasks which are to be done in sequence, and men will have specialist skills so that the completion of one task by one man will be followed by the starting of the next task by the next man. For example, the drawing of a detail by a draughtsmen may need to be followed by the stressing of that detail by a stress-man. Again, it is not likely that the work content of a job could be drawn from a negative exponential distribution, and there is limited evidence which suggests that a log-normal distribution would be more likely [3]. Nevertheless, if we use simulation to inject more realism into our calculations we still find it impossible to load any man to the point where he makes saleable use of the whole of his time.

Fig. 6.4 shows the results of simulating a number of jobs. Each job consists of three activities, and each activity is done by one man. In the first simulation

Extracts from computer simulations of loading jobs on three resources. In each case, information is given for jobs 25, 26, 27, 28, 29, and 30.

J.A. ; Times between job arrivals
TAFS ; Job arrival time from start
TAS1 (2,3) ; Required time at stage one (two, three)
STA1 (2,3) ; Time of start of stage one (two, three)
FTA1 (2,3) ; Time of finish of stage one (two, three)
FRT ; Free time of resource
K ; Job turn round time/Work content of job

Job times are log-normally distributed
Job arrival times are exponentially distributed

FIRST SIMULATION

J.A.	TAFS	TAS1	STA1	FTA1	FRT	TAS2	STA2	FTA2	FRT	TAS3	STA3	FTA3	FRT	K
69	1647	8	1647	1655	54	1	1655	1656	54	1	3181	3182	0	153.500
44	1691	2	1691	1693	36	1	1693	1694	37	6	3182	3188	0	166.333
6	1697	84	1697	1781	4	1	1781	1782	87	3	3188	3191	0	16.977
118	1815	4	1815	1819	34	6	1819	1825	37	22	3191	3213	0	43.688
19	1834	33	1834	1867	15	11	1867	1878	42	6	3213	3219	0	27.700
34	1868	11	1868	1879	1	1	1879	1880	1	372	3219	3591	0	4.487
					--------				--------				--------	
TOTAL FREE TIME					785				1349				399	
USE OF MAN					58.22%				28.24%				88.89%	

SECOND SIMULATION

J.A.	TAFS	TAS1	STA1	FTA1	FRT	TAS2	STA2	FTA2	FRT	TAS3	STA3	FTA3	FRT	K
3	839	8	991	999	0	1	1033	1034	0	1	2918	2919	0	208.000
59	898	2	999	1001	0	1	1034	1035	0	6	2919	2925	0	225.222
9	907	84	1001	1085	0	1	1085	1086	50	3	2925	2928	0	22.966
17	924	4	1085	1089	0	6	1089	1095	3	22	2928	2950	0	63.313
35	959	33	1089	1122	0	11	1122	1133	27	6	2950	2956	0	39.940
30	989	11	1122	1133	0	1	1133	1134	0	372	2956	3328	0	6.091
					--------				--------				--------	
TOTAL FREE TIME					39				603				136	
USE OF MAN					96.56%				46.83%				95.91%	

THIRD SIMULATION

J.A.	TAFS	TAS1	STA1	FTA1	FRT	TAS2	STA2	FTA2	FRT	TAS3	STA3	FTA3	FRT	K
2	609	8	956	964	0	1	998	999	0	1	2883	2884	0	227.500
43	652	2	964	966	0	1	999	1000	0	6	2884	2890	0	248.667
7	659	84	966	1050	0	1	1050	1051	50	3	2890	2893	0	25.386
12	671	4	1050	1054	0	6	1054	1060	3	22	2893	2915	0	70.125
25	696	33	1054	1087	0	11	1087	1098	27	6	2915	2921	0	44.500
22	718	11	1087	1098	0	1	1098	1099	0	372	2921	3293	0	6.706
					--------				--------				--------	
TOTAL FREE TIME					4				568				101	
USE OF MAN					99.64%				48.32%				96.93%	

Fig. 6.4

the interarrival times are about twice the task times of jobs. Each resource is employed for only a fraction of the time for which it is available, and turn-round times vary from one to a hundred and sixty six times the work content of the job. If we double the rate of arrival of jobs, we increase the percentage of the available resource time that is used but turn-round times increase. Increasing the rate of job arrival still more increases these trends towards greater occupation of resources and even later delivery. In each case the situation could have been improved by some attempt at scheduling, because the greatest delays are with the shortest jobs, and short jobs could be given some priority; but the system has only worked through thirty jobs, and already large jobs are beginning to get stuck in the system. The ideas derived from simple queueing are thus reinforced (and even made to seem optimistic) by simulation.

A short simulation of such simple jobs would give no answers from which any worthwhile numerical inferences can be drawn. Indeed, the average task times and interarrival times which obtained in this particular simulation were far from the averages of the distributions from which they were drawn. Nevertheless, the trend is clear: that the time taken to turn a job round increases unacceptably if we try to load each man with enough work to keep him usefully employed for the whole of his time on a job that will be directly paid for by the customer.

More realistic simulations also tend to confirm the ideas of simple queueing theory, and can provide a guide to the level to which we may expect to load designers with directly costed jobs. One example, (Leech, Jenkins, & Turner [4]) suggested that attempting to raise the period that the designer was fully employed from 75% of his time to 93% actually reduced the number of orders received by 20%. This was because, in that particular situation, the turn-round time of tenders was extended to such an extent that deadlines for quotations were missed.

Of course, no designer ever shows on his time sheet that he has been idle. This may be because the designer feels that advertising idleness will invite redundancy; it may be because the designer uses time in which he would otherwise be idle to extend his consideration of the jobs he has to do; or it may be that work expands to fill the time available to do it in. Whatever the cause, however, the assumption that more than about 80% of a designer's time can be saleably employed will lead to impossible turn-round times.

6.1.3 Write-off

The problem of having to work on failures in order to create some successful designs means that, of the 80% of a designer's time that might be spent on useful work, only a fraction is on jobs for which there is an eventual market. This has been discussed, to some extent, in Chapter 3, as one cost of marketing. Unsuccessful projects cannot be avoided, although the designer should, of course, aim to make every project successful. It is clear, however, that where a customer invites several possible suppliers to tender for the same job, all but one of the suppliers will be unsuccessful. Where a manufacturer sells small batches of specially designed products made to a customer's particular requirements, it is common for four or five manufacturers to be competing for one order, so that on average the competing companies will each get about 20% of the jobs for which they compete. When competing for an order of this kind a manufacturer must expect to spend some money, for before a proposal can be made to the customer or costs predicted, work must be done on the specification of the product, on a design scheme, on a demonstration of feasibility, on calculation of first costs and operating costs, and on demonstrations that the predicted performance and cost of the product may be believed. Because it is probable that work done to get an order may be wasted, the manufacturer will want to pay for as little as possible. On the other hand, if too little work is done to get the order and yet the order is won, the manufacturer will face difficulties which arise from the inadequate scheming and forecasting, which leads to underestimating costs and delivery dates and overestimating the performance. How much a company should spend trying to get an order will vary from one market to another. In the sixties and seventies, the defence industry was criticized for underestimating costs, and it might be possible to argue (from the debates which accompanied the publication of Downey [5]) that it is typical for a company to spend about 5% of the expected design and developments costs in an effort to get an order. It has also been suggested that figures as high as 10% or 15% of the expected final design and development costs should be spent to get an order if reasonably accurate costs are to be forecast.

If a company tenders for n jobs and its success rate is p, then on average, pn jobs will be won and $(1-p)n$ lost. If the design and development work content of a job is, on average, T man-hours, and if the fraction of the work done before tendering is f, then the work that must be written off for unsuccessful tendering is $f(1-p)nT$ man-hours. The total work done will be $f(1-p)nT + Tpn$, so that the proportion of design and development work that must be written off will be:

$$\frac{f(1-p)nT}{f(1-p)nT + pnT} = 1 - \frac{p}{p(1-f) + f}.$$

The significance of this is plotted in Fig. 6.5, where we see that we might well have to devote twenty or thirty per cent of our design effort to work which will never be paid for, directly, by the customer.

6.1.4 Underestimating

There is a further difficulty in allocating labour to design work, which arises because we tend to underestimate how much labour a job will require. We might write off 20% of our labour because of the difficulty of scheduling work, and a further 20% for work on unsuccessful projects, but the remaining 60% of man-hours will not all be paid for directly if they have not been allowed for in the

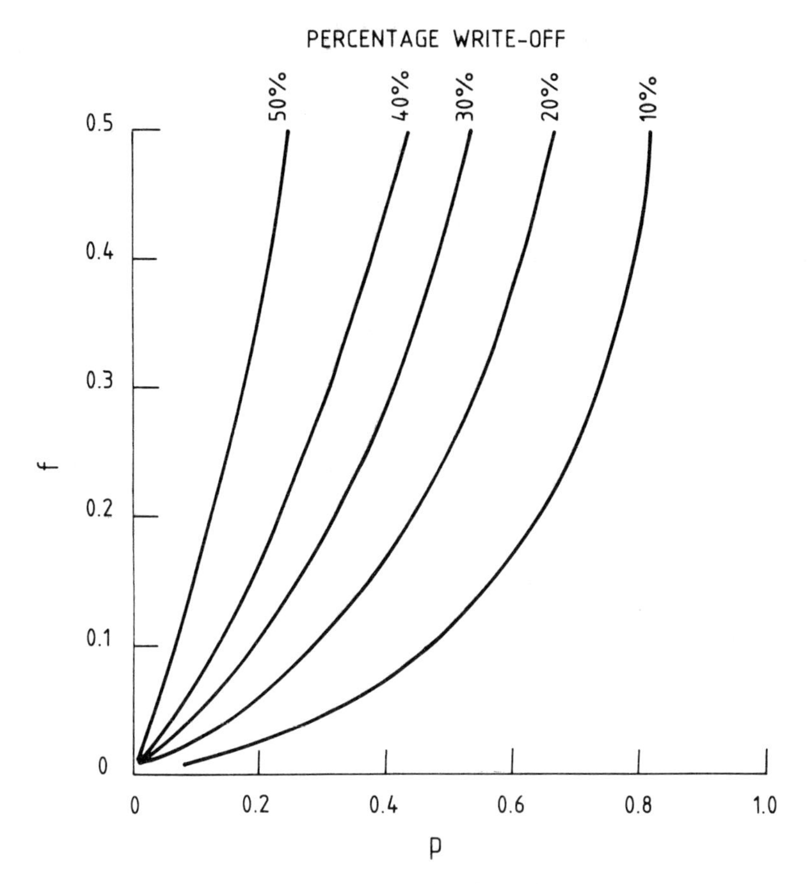

Fig. 6.5.

price that is quoted to the customer. If we are designing a special product for the customer and we estimate that design and development will require a thousand man-hours, we are likely to take this into account when quoting the price to the customer. If, subsequently, design and development acutally costs us 1500 man-hours we will find it difficult to recover the cost of the extra 500 man-hours from the customer. It is not sufficient to say that that designer should have predicted the labour costs more accurately, because no-one has yet found out how to predict design costs accurately. The best that can be done is to build yet more write-off into our design department's overheads.

Smart [6] has shown how the design process can be simulated, in any given case, to determine the write-off that must be allowed both for unsuccessful tendering and for underestimating design costs. Unfortunately, to carry out such a simulation requires information that is rarely available. How much is being spent before an order is received? How many tenders are successful? What is the distribution of job costs? What is the distribution of estimated costs? How does the success rate of tendering vary with the effort given to tendering? In some firms, where good records are kept and where a systematic attempt is made to monitor design costs, much of this information may be available, but no-one will ever know how the chances of getting an order improve as the design effort of getting the order is increased.

6.1.5 Small design costs

Problems of determining the design cost in a product are real when that cost is a large proportion of the total cost of supplying the product. Factors which increase the design and development cost as a fraction of the total cost are:

- the number of products supplied. Clearly, the more products supplied, the more thinly the design costs can be spread. If only one product is supplied, its cost must contain all design and development costs.
- the level of innovation in the product. New technology does not only mean greater risk of failure but greater cost of proving. No design is finished until it is proved that it meets the specification, and the greater the novelty, the more the proof will cost.
- the nature of the market. A product will cost more to design and develop if it is to work in a hostile environment.

Many companies, working in areas of technology that are well understood, producing products with little innovative content for markets which are large and environments which are not hostile, find it unnecessary to spend more than about one per cent of their total costs on design work. This can be significant because, if profit is five or six per cent of sales revenue, changes in design effort will be seen to affect profits, since design costs will be about 20% of profit. It may be less apparent to the accountant that reducing design costs may result in poor product design, no cost reduction, no product enhancement, and hence, high product support costs and loss of markets. The design manager must therefore choose the design resource level and quality which is a compromise between low design costs with a high risk of failing to satisfy the customer and high design costs with a complex system for predicting, justifying, and monitoring those costs.

6.2 THE USE OF INFORMATION RESOURCES

If people are the most important resource for a design office, then information is a close second. Without the relevant timely information, designs cannot be completed economically. Information implies freedom of choice; there will be a number of alternatives possible for compiling any design message. A designer gleans information from a variety of sources. Britain is particularly well endowed with information sources, for there are Government Departments and Agencies, Research Establishments, Professional Institutions, Universities and Polytechnics, Trade Associations, and individual companies as well as the British Standards Institution. All designers need to be aware of these and to cultivate contacts in order to keep up-to-date. A list of useful British Standards (BS) and Published Documents (PD) is given in Appendix 6.1. In addition, today, International Standards (for example, DIN) must be available in most design offices.

6.2.1 Range of information sources

One way of looking broadly at the spectrum of information sources is given in Fig. 6.6. At one end of the spectrum there is what might be termed refined and distilled data – the constraints of engineering design, for example value of π, e, g, λ, and so on. When these are used enough, they move further up the spectrum into the form of encapsulated tables which appear in the various engineering handbooks. Next come the text books which are more definitive and attempt to set down principles of application and the derivation of useful formulae; and yet further up the spectrum we come to more encyclopaedic publications – such as science abstracts. Finally, at the top end of the spectrum there may be review papers which set out the present state of the art in particular engineering fields, and perhaps beyond this point, in what might be termed the non-visible part of the information source spectrum, comes the current information which is mainly vested in people in specialist centres.

A survey of some 300 manufacturing firms embracing about 650 designers indicated that the following information sources listed in Table 6.1 were used. This strongly suggests that organizations producing standard data sheets have had little impact on the companies surveyed. 95% of the establishments were not subscribing to any existing design data services, but text books and handbooks were provided by the majority of firms. In general, it has been found that the larger the design department the more likely it is that a wide range of information sources will be provided.

However, self generated data from the firms concerned indicated that only 31% produced their own data sheets. Table 6.2 shows, by different industries, where abstracting services, in-house design standards, guides and procedures, and data sheets were provided.

6.2.2 How designers acquire and use information

In the translation process of moving from ideas to instructions to produce hardware, information is used to create the engineering message which then has to be encoded into a suitable form for communicating the design intent. This

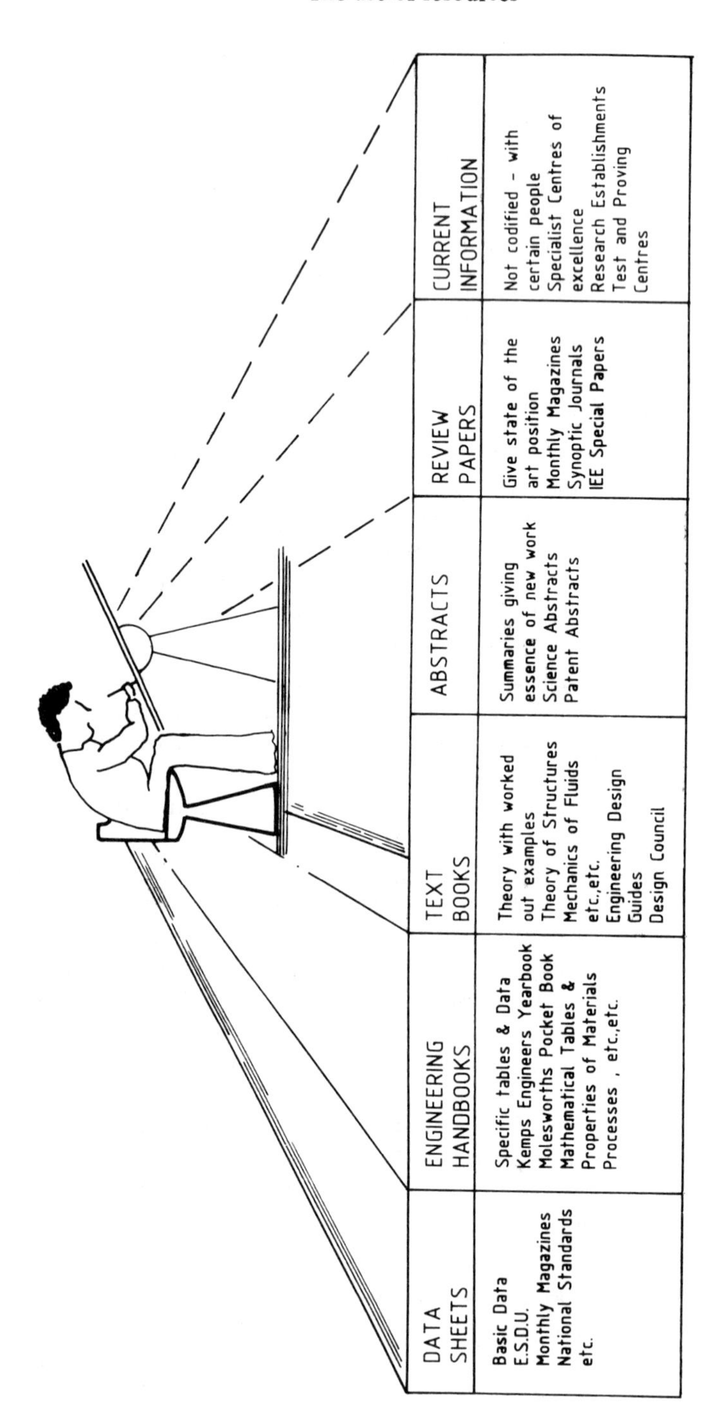

Fig. 6.6 – Spectrum of information services.

Table 6.1
Proportions of establishments providing information sources

	None %	*1–4 titles* %	*5–14 titles* %	*15–49 titles* %	*50 + titles* %	*Don't know* %
BS and international equivalent	1	14	19	23	37	6
Patents	35	7	13	14	14	8
Handbooks	28	29	29	11	2	1
Textbooks and monographs	24	12	28	12	29	5
Data sheets	95	2	2	1	0	0
Product data services	86	14	0	0	0	0
Buyers' guide	16	50	19	3	2	9
Abstracting/indexing journals	81	16	1	1	1	1
Other design journals	7	28	38	21	4	3
Research reports	35	14	14	13	15	9

Base: 310 establishments

Table 6.2

Number of establishments providing information services, analysis by sector

	All establishments	*Mechanical handling equipment*	*Other machinery*	*Industrial plant*	*Other mechanical engineering*	*Scientific instruments*
	%	%	%	%	%	%
All three types	4	6	3	4	3	7
Two types only	23	29	15	36	33	23
One type only	26	39	32	33	6	15
Not providing at all	47	26	50	27	58	55
Total	100	100	100	100	100	100
Base:	301	25	153	50	57	25

Types of information services
(1) Abstracting services
(2) House manuals of design standards of procedures
(3) Data sheets

may be depicted as indicated in Fig. 6.7 where the various inputs are available for aiding in the generation of the message which is then prepared for output in the form of specific instructions for manufacture.

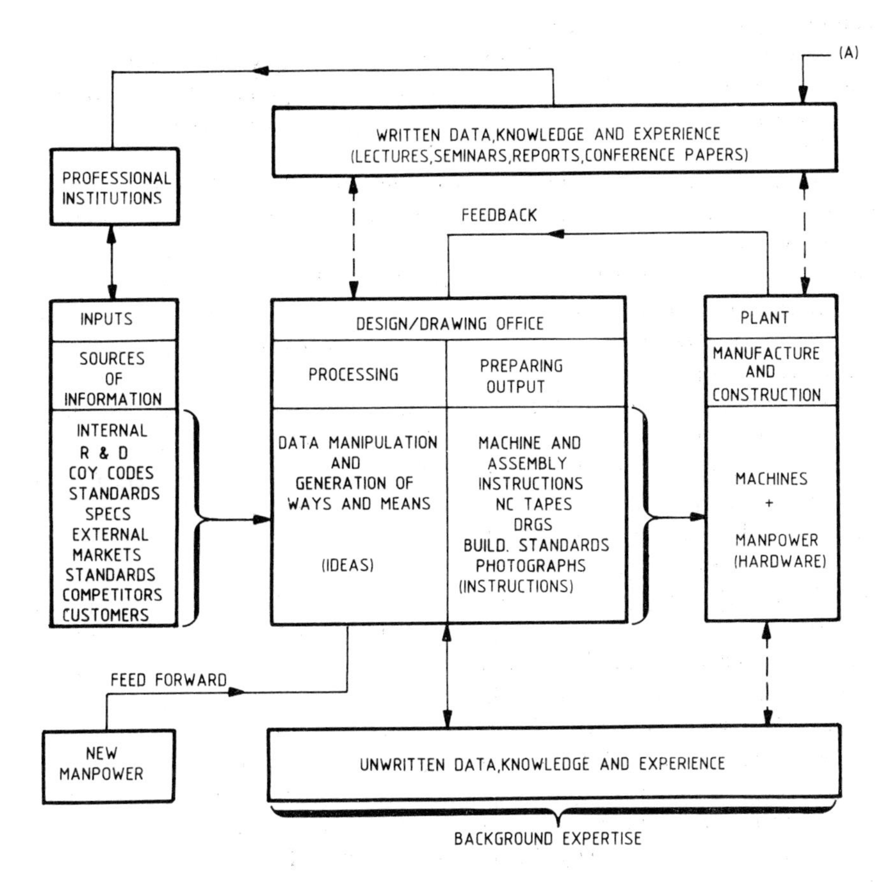

Fig. 6.7 – Information flow in the design process.

Experience has shown that with regard to information generation:

- Information at the start of design is generally incomplete, and it is often not too clear what to seek in the way of additional information.
- Time is required to find out what else to search for, and extra time is required to try to find it.
- Information obtained is often conflicting and may contain much 'noise'.
- Information costs the designer precious time and money to obtain.

Information which engineers use may be regard as 'Soft' or 'Hard'.

'Soft information' is generally nebulous, qualitative, often verbal and transient. It includes: opinions, market surveys, ideas, proposals, etc.

'Hard information' is generally verifiable, permanent, documented, often expressed numerically and includes: principles, laws, theorems, standards, contracts, physical constants, etc.

The design engineer needs information in both these areas. As progress is made through the translation process, so in-house information is generated, some of which may be formally coded in the form of data sheets, specifications, standards, lectures, etc. indicated by Box A in Fig. 6.7.

Many good designers seem to start their information search from catalogues where well-produced diagrams give clues via a high-impact visual representation. They tend to work backwards from these to deduce certain factors. There is a need to get information to the designer's elbow which is 'punchy' and easily digestible. A kind of 'handyman' approach seems best for presentation purposes. The desirable parameters for good information publications, catalogues, etc. are set out in Appendix 6.2.

Unfortunately, trade catalogues, information sheets, handbooks, standards, codes of practice, price lists, etc., although only changing slowly, do come in a variety of forms many of which do not meet many of the parameters given in Appendix 6.2. How often does one find the range of properties of materials being quoted as opposed to average figures?

Undoubtedly certain classes of engineering work lend themselves to the use of a kind of shorthand or the use of a metalanguage[†]. An obvious class here is organic chemistry in chemical engineering, and electronic and computer engineering where highly stylized languages have come into being. In mechanical engineering, however, this is much more difficult – the line, or stick diagrams, seem to break down after a certain complexity is reached.

With the ever increasing costs of publication today the problem of disseminating technical data and production information becomes more acute, and this had led to the free issue of controlled circulation journals. Here production and distribution costs are borne by the advertising revenue. Inevitably this has meant a broader approach being taken towards engineering design matters. This in turn has resulted in fewer specialist magazines being available with their in-depth content.

In this particular area there appear to be two possible remedies: synoptic publications and specialized courses, teach-ins, and seminars for designers.

6.2.3 Company-generated information

As the design process proceeds (Fig. 6.7) so more and more information is produced as the hardware end is reached. Consequently more people become concerned with the retrieval and manipulation of information about specifications,

---

[†]Metalanguage: A language that supplies terms for quick analysis of a subject.

materials, parts and their relationship with one another in the composition of products, methods of manufacture or procurement, or the make-up of various manuals. Companies set up their own systems for handling this massive information transfer.

Today, a further forcing element in the information transfer is occurring largely owing to the onerous legal requirements now being imposed upon designers through such acts as the Health and Safety Act and the Consumer Protection Act. These acts force the designer to formally and systematically codify his design information as the design process occurs. The fact that a product might be defectively or negligently designed, and the fact that such a defect of negligence is the proximate cause of an accident or injury to persons or property, now places the designer at risk. He must not only get his design right, employing a fail-safe philosophy, but also prove the design by proper testing and where possible obtain independent validation of the design.

One means of ensuring that the relevant information is recorded and checked is to conduct a design review. This is a formal, documented, systematic study of the proposed design by specialists who may not be directly concerned with the finished design's performance. This is covered more fully in Chapter 12.

All designers are concerned with availability, accessibility, authenticity, and applicability of information. The indications are that many sources of information are insensitive to the designer's work. Consequently designers often have to spend much time capturing information and translating this to their particular needs. Undoubtedly the computer can be of great assistance in this design area and this is discussed in Chapter 13.

6.3 THE USE OF MATERIAL RESOURCES

Nothing has affected the rate of technological innovation so much as the advent of new and better materials. Materials are a resource which is fundamental to so much of human activity. No matter how skillful a designer, no matter what manufacturing or construction techniques can be deployed, the ultimate hardware depends heavily on the materials used and their properties.

The development of modern spacecraft structures has only been made possible by new alloys, and the design of their engines too has only been made possible by better materials. Here we are not concerned with the properties of materials or their method of production, or ease of fabrication, *per se,* but rather with cost and availability considerations. Nevertheless some aspect of properties must be considered, for with increased economic pressures being applied to the designer's work it may be necessary to search for alternative materials. Broadly speaking, material may be classed as metals and alloys, polymers, ceramics and glasses, and composites. In comparing such materials, a designer must consider, in particular, specific elastic moduli, as indicated in Fig. 6.8. On this basis aluminium may be just as economical as steel. Most conventional materials have specific elastic moduli of the order of 4×10^6 lbf/in^2. If a design demands tougher, lighter materials, then the designer must use those with a higher specific modulus. In some circumstances their additional cost may not prohibit their use.

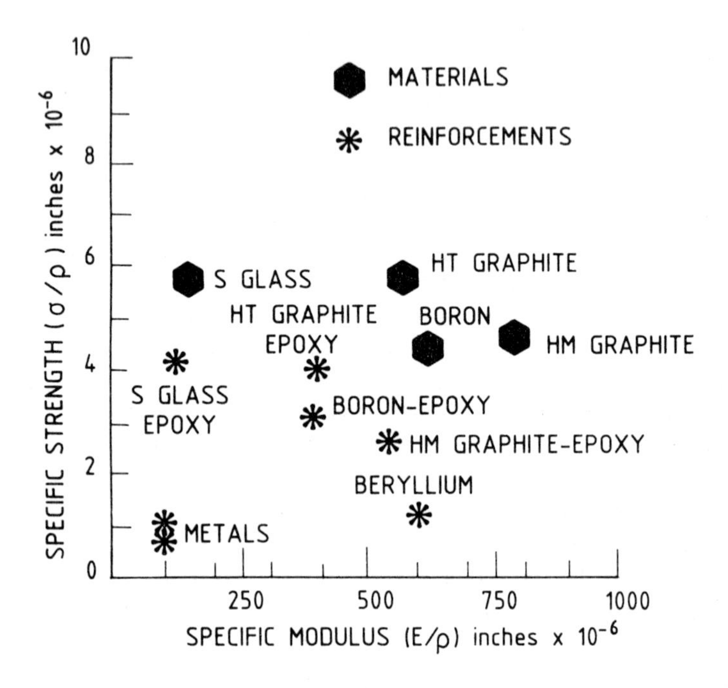

Fig. 6.8.

Today, materials of high specific modulus are being incorporated in a relatively cheap matrix as fibrous or filament reinforcement to produce 'composites' combining some of the properties of both components.

The three basic requirements for many engineering materials are: high resistance to plastic deformation, to fracture, and to high elastic stiffness. The first two properties give toughness, the last, rigidity.

The development of composites has tried to meet these design requirements. Of course nearly all materials used in engineering are to some degree composites: for example cast iron, steel, and other metal alloys; but these are isotropic. On the other hand reinforced concrete is anisotropic because of the preferred orientation of the reinforcing elements in the matrix.

In modern composites the matrix can consist of metal, ceramics, glass, concrete, or resins, and the reinforcement may be metal rods or filaments, whiskers of silicon, or carbon fibres.

Much detail design work is concerned with economically transferring loads from one element or section to another. With anisotropic materials it is possible to orient the fibres at areas of high stress-concentration so as to obtain the most effective load transfer. In structural engineering this has long been recognized as designers have used reinforced concrete in this way.

For the benefit of design and development managers a brief review will now be given of some materials development.

6.3.1 Strength of materials

The strength limits of solids depend upon the deformation and fracture-mechanisms. In all solids the theoretical strength limits are many times greater than the practical levels achieved. At low stress deformation takes place in an essentially elastic manner but as the stress increases the mechanisms of deformation differ in metals, plastics, ceramics and glasses.

6.3.1.1 Metals

As the stress in a metal is increased beyond a certain level the strain becomes plastic. The elastic limit of the metal generally occurs at strains of approximately 0.05%. In the elastic range, Hooke's Law applies; that is, stress is proportional to strain. The elastic limit is therefore widely used as a criterion in the design of safe structures.

Theoretically it can be shown that, in any perfect crystal, the maximum elastic strain should be approximately 10%, and stresses up to 100 times those used at present might be obtained. The discrepancy between theory and practice has been shown to result from the presence of small imperfections, known as dislocations, in the symmetry of the metal lattice. These faults move through the crystal at relatively low applied stress, producing plastic deformation.

Most practical techniques available for strengthening metals today; that is, work-hardening, precipitation hardening, dispersion hardening, depend on preventing the movement of dislocations.

An alternative approach would be to remove the mobile dislocations from the lattice. This has not so far proved practicable in bulk metals but has been achieved in fine whiskers of 0.1 to 3 μm diameter up to a few millimetres long. Such 'whiskers' can be strained elastically to approximately 1%, so that their strength approaches the theoretical limit.

6.3.1.2 Ceramics and glasses

In ceramics and glasses no significant plastic deformation occurs, and elastic straining continues until fracture. In bulk materials this occurs at strains of approximately 0.01%, though the reproducibility of results is poor. However, it has been found that the maximum theoretical strain of about 1% can be approached in ceramic whiskers and very thin glass fibres. It is believed that this is due to the absence of both surface and internal defects on the whiskers and fibres.

6.3.1.3 Plastics

The formation characteristics of plastics are very complex and are not yet fully understood. Their elastic moduli generally vary with strain and are low compared with metals and ceramics. Hence at this stage they do not offer much hope as ultra-strong materials.

6.3.2 The structures of composites

Since, to date, it has only been possible to approach theoretical strength limits of materials in very small dimensions, a considerable effort has been made over the last decade to utilize the properties of such materials in composite form. Two different principles have been involved:

6.3.2.1 Dispersion strengthened

Dispersion strengthening is applicable only to a matrix in which plastic deformation occurs by dislocation movement, for example a metal. Small, strong particles dispersed throughout the lattice are used to impede the movement of dislocations and so raise the elastic limit. It is important that the particles should be small, ideally about 0.001 μm diameter, spaced 0.01 μm apart. They must also have a high cleavage strength and not react with the matrix at the high temperatures generally required for fabrication. Materials which have been used include oxides, nitrides, and carbides.

The effect of adding a few per cent by volume of these to a metal is to produce an isotropic material with mechanical properties at room temperature similar to those of the heavily cold-worked metal. But these properties are retained after heating to well above the normal softening temperature. The creep properties of the dispersion-strengthened material are also very much superior to those of the base metal. However, the room temperature properties of the best materials of this type are still only of the same order as those obtained by cold work; there is still a big gap between the best of these and the theoretical limits.

6.3.2.2 Fibre strengthened (or reinforced)

The aim in fibre-strengthened materials is to utilize as much as possible of the high tensile strengths of various fine whiskers and fibres in a practical engineering bulk material. The whiskers and fibres are therefore incorporated in a continuous matrix which must transmit the applied tensile stresses to the fibres.

6.3.3 Production of composites

6.3.3.1 Dispersion-strengthened

There are various methods of producing a dispersion-strengthened material. The simplest case is undoubtedly that of pure metals which have a relatively stable oxide: aluminium, magnesium, and lead. The metal powder is oxidized and then consolidated by pressing, sintering, or, more usually, hot extrusion.

Typical properties are given in Table 6.3. The most widely available commercial dispersion-strengthened metal is of this type: S.A.P. (Sintered Aluminium Product). A variation of this technique is to oxidize one component of an alloy powder.

Table 6.3

Material	E/S.G. lbf/in^2 x 10^{-6}
Graphite	45
Silicon carbide	31
Boron	28
RAE carbon fibres	20/28
Alumina	19
Silicon nitride	18
Silica and common glasses	4
Tungsten	2.6
Molybdenum	4.8
Beryllium	25
Steel	3.8
Titanium	3.7
Aluminium	3.7
High tenacity nylon	0.6
Concrete	0.8
Wood	3.7

Another relatively simple technique is to mix and consolidate a finely powdered metal with an even more finely powdered oxide or other refractory. Practical problems, particularly in the United Kingdom, are to obtain the necessary fine powders, ideally about 1 μm grain size for the metal and 0.01 μm for the refractory. Workers in the Nelson Research Laboratories producing copper/alumina by this technique found the finest copper powder available was about 20 μm, and this was not of the purity desired for electrical applications.

Some of the best dispersion-strengthened materials have been produced by internal oxidation. A dilute alloy of, say, aluminium in copper is prepared, rolled to thin foil, and heated in an atmosphere which is oxidizing to aluminium and reducing to copper. Alumina is formed as an extremely fine, well-dispersed phase. This technique produces materials with very high tensile properties but is limited to very thin foils and wires.

The co-precipitation of hydroxides or carbonates, followed by chemical breakdown to give a metal/metal-oxide mixture, has also been investigated. The

Nelson Research Laboratories have applied for a patent for a process in which a metal powder, for example copper, is covered with a layer of oxide of controlled thickness prior to consolidation.

6.3.3.2 Fibre-strengthening and reinforcement

1) *Reinforced plastics*

Glass fibre reinforced plastics first came into use in aircraft during the 1939–45 war. Their attraction was that they had a high strength/weight ratio and were electrically nonconducting. Structures were made of either glass mats or woven glass fabrics, impregnated with a liquid thermosetting resin; they have planar or orthotropic properties, respectively. Since glass fibres of tensile strengths up to 500 000 lbf/in^2 can be drawn in the laboratory, uniaxial strengths of this order should be achievable under ideal conditions. However, commercially available glass fibres have lower strengths, and since it is rarely desirable to place all the reinforcement in one direction, the strength of the composite in practice is considerably lower than the theoretical figure.

In order to develop the optimum properties in glass-reinforced plastic structures, a considerable effort has been expended, especially in the USA, on fabrication methods which allow the reinforcement to be ideally positioned with respect to the applied stresses. These processes involve filament winding: a continuous resin-impregnated filament is wound on to a collapsible former in such a way that all the fibres are uniformly stressed.

Glass-reinforced plastics suffer from two major disadvantages: firstly, the elastic moduli of most glasses are rather low, 10^7 lbf/in^2: which leads to comparatively low strengths; and secondly the resins available for impregnating the glass have a poor temperature resistance.

Glass fibres with an elastic modulus of 20–30 $\times$ 10^6 lfb/in^2 are now available in the USA and laminates using this material have correspondingly higher strengths. Developments in chemistry have recently led to resins which are stable at up to 400°C. Whilst these are not as yet universally applicable to the production of laminates, they offer considerable promise for the future.

Some work, again principally in the USA, has also been carried out on the use of alumina silicon-nitride whiskers and silica fibres. Whilst data are sparse, there is reason to believe that these will be even more effective reinforcing materials for plastics than glass fibres.

2) *Reinforced metals*

For certain applications fibre-reinforced metallic matrices offer advantages over the equivalent fibre reinforced plastic, for example because of their better resistance to high temperatures and better thermal and electrical conductivities. To offset these advantages, they generally have a much higher density, and there are greater practical fabrication problems. Experiments designed to produce whisker-reinforced metals which, it is hoped, will approach the theoretical limits of strength, have not yet been very fruitful.

One of the most advanced materials of this type is the silica-glass reinforced aluminium being developed by Rolls Royce. They have successfully coated silica glass fibres of an average tensile strength of about 800 000 lbf/in^2 with aluminium. Hot pressed bars, containing up to a million of these per square inch, have a room-temperature tensile strength of about 110 000 lbf/in^2, dropping about 50% at 400°C.

Tensile strengths of 300 000 to 400 000 lbf/in^2 can be obtained in metal wires of 10^{-3} to 10^{-2} in. diameter and, because of the great difficulties in fabricating fibre-strengthened metals, a number of workers have investigated the properties of wire-reinforced metals, for example steel in aluminium or tungsten in copper. It has been shown that the tensile strength of a composite bar with reinforcing fibres parallel to the tension is approximately proportional to the amounts and tensile strengths of its components. The fatigue strength of the composites will also normally greatly exceed that of the matrix.

3) *Reinforced ceramics*

In the USA some work is being carried out on the development of reinforced ceramics. Because of their brittleness, ceramics have a higher compressive than tensile strength. Thus, if it is desired to subject them to high tensile loads, some form of reinforcement is necessary. However, as the ceramic matrix will normally fail at a lower strain than the reinforcement, cracking and crazing of the ceramic will occur at high tensile loads. The presence of the reinforcement is then required to maintain the integrity of the structure. The design philosophy is analogous to that used in reinforced concrete.

The application of these materials is at present limited to aerospace projects in general and re-entry vehicles in particular, where the high temperature resistance and improved thermal shock resistance and tensile strengths are utilized.

6.3.4 Economic factors

As new materials come on the market the economic laws linking price to production volume start to operate. The supplier of a new material will be able to reduce prices as output increases. But, initially, buyers are cautious and tend to try to use new materials in small quantities, with consequent high costs.

If, however, a new material has a real technical advantage, the designer knows that its costs will decrease (at least in real terms) as the output from the production unit increases. Ultimately a lower limit will be reached, but the rate at which this can take place depends on many factors.

Moreover, in the early stages it takes time for a new material to be proven and accepted in the market place. The rate at which this occurs is also unknown. Nor is the prime cost of the material the whole story.

The selling price of a product will be made up as indicated in Fig. 6.9.

THE ELEMENTS OF COST WILL BE:

	£
Direct labour	XX
Direct material	XX

Prime cost	XX
Manufacturing overhead	XX

Manufacturing cost	XXX
Selling and administration costs	XX

	£XXX

FOR A GENERAL ENGINEERING MANUFACTURING COMPANY

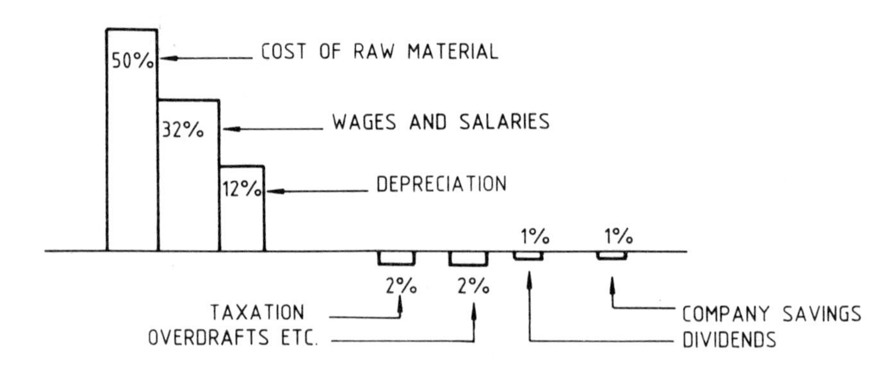

Fig. 6.9 – Cost of manufacturing a product.

In this diagram the cost of material is given as 50% of the factory cost. In some companies today this could be as high as 70%. Therefore designers must carefully assess the economic advantage gained in service (life and maintenance) or machine performance and capacity. The total costs should then be compared with those of achieving a similar result by conventional means. However, a new material may permit a result to be achieved which was hitherto altogether unattainable (for example high-temperature turbine blades made in carbon-fibre composites) so that a comparison of like with like becomes difficult.

The design manager must take care when authorising the use of new materials, to bear all these factors in mind.

6.3.4.1 New design

A new material should never be a straight substitution but should always involve redesign. The new design may be quite different from the previous one, involving new, possibly cheaper, processes; and changes in product performance.

6.3.4.2 Optimization

The balance of the various factors concerning the design, including the economic ones, will change as new materials are introduced, and criteria such as ratios of strength/weight/price, etc. need to be developed. Hence designers must keep in mind all properties of any new material, including the price. Then they must consider the overall job to be tackled from a cost point of view, as well as technically, and develop cost/performance, cost/load-carrying capacity, and cost/life parameters. Mean-time-between-failures will also usually be important. A further source of difficulty to designers is the variations in materials performance. Variations may be due to:

(i) Inherent variability of the material, however pure, owing to, for example its macrostructure. In metals, in particular, which have crystalline structure, and certainly in materials exhibiting a fibrous structure, the intrinsic physical properties may be directional. The strength and elastic properties have to be defined in relation to O_x, O_y, or O_z axes.
(ii) Lack of homogeneity due to method of process or manufacturer, nature of raw material, etc.
(iii) Change in composition or process from one time to another.

In view of these it is necessary to monitor, that is, seek and record the variations in those properties which are basic to design, and of course to take steps to make allowance for these in the actual design procedures.

6.3.5 Selecting and evaluating material suppliers

Every manager must ensure that his enterprise procures materials at the lowest cost consistent with satisfying all the other design objectives including quality, delivery, and serviceability. No longer is it a matter of accepting the lowest price offered by three different suppliers and assuming that market forces will automatically assure value for money.

Design and development have now to become active forces in creating a market. A design manager must recognize that he is involved in someone else's engineering and manufacturing efforts, so that vender capabilities become important. For development work special equipment and test devices are needed, and the purchasers are then buying capability, not mere commodities.

Competitive bidding should operate only when managers have ensured that:

(a) The item required is commercially available to acceptable standards and suitable for the application in question.
(b) All purchasing specifications are clear, unambiguous, and complete.
(c) There are enough qualified and competent suppliers willing to compete.
(d) That the market is elastic and will respond to supply and demand.

However, these conditions are often lacking, in which case design managers may have to negotiate purchases directly with suppliers. Then they must ascertain before commitment, whether the supplier can meet the requirements in terms of quality, delivery, and service. In short, vendors must be evaluated.

6.3.5.1 Vendor rating

Too often vendors are looked upon as suppliers of commodities instead of capabilities. Even when capabilities have been assessed, it is essential for buyers to evaluate suppliers in the light of actual past performance, and this requires the establishment of performance standards.

For this, some form of vendor rating becomes imperative. A standard for assessment should be set up by purchasing departments so that distinctions can be made between good and marginal suppliers.

However, rating systems should not be expected to eliminate or mimize the buyer's judgment; they are merely guides in negotiating specific purchase agreements. Clearly past quality and delivery experience are significant, but past prices are not, unless they are expressed in terms of quality, delivery, and services values. Quality must be rated in terms of cost per usable item, rather than quoted price.

Where the buyer specifies performance requirements it is necessary to measure the costs incurred in satisfying those. Similarly, where services are rendered over and above those specified, it is desirable to evaluate them in terms of costs to compare them wtih the efforts of other suppliers.

6.3.5.1.1 Quality performance

In rating the vendor, *total quality cost* has to be considered. This can only be done by accumulating for each vendor, over a reasonable period, costs under the following: *Prevention* – Vendor surveys, sample approvals, development of test and reliability specifications. *Detection* – Incoming inspection and testing. *Failure and correction* – Manufacturing losses due to vendor error, scrap, re-work, etc. customer complaints and lost sales.

When the costs under these headings are known, a quality rating summary may be drawn up based on the ratio of: Total Quality Assurance Costs/Total Value of Purchases.

6.3.5.1.2 Delivery performance

In addition to rating for quality it is necessary to rate each vendor for delivery performance. Expenditures in this area include follow-up telephone calls, or Telex expense, field expediting and surveillance costs, additional transport charges, as well as manufacturing losses due to late delivery. The total of such costs can similarly be expressed as a percentage of the value of the purchase.

6.3.5.1.3 Service performance

Over and above quality and delivery performance there may be other services given that benefit the customer, and credit should be given for them. These include the capabilities of vendors' personnel and plant, which enable them to offer assistance in emergencies, etc.

While the above rating system does not pretend to be a precise method which automatically selects and rejects vendors, it is an aid in systematic accumulation of experience. These data can then be used by the design and supply departments of a company to relate servicing costs caused by bought-out components and materials to first costs of the equipment that incorporates them.

CHAPTER SUMMARY

Human resources

(i) The uneven need for designers creates a problem of keeping a designer usefully employed for most of his time.
(ii) Only about 80% of a designer's time can be used.
(iii) An amount which varies with the environment in which the company operates must be written off to pay for the work done on unsuccessful projects.
(iv) Because it is impossible to predict which projects will be unsuccessful, it will not be possible to schedule design work far into the future.
(v) If a tendency to underestimate the cost of design is reflected in the underquoting of prices, the allowance for Design Office write-off must be increased.

Information resources

(i) Information for designers comes from various sources which all the design team should be conversant with.
(ii) Company generated information needs to be systematically compiled, coded, and stored for retrieval and re-use.
(iii) The design review can be a useful method of recording design information through the design process.

Material resources

(i) The economic use of material is of paramount importance if good designs are to be achieved.
(ii) Designers should study possible alternative materials to achieve economic designs.
(iii) Selecting material suppliers to achieve the lowest total quality cost requires a vendor rating system.

SUBJECTS FOR DISCUSSION

(1) Discuss the usefulness of critical path methods of scheduling work when resources are not taken into account.

(Any answer should consider the difficulties created by trying to do many jobs in one organization, the need to defer activity starts until resources become available, and the even greater difficulty of attempting to schedule men with specialist skills).

(2) (i) Why cannot a job be scheduled so that no man is idle and the job is never held up for lack of manpower?
(ii) If it is necessary that only 80% of a man's time be available for specified design projects, can the remaining 20% of his time be made use of?

(3) How is the cost of idle time or time spent on unsuccessful projects to be recovered?

(4) How useful would it be if the 'hiring and firing' of designers were acceptable in high-technology industries?
Is the hiring and firing of designers an available management procedure in other advanced countries?

(5) Make a list of the information that might be needed to design:

(i) mechanical linkages,
(ii) thermodynamic machinery,
(iii) electronic equipment,
(iv) medical equipment,
(v) equipment for a coal-fired power station,
(vi) equipment for an atomic power station,
(vii) military equipment.

Classify the types of information that you have listed.

(6) What methods of storing information are likely to be met in a design organization?

(7) List some national and international sources of information.

(8) Can dearer materials that are expensive to machine or mould be cheaper in some applications?

(9) What factors make for the economic purchase of materials (and bought-out sub-assemblies)?

REFERENCES

[1] Leech, *The management of engineering design,* Wiley, London, 1972.
[2] Fabrycky, Ghare, & Torgersen, *Industrial operations research,* Prentice Hall, Englewood Cliffs, N. J., 1972.
[3] Leech & Earthrowl, 'Predicting design costs', *The Aeronautical Journal,* **76** No. 741, R.Ae.S., London, 1972.
[4] Leech, Jenkins, & Turner, 'Simulating the work of a tendering technical company, *C.M.E.,* pp. 79–81, April 1979, I.Mech.E., London.

[5] Downey, W. G., *Development cost estimating,* Report of the Steering Group for the Ministry of Aviation, HMSO, 1969.
[6] Smart, P. M., 'Design budget allocation and project selection', PhD thesis, University of Wales, 1976.

Some useful references to materials

[7] *Engineering materials – An introduction to their properties and applications* Michael F. Ashby & David R. H. Jones. Pergamon Press, 1984.
[8] *Designing with composite material* I.Mech.E. book, 1973.
[9] *Engineering materials – properties and selection* Kenneth Budinski, Prentice Hall, 1979.
[10] *Materials and technology* Vol. III. Metals and Ores, Longman, 1970.
[11] *The new science of strong materials or why you don't fall through the floor* J. E. Gordon, Penguin Books. 1974.

Appendix 6.1

Some useful British Standards (BS) and published documents (PD)

1. *Management and methods*

PD 6470 The management of design for economic production.
BS 6046 Use of network techniques in project management.
BS 4335 Glossary of terms used in project network techniques.
BS 0 A standard for standards (relates to British Standards but useful in preparing standards generally).
PD 6495 IFAN Guide 1. Calculation of the profitability of (company) standardization projects.
BS 4811 Presentation of research and development reports.
PD 57 Methods of equipment reliability testing.
BS 5760 Reliability of systems, equipments and components (covering management, assessment and examples).
PD 9004 BS 9000 CECC and IECOA – UK Administration Guide. Procedures for the national implementation of quality assessment systems for electronic components.
PD 9002 BS 9000 Component selection guide (obtainable only from BSI Hemel Hempstead on annual subscription).
BS 5191 Glossary for production planning and control terms.
BS 4884 Technical manuals (covering content and presentation for operation, maintenance and repair).
BSI Handbook 22 – a complete guide to quality assurance which includes BS 5750 upon which 'assessed capability' is based.

2. *Design information*

2.1 *General*

BS 308 Engineering drawing practice.

PD 6481 Recommendations for the use of preferred numbers and preferred sizes.
BS 1916 Limits and fits for engineering.
BS 4500 ISO limits and fits.
BS 3763 The International System of units (SI).
PD 2856 The use of SI units.
BS 2856 Precise conversion of inch and metric sizes on engineering drawings.
BS 5070 Drawing practice for engineering diagrams.
BS 5536 Specification for preparation of technical drawings and diagrams for microfilming.
BS 1553 Specification for graphical symbols for general engineering.
BS 3939 Graphical symbols for electrical power, telecommunications and electric diagrams.
BS 1991 Letter symbols, signs and abbreviations.
BS 3641 Symbols for machine tools.
BS 4112 Lubrication symbols.
BS 2015 Glossary of paint terms.
BS 499 Welding terms and symbols.
BS 661 Glossary of acoustical terms.
BS 5493 Code of practice for protective coating of iron and steel structures against corrosion.
BS 4479 Recommendations for the design of metal articles that are to be coated.
BS 4675 Mechanical vibration in rotating and reciprocating machinery.
BS 5304 Code of practice for safeguarding of machinery.
BS 5497 Precision of test methods.

2.2 *Colour*

BS 381C Specifications for colours for identification, coding and special purposes.
BS 4727 Glossary of electrotechnical power, telecommunications, lighting and colour terms.
BS 5378 Safety signs and colours.
BS 4800 Specification for paint colours for building purposes.
BS 4099 Specification for colours of indicator lights, pushbuttons, annunciators and digital readouts.

2.3 *Ergonomics*

BS 4467 Anthropometric and ergonomic recommendations for dimensions in designing for the elderly.
BS 5940 Specification for design and dimensions of office workstations, desks, tables and chairs.
BS 3044 Anatomical, physiological and anthropometric principles in the design of office chairs and tables.

Appendix 6.2

Ten useful parameters to be considered in the preparation and production of design information

(1) Easy storage, retrieval and handling.
(2) Good indexing and cross-referencing.
(3) Logical order and consistent format.
(4) Clear and pleasing presentation.
(5) Appropriate degree of detail for the intended readership.
(6) Precision and accuracy of quantitative data using standard units.
(7) Inclusion of publication date and provision for updating.
(8) Indication of sources of further information.
(9) Suitability for use abroad.
(10) Effectiveness with economy.

CHAPTER 7

Planning for design

This chapter states the need for planning of Design work in the context of the business where policy is expressed in terms of a long-range plan into which short-range plans fit. How an annual tactical plan is compiled so that time and cost of design work may be monitored and then controlled is outlined.

Design, no less than any other activity, requires careful planning. Good design plans need to be flexible to take into account unforeseen eventualities. For every management objective there needs to be a plan against which subsequent decisions are made. The plan becomes the standard against which control is exercised.

7.1 BUSINESS PLANNING

We are concerned here only with business planning as it applies to engineering design. The plan will be concerned with spending and getting money. Design, will itself cost money and generate ways of spending and earning money.

There is a difference between an annual budget and a long-term plan. The formal device for expressing business plans is the budget, which amounts to a quantitative statement of plans, in financial terms, produced annually. There may be budgets of manpower, resources, outputs, and other non-monetary quantities, but these quantities are convertible directly to money.

Any business needs three types of annual budget: an operating budget, showing plans for operations for the year; a cash budget, indicating the anticipated sources and uses of cash; and a capital budget, showing planned changes in fixed assets.

On the other hand, a long-term plan seeks to map out the broad direction in which the enterprise is going to move over a number of years ahead.

Long-term planning does not mean that short-term decisions need not be considered for their long-term effects. Any enterprise must develop a set of guidelines which may be referred to, even while making day to day decisions, if

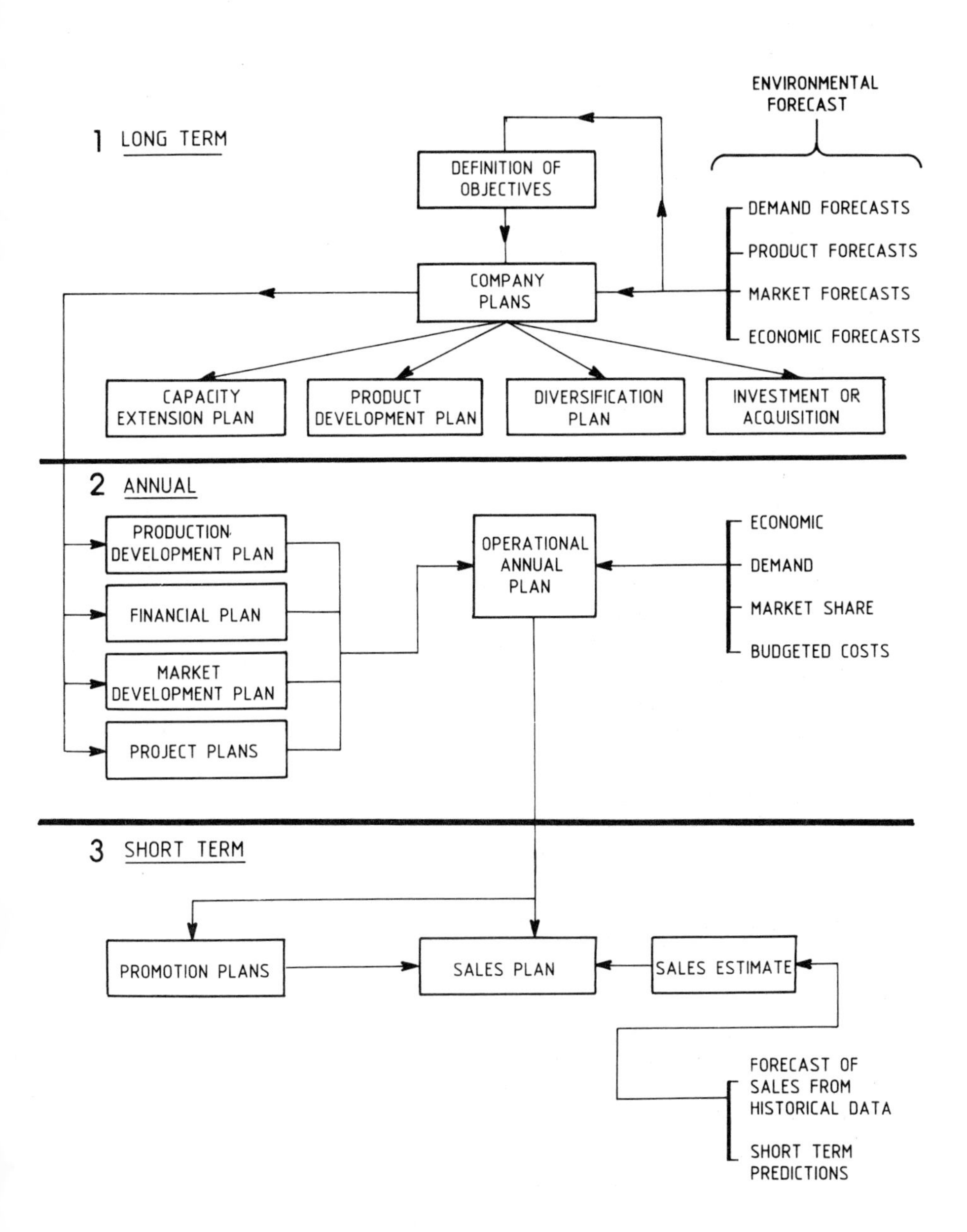

Fig. 7.1 – Planning levels.

only to ensure that no current commitment is going to militate against long-term dvelopments. Guidelines of this kind will often influence the current year's budget.

Management plans must therefore be long-term to cover overall strategy, or short-term to take care of tactics. All planning may be subjectively expressed as prediction prior to any knowledge of results, and will be objectively observable as actions initiated. All planning requires analysis and leads to control, and must express company policy.

Above all, planning involves identifying clear objectives. These must be understood and agreed by all concerned. In steering towards the set objectives, likely problems need to be anticipated and decisions taken in order to circumvent them. The planning spectrum is indicated in Fig. 7.1.

Planning, therefore, from a business point of view may be looked upon as being at three broad levels.

(1) Long-term, strategic planning.
(2) Annual, tactical planning.
(3) Operational, short-term planning.

The time-span varies at level one from five to ten years, depending on the nature of the business, and at level three may be as short as five to ten weeks. It is this level which describes what a firm is doing now, while level two allows current operations to be measured against potential and adjustments to be made in the light of the strategic plans.

All planning must initiate a cycle of planning, doing, and reviewing so that control may be exercised.

7.2 LONG-RANGE PLANNING

First, then, let us look at long-range planning. Terms like 'corporate planning' or 'technological forecasting' are often invoked to cover this work, but we shall not be concerned here with specialized techniques so much as to state the main benefits of long-range planning.

The essence of long-range planning is an attempt to predict competitive threats and market opportunities arising in the trading environment, and to develop a set of guidelines for management decision making to meet them.

These guidelines should show managers the direction their efforts should take and the areas in which resources should be deployed to take advantage of the opportunities or to parry the threats. Note that we are not trying to propose a system which caters only for a threat that might arise in the distant future, but one which will also cope with a current threat.

A strategic plan must delineate a course of action which can cope with changed circumstances and ensure that facilities and skills are flexible enough to change direction as required for survival. In short it enables priority objectives to be kept clearly in view while allowing latitude in the way they are met.

7.2.1 The structure of long-range planning

The first step in business planning is to assess, under various headings, the firm's potential. There will be the potential of the internal resources under such headings as Marketing, Production, Finance, Design, Research and Development. These internal resources will need to be considered in the light of the external influence of the actual trading conditions. The balancing of these two factors will lead to the overall business objectives depicted in Fig. 7.1. This in turn will enable the overall objectives to be broken down into individual targets for each specialist function, including design.

The procedure must be iterative, and a very simplified description of a long-range plan is that:

(1) the managing director determines what would be a reasonable return on the capital employed;
(2) the finance and marketing directors determine the volume of sales that would be needed to generate the stated profit in the context of the business that the firm is in, and the state of the market;
(3) the marketing director determines the product mix and delivery times for the required volume of sales;
(4) the production director agrees that the required product mix can be made in the time available. In doing so he will specify what money he needs to meet the target;
(5) the finance director agrees that the capital required by the programme can be obtained at an acceptable cost;
(6) the technical director agrees that the drawings of the developed products will be made available in time to meet the programme.

It is clearly unlikely that all the directors will immediately find common ground; that the production director will be able to make what the marketing director wants to sell; that the technical director will be able to develop a new product by the time it is wanted; that the finance director will be able to raise enough capital at an acceptable cost; etc. There will be gaps between what is desired and what can be achieved. These gaps will be analysed and plans modified.

Eventually a plan will be formulated which all managers will agree to be acceptable and feasible. Each department manager in the company will then have long-range objectives and a knowledge of the money that will be available to meet those objectives. In particular the technical director will know the long-term design needs and the money that he will be permitted to spend to meet those needs. The process will be reviewed at least annually to make sure that new factors are taken into account as they arise. This type of planning will set targets in four main areas of business:

(1) Capacity extension plan.
(2) Product or project development plan.

(3) Diversification plan.
and, if necessary,
(4) Divestment and acquisition plan.

The discipline of such an analysis provides top management with the following advantages:

- it makes everyone aware of the cost targets and anticipated profits,
- it provides a set of standards for control purposes,
- it gives a clear indication of the priorities that should be set at any time, and
- it shows where management development is necessary.

7.3 SHORT-RANGE PLANNING

In any business it is just as important to keep track of the short-term events as of the long-term plans. It is often desirable to turn big unwieldy jobs into small management tasks so that proper supervision can be exercised.

By and large, any supervisory task is concerned with carrying out instructions for operating and organizing, by breaking down the task into individual assignments, and ensuring their satisfactory completion by the imposition of some form of discipline and control. It is in this area of work that it becomes essential for proper short-term plans to be prepared and progressed carefully.

7.3.1 Annual tactical planning

Partly as a consequence of, and certainly complementary with, the long-term plan, the annual plan will be broken down into a budget and a responsibility structure. The former will lay down the anticipated revenue and cost associated with each product line, while the latter will assign responsibility for these to groups or individuals within the organization, and consequently becomes a way of delegating work. The budget therefore primarily becomes a control and communicaton device.

It is easy to see that, in the Production Department, the annual tactical plan will state how much money may be spent on capital equipment and how much money may be spent in each accounting period on wages, materials, and other cash operating costs. The plan will also state what finished products must leave the factory in each accounting period. The questions, 'Have we spent no more than we predicted in making each of these products?' 'Have we made the products demanded, to the value demanded and by the time demanded?', clearly form a basis for the management of the department.

The costs must be predicted in the context of the required delivery programme. Then the costs incurred must be monitored, and the attainment of delivery dates must be measured. Although, in the case of production, the management procedure is clear and clearly derived from the long- and short-term plans, it is not often that plans are adhered to without difficulty. Commonly, materials and manufacturing processes cost more than were allowed for, and, either through poor prediction or competition, revenue from sales is less than had been predicted. Cost reduction or product enhancement exercises will then be generated. These exercises will, in turn, generate design projects.

7.4 CONTROL OF DESIGN

In the design department, the management process is less easy to equate with the control of money. This is partly because most of the costs are labour costs, and the total cost of labour will not change dramatically from one accounting period to another. Also, since the finished product which emerges from the design department is a set of drawings and other documents for use by other departments within the company, the monetary value of the work done in an accounting period is not obvious. Nevertheless, if the drawings are not produced on time, the production department will not meet its target, and if the development is not done on time, the drawings will not be done on time.

The problem of the control of design becomes one of

- forecasting costs,
- allocating resources (costs) to each project,
- setting the programme for each project so that the drawings will be delivered on time,
- monitoring the use of resources by each project, and
- monitoring the course of each project against its programme.

The role of the designer in all this is crucial. In formulating the long-term plan the prediction will not be detailed, and some approximation may be permissible, but in formulating the annual plan detailed costs must be predicted. In the first instance, the prediction will be necessary if management are to decide whether a project is worth undertaking, and as we shall see, it is as important to know the timing of costs as the costs themselves. Secondly, if costs are to be monitored their magnitude and timing must be predicted. As these predictions must be made when only the designer has any knowledge of what will be built, he cannot be left out of the prediction process. The designer therefore has a part to play in setting up cost data against which the performance of every department in the company will be monitored.

The design department itself is a place where money is spent, and so some of the costs of any project, which must be predicted, are those incurred in design and development and against which the designer's own performance will be monitored.

Generally, we think of a design project as the design, manufacture, and marketing of a new product, but, as we have seen, a project may be the reduction of the cost of an existing product or the enhancement of an existing product to make it easier to sell. Cost reduction and product enhancement may even form the major part of the design department's work, and such projects require no less creativity and technical skill than the invention of totally new products. The fact that such projects are sometimes forced on a company does, however, mean that the annual plan must be flexible enough to stand modification.

Whatever the nature of the design work may be, the first step in planning is to define the scope and divide up the work into convenient packages specifying tasks which have to be completed. This amounts to preparing a logical breakdown of what has to be done, which, incidentally, can be used to provide a cost structure for the complete design project work. See Fig. 7.2.

Whatever may be the size or outcome of a project, we must consider the cash flows that will result from undertaking it. Without those cash flows we can neither assess the desirability of the project nor manage it when we have undertaken it.

7.5 CHAPTER SUMMARY

7.5.1 Engineering design must be considered in the formulation of a company's long-range business plan. One outcome of planning is the company's annual design plans.

7.5.2 Long and short range planning leads to the formulation of a detailed programme of design work.

7.5.3 The programme of design work provides cost and time data against which actual design costs and times may be monitored. The programme is thus the basis of any system of design management.

FURTHER READING

Turner, B. T., Design Management and Product Strategy, read at the conference 'Efficiency in the Design Office', June 1983, London, and published by Mechanical Engineering Publications, Worthing.
This paper contains further references.

SUBJECTS FOR DISCUSSION

(1) Give examples from history of major changes in an industry which might have been taken into account by long-range planning.
(2) How often should a long-range plan be formulated?
(3) What procedure could be adopted for formulating a long-range plan?
(4) What costs are involved in the formulation of a long-range plan?
(5) What costs might be generated by failure to formulate a long-range plan?
(6) How could a long-range plan be published?
(7) Why not let a company die?

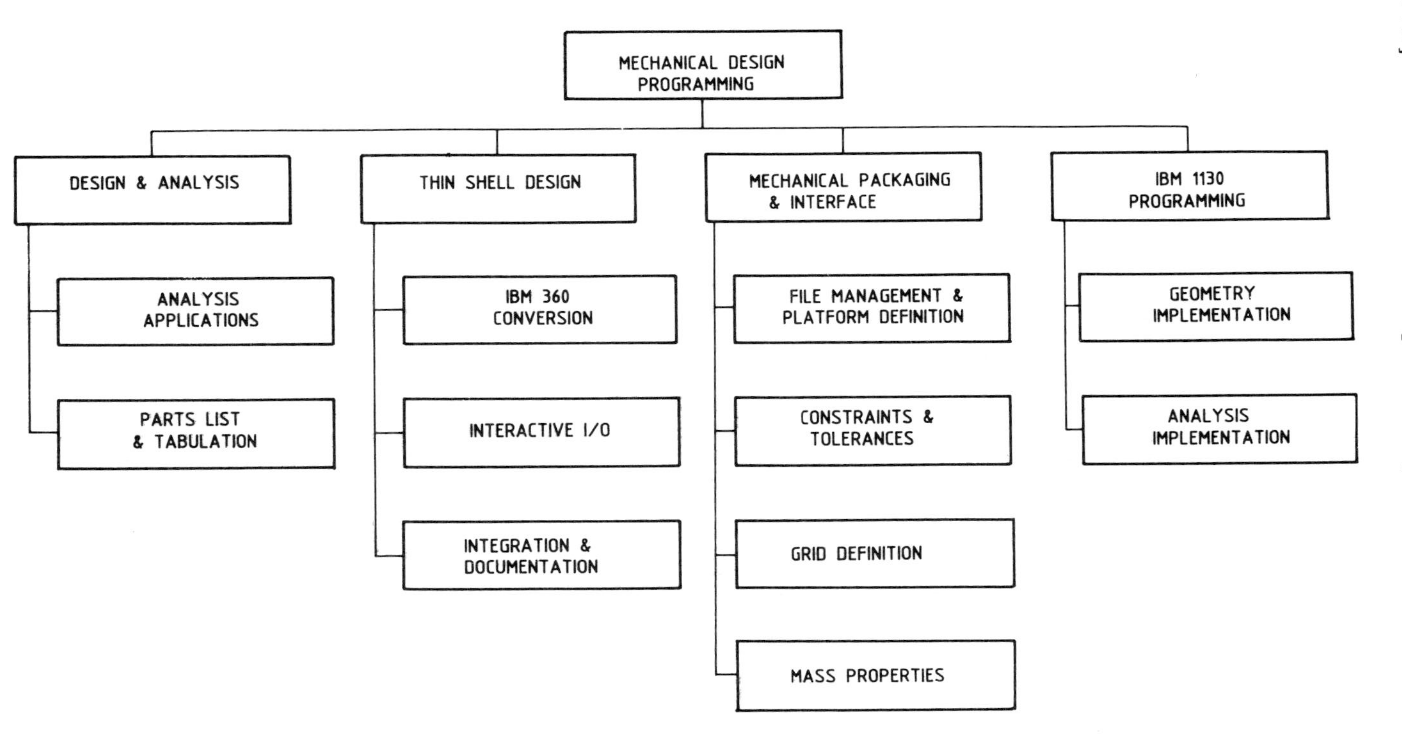

Fig. 7.2 – A work breakdown structure for a CAD project.

(8) Give examples of changes in technology which needed to be taken into account by a company wishing to stay in its established business.
Examples could be:

(a) aeroplane engine manufacturers and the change from propellors to jets;
(b) aeroplane engine manufacturers and the development of carbon fibre;
(c) the change from black and white to colour television;
(d) the reduction in the cost of computers that can be used in the management of small business;
(e) . . .
(f) . . .

Discuss one of your examples.

(9) What procedure could be adopted for formulating an annual plan, and who would be involved?
(10) How is the annual plan related to (or constrained by) the long-range plan?
(11) What is the relation between the plan and the budget as seen by the Senior Designer?
(12) How does the plan provide a basis for managing design?

CHAPTER 8

The design project as a cash flow stream

This chapter shows that accepting a design project will generate a stream of cash flows. Discounted Cash Flow methods are offered for the calculation of the Net Present Value or the Rate of Return of the cash flow stream that the project generates.

8.1 THE CASH FLOW STREAM

Is a project worth doing? Is it worth spending money on the design of a new product?

Many people believe that it would have been better never to have started work on the Concorde or Advanced Gas Cooled Reactors or Tracked Hovercraft. This is not because they believed that the designers would fail to solve the technical problems, but that when the problems were solved, the products had little or no chance of being profitable. The designer often sees an exciting technical problem which he solves with a few more hours of drawing or a few more hours of analysis on the computer. In fact, the inventive work of the designer is only the first, small cost in a project which will, inevitably, require much greater expenditure before the product can be successful. A manager should not, generally, allow a designer to spend a few hundred hours solving technical problems if he is not prepared to spend much more money (perhaps hundreds of thousands of pounds) later on development and tooling.

One way of looking at design work is to examine the cash flow stream that it is expected to generate. We may then decide that it would be better not to start the project. If we do start the project, the predicted cash flows and their timing will provide benchmarks against which the project may be monitored and managed. Fig. 8.1 shows the shape of the cash flow stream generated by a project to design and build a new product. This curve is sometimes referred to as the 'White Elephant Curve'.

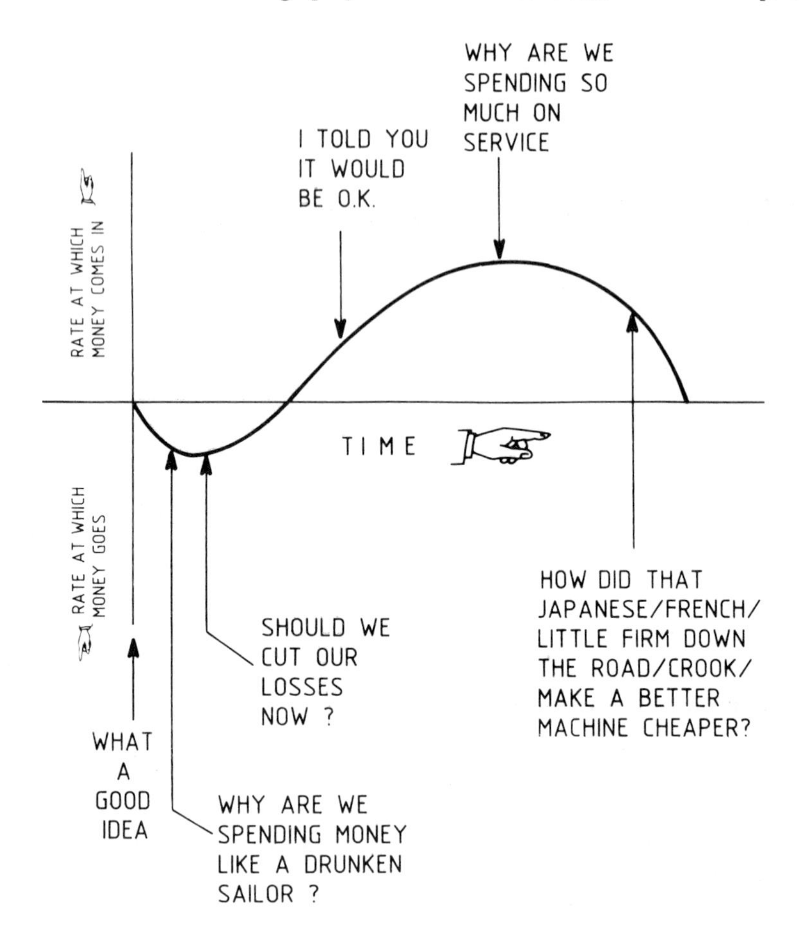

Fig. 8.1 – White elephant curve.

The early cash flows are negative because they are associated with costs to the manufacturer. They are costs of design and development, the costs of setting up appropriate manufacturing facilities, the costs of setting up product support systems and other costs which will be discussed in more detail later. Even when the product is being sold and is generating a positive cash flow, the net cash flow will have negative components which result from servicing costs, design changes, the rectification of faults and perhaps from extending the life of the product.

Clearly the project cannot be profitable unless later positive cash flows more than compensate for early costs. All we are saying, here, is that by the end of the project, more money must have flowed into the company than out. One suspects that even this simple criterion is ignored by some designers, but, as we shall see, even an apparent profit over the life of a project may be insufficient to justify it. Capital must usually be borrowed, so that apparent profit needs to be great enough to pay back interest on debts before it can be considered as a real profit. One reason for calling the cash flow profile a white elephant curve is that, early in the project life, only negative cash flows will be apparent. Before the curve starts turning up, the managers will be asking if the project is a white elephant. Is development a bottomless pit into which money is being poured with little chance of technical success? Will the investment in tooling ever be paid for by sales of the product?

8.2 DISCOUNTED CASH FLOWS

8.2.1 Present values

Consider that at the end of year 0 (that is, at the beginning of the project) a sum of £P is borrowed at an interest rate of $100 \times i\%$; consider that at the end of year n the debt is repaid by a single payment of £$F(n)$; then at the end of year n, immediately before repayment, the sum owed is

$$£P(1 + i)^n$$

Therefore

$$F(n) = P(1 + i)^n$$

P is called the Present Value of $F(n)$.
In other words, the present value of $F(n)$ is

$$F(n)/(1 + i)^n$$

Suppose the project generates the Cash Flow Stream,

$$F(0), F(1) \ldots\ldots\ldots, F(n), \ldots\ldots\ldots\ldots F(N);$$

where $F(n)$ is the cash flow generated at the end of year n, and the project lasts N years. It is likely that the early cash flows, $F(0), F(1), \ldots$, will be negative and will represent borrowing, while the later cash flows will be repayments.

All loans will be cleared, however, if the Net Present Value (NPV) of the project which is the sum of the Present Values of all the cash flows, $F(n)$, is zero.

If the NPV of the project is less than zero, then the project will not pay back those who have invested in it.

Note that a cash flow cannot always be identified with an accounting entry. Fig. 8.2 shows the cash flows and gives some definitions. In discounted cash flow calculations, note that tax is one of the cash flows that must be taken into account. Interest and dividends are allowed for by using the weighted average, tax-adjusted cost of capital.

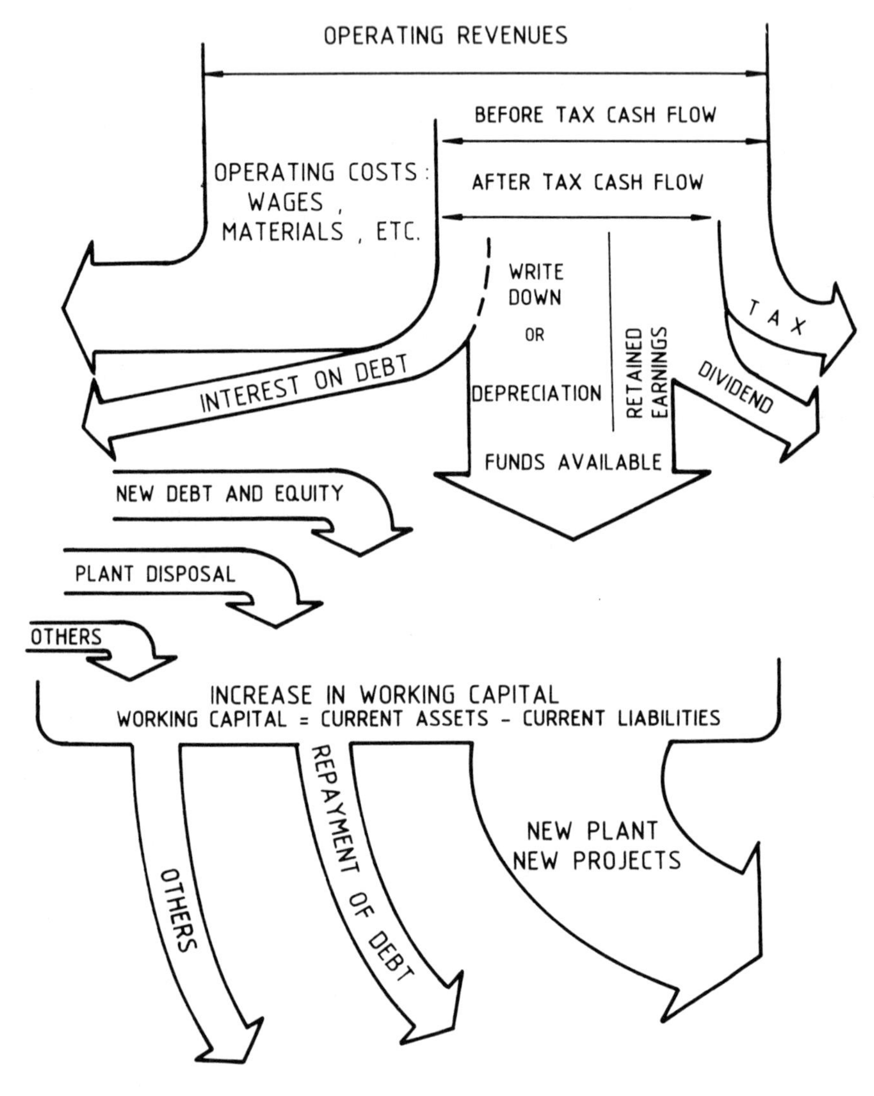

Fig. 8.2 – The owner's firm.

Fig. 8.3 lists some formulae that are useful in calculating the Net Present Values of projects.

P = Present Value; $F(n)$ = cash flow at end of year n;
i = discount rate;
N = life of project in years.

If $F(n)$ is the only future cash flow,

$$P = \frac{F(n)}{(1+i)^n}$$

otherwise

$$P = \frac{F(1)}{1+i} + \frac{F(2)}{(1+i)^2} \quad \ldots \quad \frac{F(N)}{(1+i)} N.$$

If $F(1) = F(2) = F(3) \ldots = F(N) = A$

$$P = \frac{A}{1+i} + \frac{A}{(1+i)^2} \ldots \frac{A}{(1+i)} N$$

$$= \frac{A}{i} \frac{(1+i)^N - 1}{(1+i)^N}.$$

If i is the nominal annual interest rate but discounting is done m times a year,

$$P = \frac{F(n)}{(1 + i/m)^{mn}}.$$

As m increases, discounting becomes continuous and

$$P = F(n)\, e^{-in}$$

Some values of $P/F(n)$ and P/A are tabulated on Fig. 8.4.

Fig. 8.3 – Calculating net present values

Fig. 8.4 tabulates present value factors for different project lives.

Fig. 8.4 – Present worth of a future payment (P/F)I, N

YEARS	0.5%	1.0%	1.5%	2.0%	2.5%	3.0%	4.0%	5.0%	6.0%	8.0%	10.0%	12.0%	14.0%	15.0%	16.0%	18.0%
1	0.995	0.990	0.985	0.980	0.976	0.971	0.962	0.952	0.943	0.926	0.909	0.893	0.877	0.870	0.862	0.847
2	0.990	0.980	0.971	0.961	0.952	0.943	0.925	0.907	0.890	0.857	0.826	0.797	0.769	0.756	0.743	0.718
3	0.985	0.971	0.956	0.942	0.929	0.915	0.889	0.864	0.840	0.794	0.751	0.712	0.675	0.658	0.641	0.609
4	0.980	0.961	0.942	0.924	0.906	0.888	0.855	0.823	0.792	0.735	0.683	0.636	0.592	0.572	0.552	0.516
5	0.975	0.951	0.928	0.906	0.884	0.863	0.822	0.784	0.747	0.681	0.621	0.567	0.519	0.497	0.476	0.437
6	0.971	0.942	0.915	0.888	0.862	0.837	0.790	0.746	0.705	0.630	0.564	0.507	0.456	0.432	0.410	0.370
7	0.966	0.933	0.901	0.871	0.841	0.813	0.760	0.711	0.665	0.583	0.513	0.452	0.400	0.376	0.354	0.314
8	0.961	0.923	0.888	0.853	0.821	0.789	0.731	0.677	0.627	0.540	0.467	0.404	0.351	0.327	0.305	0.266
9	0.956	0.914	0.875	0.837	0.801	0.766	0.703	0.645	0.592	0.500	0.424	0.361	0.308	0.284	0.263	0.225
10	0.951	0.905	0.862	0.820	0.781	0.744	0.676	0.614	0.558	0.463	0.386	0.322	0.270	0.247	0.227	0.191
11	0.947	0.896	0.849	0.804	0.762	0.722	0.650	0.585	0.527	0.429	0.350	0.287	0.237	0.215	0.195	0.162
12	0.942	0.887	0.836	0.788	0.744	0.701	0.625	0.557	0.497	0.397	0.319	0.257	0.208	0.187	0.168	0.137
13	0.937	0.879	0.824	0.773	0.725	0.681	0.601	0.530	0.469	0.368	0.290	0.229	0.182	0.163	0.145	0.116
14	0.933	0.870	0.812	0.758	0.708	0.661	0.577	0.505	0.442	0.340	0.263	0.205	0.160	0.141	0.125	0.099
15	0.928	0.861	0.800	0.743	0.690	0.642	0.555	0.481	0.417	0.315	0.239	0.183	0.140	0.123	0.108	0.084
16	0.923	0.853	0.788	0.728	0.674	0.623	0.534	0.458	0.394	0.292	0.218	0.163	0.123	0.107	0.093	0.071
17	0.919	0.844	0.776	0.714	0.657	0.605	0.513	0.436	0.371	0.270	0.198	0.146	0.108	0.093	0.080	0.060
18	0.914	0.836	0.765	0.700	0.641	0.587	0.494	0.416	0.350	0.250	0.180	0.130	0.095	0.081	0.069	0.051
19	0.910	0.828	0.754	0.686	0.626	0.570	0.475	0.396	0.331	0.232	0.164	0.116	0.083	0.070	0.060	0.043
20	0.905	0.820	0.742	0.673	0.610	0.554	0.456	0.377	0.312	0.215	0.149	0.104	0.073	0.061	0.051	0.037
21	0.901	0.811	0.731	0.660	0.595	0.538	0.439	0.359	0.294	0.199	0.135	0.093	0.064	0.053	0.044	0.031
22	0.896	0.803	0.721	0.647	0.581	0.522	0.422	0.342	0.278	0.184	0.123	0.083	0.056	0.046	0.038	0.026
23	0.892	0.795	0.710	0.634	0.567	0.507	0.406	0.326	0.262	0.170	0.112	0.074	0.049	0.040	0.033	0.022
24	0.887	0.788	0.700	0.622	0.553	0.492	0.390	0.310	0.247	0.158	0.102	0.066	0.043	0.035	0.028	0.019
25	0.883	0.780	0.689	0.610	0.539	0.478	0.375	0.295	0.233	0.146	0.092	0.059	0.038	0.030	0.024	0.016
26	0.878	0.772	0.679	0.598	0.526	0.464	0.361	0.281	0.220	0.135	0.084	0.053	0.033	0.026	0.021	0.014
27	0.874	0.764	0.669	0.586	0.513	0.450	0.347	0.268	0.207	0.125	0.076	0.047	0.029	0.023	0.018	0.011
28	0.870	0.757	0.659	0.574	0.501	0.437	0.333	0.255	0.196	0.116	0.069	0.042	0.026	0.020	0.016	0.010
29	0.865	0.749	0.649	0.563	0.489	0.424	0.321	0.243	0.185	0.107	0.063	0.037	0.022	0.017	0.014	0.008
30	0.861	0.742	0.640	0.552	0.477	0.412	0.308	0.231	0.174	0.099	0.057	0.033	0.020	0.015	0.012	0.007
31	0.857	0.735	0.630	0.541	0.465	0.400	0.296	0.220	0.164	0.092	0.052	0.030	0.017	0.013	0.010	0.006
32	0.852	0.727	0.621	0.531	0.454	0.388	0.285	0.210	0.155	0.085	0.047	0.027	0.015	0.011	0.009	0.005
33	0.848	0.720	0.612	0.520	0.443	0.377	0.274	0.200	0.146	0.079	0.043	0.024	0.013	0.010	0.007	0.004
34	0.844	0.713	0.603	0.510	0.432	0.366	0.264	0.190	0.138	0.073	0.039	0.021	0.012	0.009	0.006	0.004
35	0.840	0.706	0.594	0.500	0.421	0.355	0.253	0.181	0.130	0.068	0.036	0.019	0.010	0.008	0.006	0.003
36	0.836	0.699	0.585	0.490	0.411	0.345	0.244	0.173	0.123	0.063	0.032	0.017	0.009	0.007	0.005	0.003
37	0.831	0.692	0.576	0.481	0.401	0.335	0.234	0.164	0.116	0.058	0.029	0.015	0.008	0.006	0.004	0.002
38	0.827	0.685	0.568	0.471	0.391	0.325	0.225	0.157	0.109	0.054	0.027	0.013	0.007	0.005	0.004	0.002
39	0.823	0.678	0.560	0.462	0.382	0.316	0.217	0.149	0.103	0.050	0.024	0.012	0.006	0.004	0.003	0.002
40	0.819	0.672	0.551	0.453	0.372	0.307	0.208	0.142	0.097	0.046	0.022	0.011	0.005	0.004	0.003	0.001
41	0.815	0.665	0.543	0.444	0.363	0.298	0.200	0.135	0.092	0.043	0.020	0.010	0.005	0.003	0.002	0.001
42	0.811	0.658	0.535	0.435	0.354	0.289	0.193	0.129	0.087	0.039	0.018	0.009	0.004	0.003	0.002	0.001
43	0.807	0.652	0.527	0.427	0.346	0.281	0.185	0.123	0.082	0.037	0.017	0.008	0.004	0.002	0.002	0.001
44	0.803	0.645	0.519	0.418	0.337	0.272	0.178	0.117	0.077	0.034	0.015	0.007	0.003	0.002	0.001	0.001
45	0.799	0.639	0.512	0.410	0.329	0.264	0.171	0.111	0.073	0.031	0.014	0.006	0.003	0.002	0.001	0.001
46	0.795	0.633	0.504	0.402	0.321	0.257	0.165	0.106	0.069	0.029	0.012	0.005	0.002	0.002	0.001	0.000
47	0.791	0.626	0.497	0.394	0.313	0.249	0.158	0.101	0.065	0.027	0.011	0.005	0.002	0.001	0.001	0.000
48	0.787	0.620	0.489	0.387	0.306	0.242	0.152	0.096	0.061	0.025	0.010	0.004	0.002	0.001	0.001	0.000
49	0.783	0.614	0.482	0.379	0.298	0.235	0.146	0.092	0.058	0.023	0.009	0.004	0.002	0.001	0.001	0.000
50	0.779	0.608	0.475	0.372	0.291	0.228	0.141	0.087	0.054	0.021	0.009	0.003	0.001	0.001	0.001	0.000

Fig. 8.4 – continued

YEARS	20.0%	25.0%	30.0%	35.0%	40.0%	50.0%
1	0.833	0.800	0.769	0.741	0.714	0.667
2	0.694	0.640	0.592	0.549	0.510	0.444
3	0.579	0.512	0.455	0.406	0.364	0.296
4	0.482	0.410	0.350	0.301	0.260	0.198
5	0.402	0.328	0.269	0.223	0.186	0.132
6	0.335	0.262	0.207	0.165	0.133	0.088
7	0.279	0.210	0.159	0.122	0.095	0.059
8	0.233	0.168	0.123	0.091	0.068	0.039
9	0.194	0.134	0.094	0.067	0.048	0.026
10	0.162	0.107	0.073	0.050	0.035	0.017
11	0.135	0.086	0.056	0.037	0.025	0.012
12	0.112	0.069	0.043	0.027	0.018	0.008
13	0.093	0.055	0.033	0.020	0.013	0.005
14	0.078	0.044	0.025	0.015	0.009	0.003
15	0.065	0.035	0.020	0.011	0.006	0.002
16	0.054	0.028	0.015	0.008	0.005	0.002
17	0.045	0.023	0.012	0.006	0.003	0.001
18	0.038	0.018	0.009	0.005	0.002	0.001
19	0.031	0.014	0.007	0.003	0.002	0.000
20	0.026	0.012	0.005	0.002	0.001	0.000
21	0.022	0.009	0.004	0.002	0.001	0.000
22	0.018	0.007	0.003	0.001	0.001	0.000
23	0.015	0.006	0.002	0.001	0.000	0.000
24	0.013	0.005	0.002	0.001	0.000	0.000
25	0.010	0.004	0.001	0.001	0.000	0.000

Fig. 8.4 – continued – Present worth of a uniform series (Annuities) (P/A)I, N

YEARS	0.5%	1.0%	1.5%	2.0%	2.5%	3.0%	4.0%	5.0%	6.0%	8.0%	10.0%	12.0%	14.0%	15.0%	16.0%	18.0%
1	0.995	0.990	0.985	0.980	0.976	0.971	0.962	0.952	0.943	0.926	0.909	0.893	0.877	0.870	0.862	0.847
2	1.985	1.970	1.956	1.942	1.927	1.913	1.886	1.859	1.833	1.783	1.736	1.690	1.647	1.626	1.605	1.566
3	2.970	2.941	2.912	2.884	2.856	2.829	2.775	2.723	2.673	2.577	2.487	2.402	2.322	2.283	2.246	2.174
4	3.950	3.902	3.854	3.808	3.762	3.717	3.630	3.546	3.465	3.312	3.170	3.037	2.914	2.855	2.798	2.690
5	4.926	4.853	4.783	4.713	4.646	4.580	4.452	4.329	4.212	3.993	3.791	3.605	3.433	3.352	3.274	3.127
6	5.896	5.795	5.697	5.601	5.508	5.417	5.242	5.076	4.917	4.623	4.355	4.111	3.889	3.784	3.685	3.498
7	6.862	6.728	6.598	6.472	6.349	6.230	6.002	5.786	5.582	5.206	4.868	4.564	4.288	4.160	4.039	3.812
8	7.823	7.652	7.486	7.325	7.170	7.020	6.733	6.463	6.210	5.747	5.335	4.968	4.639	4.487	4.344	4.078
9	8.779	8.566	8.361	8.162	7.971	7.786	7.435	7.108	6.802	6.247	5.759	5.328	4.946	4.772	4.607	4.303
10	9.730	9.471	9.222	8.983	8.752	8.530	8.111	7.722	7.360	6.710	6.145	5.650	5.216	5.019	4.833	4.494
11	10.677	10.368	10.071	9.787	9.514	9.253	8.760	8.306	7.887	7.139	6.495	5.938	5.453	5.234	5.029	4.656
12	11.619	11.255	10.908	10.575	10.258	9.954	9.385	8.863	8.384	7.536	6.814	6.194	5.660	5.421	5.197	4.793
13	12.556	12.134	11.732	11.348	10.983	10.635	9.986	9.394	8.853	7.904	7.103	6.424	5.842	5.583	5.342	4.910
14	13.489	13.004	12.543	12.106	11.691	11.296	10.563	9.899	9.295	8.244	7.367	6.628	6.002	5.724	5.468	5.008
15	14.417	13.865	13.343	12.849	12.381	11.938	11.118	10.380	9.712	8.559	7.606	6.811	6.142	5.847	5.575	5.092
16	15.340	14.718	14.131	13.578	13.055	12.561	11.652	10.838	10.106	8.851	7.824	6.974	6.265	5.954	5.668	5.162
17	16.259	15.562	14.908	14.292	13.712	13.166	12.166	11.274	10.477	9.122	8.022	7.120	6.373	6.047	5.749	5.222
18	17.173	16.398	15.673	14.992	14.353	13.754	12.659	11.690	10.828	9.372	8.201	7.250	6.467	6.128	5.818	5.273
19	18.082	17.226	16.426	15.678	14.979	14.324	13.134	12.085	11.158	9.604	8.365	7.366	6.550	6.198	5.877	5.316
20	18.987	18.046	17.169	16.351	15.589	14.877	13.590	12.462	11.470	9.818	8.514	7.469	6.623	6.259	5.929	5.353
21	19.888	18.857	17.900	17.011	16.185	15.415	14.029	12.821	11.764	10.017	8.649	7.562	6.687	6.312	5.973	5.384
22	20.784	19.660	18.621	17.658	16.765	15.937	14.451	13.163	12.042	10.201	8.772	7.645	6.743	6.359	6.011	5.410
23	21.676	20.456	19.331	18.292	17.332	16.444	14.857	13.489	12.303	10.371	8.883	7.718	6.792	6.399	6.044	5.432
24	22.563	21.243	20.030	18.914	17.885	16.936	15.247	13.799	12.550	10.529	8.985	7.784	6.835	6.434	6.073	5.451
25	23.446	22.023	20.720	19.523	18.424	17.413	15.622	14.094	12.783	10.675	9.077	7.843	6.873	6.464	6.097	5.467
26	24.324	22.795	21.399	20.121	18.951	17.877	15.983	14.375	13.003	10.810	9.161	7.896	6.906	6.491	6.118	5.480
27	25.198	23.560	22.068	20.707	19.464	18.327	16.330	14.643	13.211	10.935	9.237	7.943	6.935	6.514	6.136	5.492
28	26.068	24.316	22.727	21.281	19.965	18.764	16.663	14.898	13.406	11.051	9.307	7.984	6.961	6.534	6.152	5.502
29	26.933	25.066	23.376	21.844	20.454	19.188	16.984	15.141	13.591	11.158	9.370	8.022	6.983	6.551	6.166	5.510
30	27.794	25.808	24.016	22.396	20.930	19.600	17.292	15.372	13.765	11.258	9.427	8.055	7.003	6.566	6.177	5.517
31	28.651	26.542	24.646	22.938	21.395	20.000	17.588	15.593	13.929	11.350	9.479	8.085	7.020	6.579	6.187	5.523
32	29.503	27.270	25.267	23.468	21.849	20.389	17.874	15.803	14.084	11.435	9.526	8.112	7.035	6.591	6.196	5.528
33	30.352	27.990	25.879	23.989	22.292	20.766	18.148	16.003	14.230	11.514	9.569	8.135	7.048	6.600	6.203	5.532
34	31.196	28.703	26.482	24.499	22.724	21.132	18.411	16.193	14.368	11.587	9.609	8.157	7.060	6.609	6.210	5.536
35	32.035	29.409	27.076	24.999	23.145	21.487	18.665	16.374	14.498	11.655	9.644	8.176	7.070	6.617	6.215	5.539
36	32.871	30.108	27.661	25.489	23.556	21.832	18.908	16.547	14.621	11.717	9.677	8.192	7.079	6.623	6.220	5.541
37	33.703	30.800	28.237	25.969	23.957	22.167	19.143	16.711	14.737	11.775	9.706	8.208	7.087	6.629	6.224	5.543
38	34.530	31.485	28.805	26.441	24.349	22.492	19.368	16.868	14.846	11.829	9.733	8.221	7.094	6.634	6.228	5.545
39	35.353	32.163	29.365	26.903	24.730	22.808	19.584	17.017	14.949	11.879	9.757	8.233	7.100	6.638	6.231	5.547
40	36.172	32.835	29.916	27.355	25.103	23.115	19.793	17.159	15.046	11.925	9.779	8.244	7.105	6.642	6.233	5.548
41	36.987	33.500	30.459	27.799	25.466	23.412	19.993	17.294	15.138	11.967	9.799	8.253	7.110	6.645	6.236	5.549
42	37.798	34.158	30.994	28.235	25.821	23.701	20.186	17.423	15.225	12.007	9.817	8.262	7.114	6.648	6.238	5.550
43	38.605	34.810	31.521	28.662	26.166	23.982	20.371	17.546	15.306	12.043	9.834	8.270	7.117	6.650	6.239	5.551
44	39.408	35.455	32.041	29.080	26.504	24.254	20.549	17.663	15.383	12.077	9.849	8.276	7.120	6.652	6.241	5.552
45	40.207	36.095	32.552	29.490	26.833	24.519	20.720	17.774	15.456	12.108	9.863	8.283	7.123	6.654	6.242	5.552
46	41.002	36.727	33.056	29.892	27.154	24.775	20.885	17.880	15.524	12.137	9.875	8.288	7.126	6.656	6.243	5.553
47	41.793	37.354	33.553	30.287	27.467	25.025	21.043	17.981	15.589	12.164	9.887	8.293	7.128	6.657	6.244	5.553
48	42.580	37.974	34.043	30.673	27.773	25.267	21.195	18.077	15.650	12.189	9.897	8.297	7.130	6.659	6.245	5.554
49	43.364	38.588	34.525	31.052	28.071	25.502	21.341	18.169	15.708	12.212	9.906	8.301	7.131	6.660	6.246	5.554
50	44.143	39.196	35.000	31.424	28.362	25.730	21.482	18.256	15.762	12.233	9.915	8.304	7.133	6.661	6.246	5.554

Fig. 8.4 – continued

YEARS	20.0%	25.0%	30.0%	35.0%	40.0%	50.0%
1	0.833	0.800	0.769	0.741	0.714	0.667
2	1.528	1.440	1.361	1.289	1.224	1.111
3	2.106	1.952	1.816	1.696	1.589	1.407
4	2.589	2.362	2.166	1.997	1.849	1.605
5	2.991	2.689	2.436	2.220	2.035	1.737
6	3.326	2.951	2.643	2.385	2.168	1.824
7	3.605	3.161	2.802	2.508	2.263	1.883
8	3.837	3.329	2.925	2.598	2.331	1.922
9	4.031	3.463	3.019	2.665	2.379	1.948
10	4.192	3.571	3.092	2.715	2.414	1.965
11	4.327	3.656	3.147	2.752	2.438	1.977
12	4.439	3.725	3.190	2.779	2.456	1.985
13	4.533	3.780	3.223	2.799	2.469	1.990
14	4.611	3.824	3.249	2.814	2.478	1.993
15	4.675	3.859	3.268	2.825	2.484	1.995
16	4.730	3.887	3.283	2.834	2.489	1.997
17	4.775	3.910	3.295	2.840	2.492	1.998
18	4.812	3.928	3.304	2.844	2.494	1.999
19	4.843	3.942	3.311	2.848	2.496	1.999
20	4.870	3.954	3.316	2.850	2.497	1.999
21	4.891	3.963	3.320	2.852	2.498	2.000
22	4.909	3.970	3.323	2.853	2.498	2.000
23	4.925	3.976	3.325	2.854	2.499	2.000
24	4.937	3.981	3.327	2.855	2.499	2.000
25	4.948	3.985	3.329	2.856	2.499	2.000

The following is an example of the calculation of the net present value of a cash flow stream. This example is, of course, trivial. A real design project would have a much more complicated cash flow stream and probably a much longer life. The cash flows are –£6000; £700; £1500; £4200; and £2100 at the end of years 0, 1, 2, 3, and 4, respectively. The discount rate is 10%.

	Cash flow	$P/F(n)$	P
End year 0	–£6000	1.000	–6000
1	700	0.909	636
2	1500	0.826	1239
3	4200	0.751	3154
4	2100	0.683	1434
		NPV =	463

Good. Borrow £6463
Invest £6000 in the project
Spend £463 on a holiday (or use it to increase the value of the company).

8.2.2 Rate of return

NPV value, when used as a means of judging a project, assumes that the project is borrowing money from the company and then repaying the loan as it generates positive cash flows.

We may, however, prefer to ask the following question:

'If the company lends money to the project, what is the effective interest rate that the company is getting?'

Consider a cash flow stream:

$$£f_0; £f_1; \ldots £f_i; \ldots £f_n,$$

where $£f_i$ is the cash flow at the end year i.

Any debts will be cleared if the NPV of the cash flow stream is zero.

Our problem, then, is to determine the value of i for which the NPV of the cash flow stream is zero. This value of i is called the Rate of Return of the project or, sometimes, the Internal Rate of Return.

Usually, we have to find i by trial and error.

The following is an example of the calculation of the rate of return of a project which generates the following cash flows:

End of year	Cash flow (£)
0	–10 000
1	1 500
2	4 000
3	5 000
4	3 000
5	3 000

The net present value of the cash flow stream is

$$-10\,000 + \frac{1500}{1+i} + \frac{4000}{(1+i)^2} + \frac{5000}{(1+i)^3} + \frac{3000}{(1+i)^4} + \frac{3000}{(1+i)^5}$$

and, to find the rate of return of the project, we must determine that value of i which yields a NPV of zero.

If $i = 10\% = 0.1$

$$\frac{1}{1+i} = \frac{1}{1.1} = 0.909$$

$$\frac{1}{(1+i)^2} = \frac{1}{1.1^2} = 0.826$$

$$\frac{1}{(1+i)^3} = \frac{1}{1.1^3} = 0.751$$

$$\frac{1}{(1+i)^4} = \frac{1}{1.1^4} = 0.683$$

$$\frac{1}{(1+i)^5} = \frac{1}{1.1^5} = 0.621$$

The net present value of the cash flow stream is thus

$$-10\,000 + 15(0.909) + 4000(0.826) + 5000(0.751) + 3000(0.683) + 3000(0.621) = 2337$$

Luckily we do not have to calculate $1/(1+i)^n$. The values of $1/(1+i)^n$ are given in the tables of Fig. 8.4.

Alternatively, and more easily, we may use an existing computer program, such as Supercalc, to calculate the NPV of a cash flow stream. We may even write a simple program in BASIC, to calculate NPVs on a home computer.

As we are trying to find the value of i for which the NPV is zero, we must try again. Usually, if we increase the value of i, the NPV reduces.

Try a value of i = $20\% = 0.2$
NPV = −426
Try again, with i = $15\% = 0.15$
NPV = 823
Try again, with i = $18\% = 0.18$
NPV = 46
Try again, with i = $19\% = 0.19$
NPV = −195 .

Clearly, the value of i for which the NPV = 0, lies between 18% and 19% and is much nearer 18% than 19%.

As we cannot trust our predictions of cash flows to be very accurate, we could accept 18% as a near enough value of the rate of return of the project.

The company will have available a number of projects from which to select those in which it will invest. There will be a number of internal projects, that is, projects which involve the design of new products or the modification of existing products or procedures. But there is no reason why the company should not invest its money externally if external projects will give a better rate of return.

The criterion for accepting an internal project will therefore be whether its rate of return is at least as great as some 'cut off rate' determined by the rate of return of available external projects. If the investment is small, external opportunities may range from building societies to local government stock, while larger funds may be invested in other companies. Generally, an external investment will be fairly safe (after all, there will be no shortage of high risk, internal design projects which offer high return) and will permit disengagement of funds in case a good internal project offers itself.

The situation in which a design project is judged by whether its rate of return is greater than some 'cut-off' figure is really one in which the designer is asked to show that he can offer a better return on the shareholder's money than some other investment such as a building society.

8.2.3 The cost of capital

What is the value of i that we should use in discounted cash flow calculations?

Capital is raised from several sources, and its cost is the weighted average of the costs which apply to the different sources. Interest on debt, however, is less than the nominal value would suggest because interest is a tax allowable expense.

Consider borrowing £100 at 12% interest. Nominally, the cost of using this £100 would be £12 a year. Suppose, however, that the company is paying tax at 52% on taxable profits. Because the £12 a year is a tax allowable expense, the company will pay (0.52) × (£12) less tax than had it not borrowed the money, and the true cost of using the £100 for a year will be (£12) × (1 − 0.52) = £5.76.

Generally, if the interest on debt is $100 \times I(d)\%$ and the tax rate is T, the after tax cost of debt will be $(1 - T) \times I(d)$.

Consider £100 of shareholder's money. The shareholder might expect £10 a year as dividend, but he might also expect the value of his shares to rise by £10. The cost of this shareholder's capital is thus £20 a year because there must be sufficient earnings per share to pay the dividend and to retain enough money in the company to increase the company's value. The cost of equity is not only the dividend.

A guide to the cost of equity can be the P/E ratio ratio given in the financial columns of *The Times* and of the *Financial Times.* The value of E (earnings) includes both the dividends paid and retained earnings. P (the price of the share) is the price that the share could be bought for, and is not a measure of the amount of capital that the company obtained from its issue.

The simple formula for the cost of capital is

$$I = [1 - R(d)] \times I(e) + R(d) \times (1 - T) \times I(d), \text{ where}$$

$R(d)$ = the debt ratio
$I(d)$ = the interest on debt
$I(e)$ = the cost of equity, and
T = the tax rate.

As an example, consider a company in which 30% of the capital is debt at an interest of 17%; the shareholders expect 20% return on their capital, and the rate of corporation tax is 52%. Effectively, the cost of capital is 16.45%.

The approach here has simplified the problem because, as a glance at any company accounts will show, debt comes from several sources, and different types of shares carry different rates of return. The effective cost of capital is still a weighted, tax adjusted average, but more classes of capital must be weighted.

8.2.4 The calculation of net present value

Whether we are using NPV as a measure of the value of a project or using rate of return, we need to calculate the net present value. In the examples given above, formulae have been listed and use has been made of tables. While it is necessary to understand these methods of calculating NPVs or rates of return if we are to use them as measures of the value of a project, programs are available on the simplest of computers for making discounted cash flow calculations. Figs. 8.5, 8.6, and 8.7 are respectively a calculation of the NPV of a project using a simple computer program, a calculation of the rate of return of a project using a simple program and calculating various NPVs of a project, and its rate of return, using a spreadsheet program.

```
        i=?

!. 1

                              CASH FLOW
YEAR    [ 0 ]                                  -6000
YEAR    [ 1 ]                                  700
YEAR    [ 2 ]                                  1500
YEAR    [ 3 ]                                  4200
YEAR    [ 4 ]                                  2100


THE INTEREST RATE                                                          i=. 1

THE NET PRESENT WORTH FOR THE GIVEN CASH FLOW STREAM         PW=465. 8834779046

IF YOU WANT TO TRY ANOTHER INTEREST RATE ,
ENTER  YES , OTHERWISE ENTER  NO .
```

Fig. 8.5 – Computer calculation of NPV.

```
!5
ENTER THE AMOUNT FOR YEAR      [ 0 ] .
!-10000
ENTER THE AMOUNT FOR YEAR      [ 1 ] .
!1500
ENTER THE AMOUNT FOR YEAR      [ 2 ] .
!4000
ENTER THE AMOUNT FOR YEAR      [ 3 ] .
!5000
ENTER THE AMOUNT FOR YEAR      [ 4 ] .
!3000
ENTER THE AMOUNT FOR YEAR      [ 5 ] .
!3000
                                        CASH FLOW
YEAR    [ 0 ]                                        -10000
YEAR    [ 1 ]                                        1500
YEAR    [ 2 ]                                        4000
YEAR    [ 3 ]                                        5000
YEAR    [ 4 ]                                        3000
YEAR    [ 5 ]                                        3000

THE RATE OF RETURN FOR THE GIVEN CASH FLOW STREAM                   ROR=. 181640625
```

Fig. 8.6 – Computer calculation of rate of return.

	A	B	C	D	E	F	G
1	-6000	1	-6000	1	-6000	1	-6000
2	700	.9090909	636.3636	.8695652	608.6957	.8849558	619.4690
3	1500	.8264463	1239.669	.7561437	1134.216	.7831467	1174.720
4	4200	.7513148	3155.522	.6575162	2761.568	.6930502	2910.811
5	2100	.6830135	1434.328	.5717532	1200.682	.6133187	1287.969
6			465.8835		-294.839		-7.03094
7							

Fig. 8.7 – Calculation of NPVs (and ROR) using SUPERCALC. Column A gives the annual cash flows. Columns B, D, and F give the P/F ratios at 10%, 15%, and 13%. Row 6 give the NPVs at the various discount rates. Entry G, 7 is near enough to zero for us to accept 13% as the rate of return of the given cash flow stream.

The simple programs are written in basic and can be made available to run on most microcomputers. The spreadsheet program was Supercalc, a package thay may be used on most microcomputers with CP/M operating systems, but similar packages (for example, VISUCALC on the BBC Micro computer) are equally useful and readily available.

An advantage of using a computer for these calculations is that the user will easily be able to determine the sensitivity of the project to changes, simply by repeated calculation with modified data.

8.3 PAY-BACK PERIOD

Providing an estimate of the cost of capital which could be used as the discount rate in determining a project's net present value requires some guesswork. Interest rates change, the shareholders' expectations change, and capital structure changes. Moreover, some guesswork is also required when predicting the cash flows that a project will generate in the future.

If the decision maker has little faith in his estimates or predictions he may use 'Pay-Back Period' as a criterion by which to judge a project.

The pay-back period is simply the time that it will take for the sum of the positive cash flows to exceed the sum of the negative cash flows. A complex project like the design and build of an aircraft is bound to have a long pay-back period because it will take several years of expensive design, development, tooling, and proving before any products can be sold, and hence before there will be any income even to start cancelling out the early costs. A simpler project such as the modification of a plastic storage bin could be expected to have a short, inexpensive period of design and development and tooling costing only tens of thousands of pounds with a lead time of a few months. In such a case, the project might start paying for itself in less than eighteen months. The point about pay-back period is that the decision maker is only asked to look into the future for the length of the pay-back period and to believe that cash flows will be predominantly positive thereafter. Pay-back period as a criterion is crude but may be all that the quality of the information about some projects will justify.

8.4 CHAPTER SUMMARY

8.4.1 Accepting a design project generates a cash flow stream. If the project is more than trivial, early cash flows will be large costs to the company. The project will be worth accepting only if it ultimately generates enough profit to pay for the early project costs.

8.4.2 If we regard the project as borrowing money from the company, it will be acceptable if it has a positive net present value at a discount rate which is the weighted average, tax adjusted, cost of capital.

8.4.3 The shareholder sees himself as lending money to the project and will regard it as acceptable if it has a rate of return that is greater than the return on risk-free, external projects.

8.4.4 If it is difficult to forecast the cash flow streams that a project will generate, it may be sensible to use 'pay-back period' as a criterion for accepting or rejecting a project.

8.4.5 Simple, user friendly programs are available for the calculation of net present values and rates of return.

FURTHER READING

For a more extensive but elementary treatment of discounted cash flow methods:

Leech, D. J., *Economics and financial studies for engineers,* Ellis Horwood, Chichester, 1982.

or for a more comprehensive treatment:

Smith, G., *Engineering economy,* Iowa State University Press, Ames, Iowa, 1979.

SUBJECTS FOR DISCUSSION, AND PROBLEMS

(1) Approximately what cash flows will be generated by the following project, and when, during the life of the project, would you expect those cash flows to occur?

(i) Design, develop, and build a nuclear power station.

(ii) Eliminate a source of unreliability in a washing machine that you already make and sell.

(iii) . . .

(2) The cost of design is almost the smallest cash flow that is generated when a project is accepted. Does this mean that design is the least important activity?

(3) The Rubic Engineering Company are in the process of purchasing a new lathe which will cost £7000. It is expected that this will have a useful life of 4 years, at the end of which it will be traded in for an expected scrap value of £900. During its life it is expected to generate a cash flow at the end of each year as follows:

year 1	year 2	year 3	year 4
£700	£1500	£4200	£1200

(a) Estimate the net present value of the project for discount rates of 10% and 3%, and show that the internal rate of return is about 6.97%.

(b) The company may also make a once-off maintenance payment of £1000, payable at the time of purchase, which although not altering the cash flow stream, will result in a higher scrap price at the end of year 4. If the increased price is £2300, determine the internal rate of return, and hence deduce whether such a payment should be made.

(4) A company is evaluating the worth of a maintenance agreement on a new moulding machine. The machine costs £8000, and the cash flows expected at the end of each year are expected to be, ignoring maintenance, as follows:

year 1	year 2	year 3	year 4
£580	£1200	£4870	£4980

Assuming the mcahine has nil value at the end of year 4, evaluate and illustrate graphically the net present value for various interest rates. From this estimate the internal rate of return (IRR).

If a maintenance contract costing £500 (in each of years one to four) is taken out, the cash flow (ignoring the contract cost) is expected to be:

year 1	year 2	year 3	year 4
£1200	£2000	£4700	£5400

Determine the net present value at rates between 0% and 15%, and sketch your result. If the company evaluates whether the contract is worthwhile on the basis of IRR, what will they conclude?

(5) The firm of Eastland Ltd are considering the purchase of a numerically controlled machine to replace an existing mill. The new machine will cost £120 000 and is expected to have a life of ten years. The revenue expected to be earned by the machine over its life is:

year 1	£45 000
year 2	£60 000
year 3	£60 000
year 4	£60 000
year 5	£60 000
year 6	£60 000
year 7	£54 000
year 8	£48 000
year 9	£30 000
year 10	£15 000

Overhauls costing £60 000 and £40 000 will be required in years 5 and 8. The salvage value of the machine at the end of ten years will just balance the cost of taking it down.

The existing mill has already been written off, but it is mechanically sound although not tape controlled. With annual maintenance costing £20 000, it will perform satisfactority for seven years. At the end of seven years its disposal will bring in £15 000. Because the old mill is less easily set up than the new, it will be more expensive to operate and, effectively its revenues will be

year 1	£45 000
year 2	£45 000
year 3	£45 000
year 4	£45 000
year 5	£40 000
year 6	£36 000
year 7	£22 000

If the company evaluates investments at a 10% discount rate, should the existing machine be replaced?

(6) The following costs have been estimated for several insulation design alternatives in an office block that is being built.

Thickness of insulation	Cost of the insulation and its installation	Annual cost of heating the building
0 cm	£ 0	£7500
2 cm	2500	4250
4 cm	2900	3750
6 cm	3300	3450
8 cm	3700	3250
10 cm	4100	3100
12 cm	4500	3000
14 cm	4900	2950
16 cm	5300	2910

The building is designed for a life of 20 years, and the cost of capital is 14%.

What is the optimum insulation?

How is our answer different if the cost of capital is 20% but the life of building is to be only 10 years?

(7) Find the net present values of the following cash flow streams at discount rates of

(i) 5%, (ii) 10%, (iii) 15%, (iv) 20%, and (v) 25%

Cash flow stream (£)	a	b	c	d	e
Year					
0	–8 000	–40 000	–15 000	–60 000	–100 000
1	3 500	–40 000	–10 000	5 000	50 000
2	2 600	10 000	– 5 000	8 000	– 90 000
3	2 100	12 000	10 000	11 000	70 000
4	2 000	20 000	12 000	15 000	– 20 000
5	1 500	30 000	15 000	18 000	45 000
6	1 000	40 000	12 000	20 000	60 000
7	600	32 000	10 000	25 000	48 000
8	600	42 000	8 000	27 000	40 000
9	600	28 000	6 000	20 000	36 000
10	600	14 000	3 000	15 000	32 000

ANOTHER WORKED EXAMPLE

Consider the cash flow stream generated by the purchase of an asset during the first year of its life, when tax is taken into account. Corporation tax, at present, is 52% of profits. Writing down allowance on the asset is 100% in the first year of its life.

A machine costs £10 000, it makes products which are sold for £4000 a year, the running expenses (labour costs, raw material costs, maintenance, etc.) are £1000 a year, and the project lasts for 10 years, with no salvage at the end. The assumed discount rate is 12%.

At the beginning of the project (at the end of year 0) the company spends £10 000 to buy the machine.

During the first year (computed at the end of year 1) the company has a revenue of £4000 and expenses of £1000. Tax would normally be paid on the £3000 profit, but the writing down allowance for the purchase of the machine is £10 000. For tax purposes, then, the profit is assumed to be

$$\pounds 4000 - \pounds 1000 - \pounds 10\,000 = -\ \pounds 7000$$

The tax paid on this loss will be

$$-(0.52) \times \pounds 7000 = -\ \pounds 3640\ .$$

The cash flow computed at the end of year 1 will therefore be

$$\pounds 4000 - \pounds 1000 + \pounds 3640 = \pounds 6640\ .$$

This calculation hides some complication by assuming that there is such a thing as negative tax in implying that the government pays the company instead of the company paying the government. The real situation is more complicated than this. What we have calculated is really a tax reduction and not a negative tax; but if, as is often the case, the project we are discussing is only one of many which the company has undertaken, it is probable that the company is making enough profit from other projects for the negative tax to be a genuine reduction in the amount that the company would otherwise pay. If this is not the case – if the company would not otherwise pay sufficient tax in the year to take advantage of the tax reduction – then that reduction may be held over until use may be made of it.

This does, of course, require further calculation because taking advantage of tax reduction in later years will change the present value of that reduction. Generally we will ignore these complications and assume that a company can take immediate advantage of any tax reduction.

A further complication arises because tax is not paid immediately at the end of the year in which the company makes the profit that is taxed. Tax is paid about a year after this (not less than eight months after). This again alters the timing of the cash flows so that our calculations are approximate.

Still more complication is introduced when the company pays a dividend since it becomes liable for immediate payment of Advanced Corporation Tax (based on the income tax that shareholders might be expected to pay). This does not increase the company's costs, but again it alters the timings of the cash flows and, once again, our calculations must be assumed to be approximate.

Complete the above example (assuming that immediate advantage may be taken of tax allowances).

Year 0	Cash flows	=	– £10 000 (cost of machine)
Year 1	Revenue	=	4 000
	Tax allowable expenses	=	1 000 (running costs)
		plus	10 000 (writing down allowance)
	Profit for tax purposes	=	– 7 000
	Tax	=	– 3 640

Net cash flow for the year is

Revenue – Running costs – Tax = 6 640 .

Note that the writing down allowance is not a cash flow because it is merely a book transaction and no money changes hands.

Year 2	Revenue	= 4 000
	Running costs	= 1 000
	Profit before tax	= 3 000
	Tax	= 1 560
	Net cash flow	= 1 440

Note that once the value of the assets have been completely written down they cannot again be claimed as an expense.

Years 3–10 will be as for year 2, so that the cash flow stream will be

year	£
0	– 10 000
1	6 640
2	1 440
3	1 440
4	1 440
5	1 440
6	1 440
7	1 440
8	1 440
9	1 440
10	1 440

The student may verify that, at 12%, the Present Value of this cash flow stream is £2 779.

What is the rate of return of the cash flow stream in the above Example

The student may verify that this is about 21.5%.

Notice that we have now broken new ground. What we have calculated is the rate of return of the project AFTER TAX.

This, of course, is what ultimately interests the investor.

CHAPTER 9

Selecting or rejecting the project

Although the ranking of projects by D.C.F. methods (according to their net present values or their rates of return) is logical, we do not usually have the information needed to make these calculations early in the life of any project. Most companies expect to work on a project for some time before any decision can be taken about it. Usually a project is reviewed several times to justify further expenditure or to kill it. If too little effort is spent studying a project before the company commits itself to it fully, project costs are likely to be underestimated.

9.1 CAPITAL RATIONING

The discussion in chapter eight, of discounted cash flow methods assumes that we are determining whether a project should be accepted or not. In some cases, however, we must accept a project even if it is not profitable. The Health and Safety at Work Act, local government planners, foreign environment protection laws, or a hundred other causes may make a project mandatory. If a project is mandatory and we cannot find a profitable way of doing it we will have to use discounted cash flow methods to choose the least costly method.

In other cases a project may, itself, be a loser but may permit the acceptance of other profitable projects (as, for example, in the designing and building of an environmental test rig which will provide no direct profit but will enable other, profitable, projects to be undertaken).

Even allowing for such exceptions, a project with a positive net present value or a rate of return above some specified cut-off point will not necessarily be accepted; it merely becomes a candidate for further consideration. The difficulty is created by the limited availability of capital or of such other resources as design capacity, manufacturing skills, manufacturing capacity, or time. A project which is apparently desirable could be rejected for lack of resources, although this is only another way of saying that a project must fit in with the corporate plan.

It is easy to show that, if resources are limited, we may have to reject better projects than we accept. Consider projects A, B, and C, where A has a NPV of £2500, B of 2000, and C of £1500. A requires a negative cash flow of £20 000 at the end of year 0, B of £16 000, and C of £14 000, although only £30 000 is available at the end of year 0. If we ranked these projects in order of NPV and then chose A first, we would have to reject B and C for lack of resources. Clearly we would do better to choose B and C (which would force us to reject A).

In fact, lack of capital is unlikely to be a good reason for rejecting a profitable project. If a project has a positive NPV, calculated at a realistic cost of capital, it should be quite easy to raise the capital needed. A problem could arise if the company is asking for a loan of such a size that the debt ratio ($R(d)$) will be raised to a level at which the creditor will be taking more risk than he is prepared to, but the argument must apply that if a realistic figure has been used for the cost of capital, capital should be obtainable at that price. If other resources, such as manpower, test or manufacturing facilities prevent us from accepting a profitable project then we probably will have to reject it because such resources cannot often be obtained quickly.

If profitable projects are frequently rejected for lack of capital, then potential investors are presumably unconvinced by the designer's calculations, and we must ask why. Sometimes the designer has been over optimistic because he wants to justify himself, but sometimes he has not bothered to make a convincing economic case for the project he is proposing.

If projects are frequently rejected through lack of non-monetary resources, then we must ask whether those resources should be acquired.

While it is generally true that a project is not worth doing unless it has a positive NPV, we cannot simply extend this method to the selection of a large number of projects. The methods of project selection discussed in the literature and which rely, basically, on project ranking are rarely of direct use in industry, although we can learn from their study. At the simplest extreme there are scoring methods (see, for example, Hart [1]) which do little more than put numbers to subjective views and to which many firms pay lip service without requiring guidance from academic authors. There are ranking methods (see, for example, Smith [2]) which can work if no large projects claim more than a small share of the resources (Leech & Smart [3]) but which are more suitable for capital investment projects than for design projects. More complicated are optimization methods (see, for example, Weingartner [4]), but these are not really suitable for decision making in industry because they require more information about the projects than can be provided when the decisions have to be taken.

One of the difficulties of project selection when resources are rationed is that many of the models used assume that a large number of projects is generated and then the selection is made from them. The real situation is often very different. A company usually knows what business it is in and will not often reject a project in that business. Projects may be rejected for simple reasons, such as a belief that the customer will not pay or that he will not be allowed the necessary foreign currency by his government, but such decisions can be made very early and without appealing to any formal management procedure. Again, it is often argued that there needs to be a long period of study to determine whether a project is physically feasible or likely to be of interest to a possible customer, but in an area with which the designer is familiar (that is, the business that the firm is in) it is not often that much time has to be given to these problems.

Sometimes the company is not so much selecting or rejecting projects from a virtually endless list as trying to find a project which will fit the resources available. This situation can arise when a company is engaged in a major project which makes uneven use of manpower. It is not usually possible or even desirable to hire and fire men as the demand for them changes, and it can be worth introducing a project which will make some use of the resources which would otherwise be idle.

9.2 THE STAGES OF PROJECT EVALUATION

With clear exceptions, unless a project is expected to have a positive NPV, it should not be started. An obvious difficulty in using such a criterion is that we do not know what cash flow stream a project will generate when we first consider it. Most companies would not, therefore, expect to accept or reject a project before some money (man-hours, at least) has been spent studying it. Usually there are several stages in which a project is assessed, and at the end of any stage the project may be rejected. If a project is not rejected, we assume that the work done at each stage gives us a more accurate prediction than the last, of the cash flows that the project will generate.

A project is born when a need is perceived. There are many reasons and procedures which generate projects, and these will be listed and discussed at greater length in Chapter 11. Most ways of generating a project do not cost much and may not even be monitored. If a potential customer invites the manufacturer to offer a product which has yet to be designed, the project will have been generated at almost no cost to the manufacturer. If the project is initiated by the sales engineer, a service engineer, a production engineer, or a designer, the business of suggesting it will be almost incidental to his main task. Often the generator of a project will be senior and not required to account for his time. In almost all cases, the generation of a project, by which we mean a first brief statement of what is required, costs little, and those costs are taken from a blanket budget, that is they are not budgeted project by project.

The first decision which has to be made is not whether the project should be accepted or not but how much should be spent on the next stage, the feasibility study. The term 'feasibility study' suggests that the physical feasibility is being studied, but, as has been described, designers do not usefully spend much of their time deciding whether a proposal breaks natural laws, although it is not unknown for perpetual motion machines to be proposed. A feasibility study has three main parts. The first is a study of the market to predict the expected sales, the acceptable selling price of the product, the chances of success, the expected life of the project, and generally its commercial desirability. The second requires the production of a design scheme which will meet the market requirements so that design and production difficulties may be predicted. The third will be the prediction of the cost of the next stage of the design work. Considerable approximation is to be expected from the first two parts of the study, but the third should be accomplished with some accuracy. If the project is derived from an invitation by a customer to tender for a job, the feasibility study should provide approximate indications of the value of the job and the magnitude of any difficulties to be expected, but it should provide an accurate forecast of the cost of tendering. If the project has been generated within the designer's company the requirements are essentially the same, that is approximate estimates of the value of the job and its probable difficulties and an accurate forecast of putting a proposal together so that the decision makers (this time within the company) will be able to judge a formal proposal.

The time allowed for the feasibility study will depend on the size of the project, but is unlikely to be less than about 80 man-hours. For a very large project the feasibility study could take hundreds or even thousands of man-hours, but the less time that needs to be allocated, the better. It must be remembered that no significant time has been spent on the project before the feasibility study, and no serious effort will have been made even to define the requirements. What must be spent during the feasibility study is just enough time to enable a second decision to be made, and the second decision will be whether to embark on the more expensive 'project definition' stage of design. Some companies accept that it is not really possible to forecast the cost of a feasibility study and simply allocate a fixed amount against the budget, whatever the project.

There are four objectives to the project definition stage of design. The first, on which the others depend, is the production of the design scheme, the second is an analysis of the scheme to demonstrate that the customer's requirements will be met, the third is a list of what will have to be done in the final stage of design, and the fourth is the calculation of the cost and effort that will be required to complete the design. If the feasibility study has taken 80 man-hours, the project definition stage will take at least several hundred, but the work will be done against a forecast cost.

Where the manufacturer designs products to the particular needs of particular customers, he is selling the company's ability to design a product to the customer's satisfaction, and he will hope to get an order on the strength of the scheme drawings, the analysis, and the predicted prices. Under these conditions it is not difficult to define the point at which the project definition stage is finished, because it is the point at which the manufacturer is prepared to make a commiting proposal to the customer. Also, it is the customer who decides whether design should proceed further by ordering the product or rejecting it.

Where the product is being designed for sale on the open market and has to be built before any one can be expected to buy, the decision to proceed from one design stage to the next must be taken inside the manufacturer's company, but there is no reason to expect the decision to be made with less information than a customer would require.

If, as a result of project definition, the decision is taken to complete the design of the product, it is still possible that the project will be abandoned when design is complete but before the next, even more expensive, stage of tooling up for manufacture.

Although three main decision points have been described, most companies review the progress of projects at regular (perhaps monthly) meetings, and so there will be opportunities to abandon a project between the decision points described above.

The above discussion of the periodic review of design projects considers that there are three stages of design: the feasibility study, project definition, and the complete detail design and development for manufacture. In some industries, where the designs are simple enough for outcomes to be predicted with reasonable accuracy, two stages may be enough. In high technology, more stages may be desirable. The terminology used has been derived from the Downey Report [5], which suggests that four stages of design should be used (two stages of project definition) and was intended to offer a procedure for assessing large defence projects.

A way of looking at the project appraisal process is shown in Fig. 9.1. This figure assumes that there are four design stages after project generation, and after each stage the decision may be made to reject the project or to carry on. In fact, the number of decision points may be multiplied to as many as the company thinks desirable.

If we accept the decision tree as representing at least some design situations, we can determine the sort of gamble that the decision maker sees himself taking at each stage of the work. If, for example, he decides that a feasibility study should be done, he is spending the cost of the study in the hope of an ultimate pay-off. The size of the pay-off will be predictable in only the vaguest of terms, and the probability of pay-off $[p(1) \times p(2) \times p(3) \times p(4)]$ will be known only to the extent that experience with other projects will justify a prediction. Deciding to embark on the first stage of project definition means gambling the cost of PD1 with a higher probability $[p(2) \times p(3) \times p(4)]$ of pay-off. If the decision maker is making the same sort of gamble at each stage we see that the amount he is prepared to spend on each stage increases inversely with the probability of carrying on with the project.

The stages of decision making shown in Fig. 9.1 correspond with the early stages of design in Fig. 9.2.

DEFENCE PROJECTS

DOWNEY 1968

FEASIBILITY STUDY (FS)

PROJECT DEFINITION 1 (PD1)

PROJECT DEFINITION 2 (PD2)
(FOR SMALL PROJECTS PD2 IS ELIMINATED)

DESIGN

SUGGESTED THAT SPEND
BEFORE COMMITMENT IS 5%
COULD BE 20%

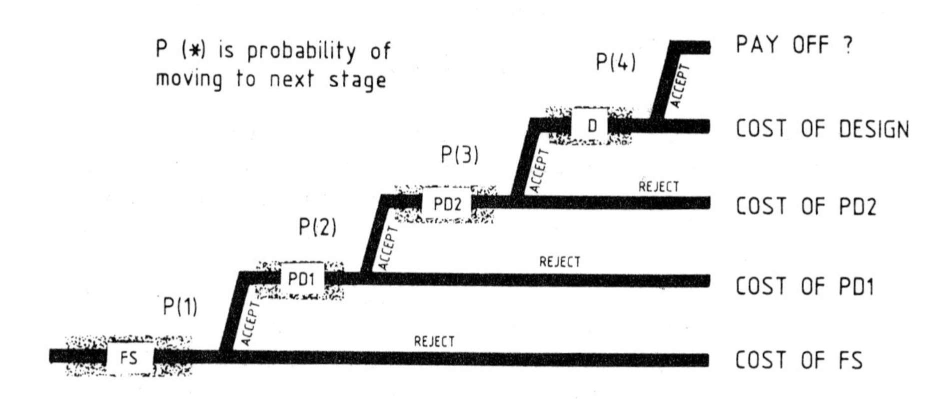

COSTS PROBABLE RISE EXPONENTIALLY

Fig. 9.1 – Stagewise project appraisal.

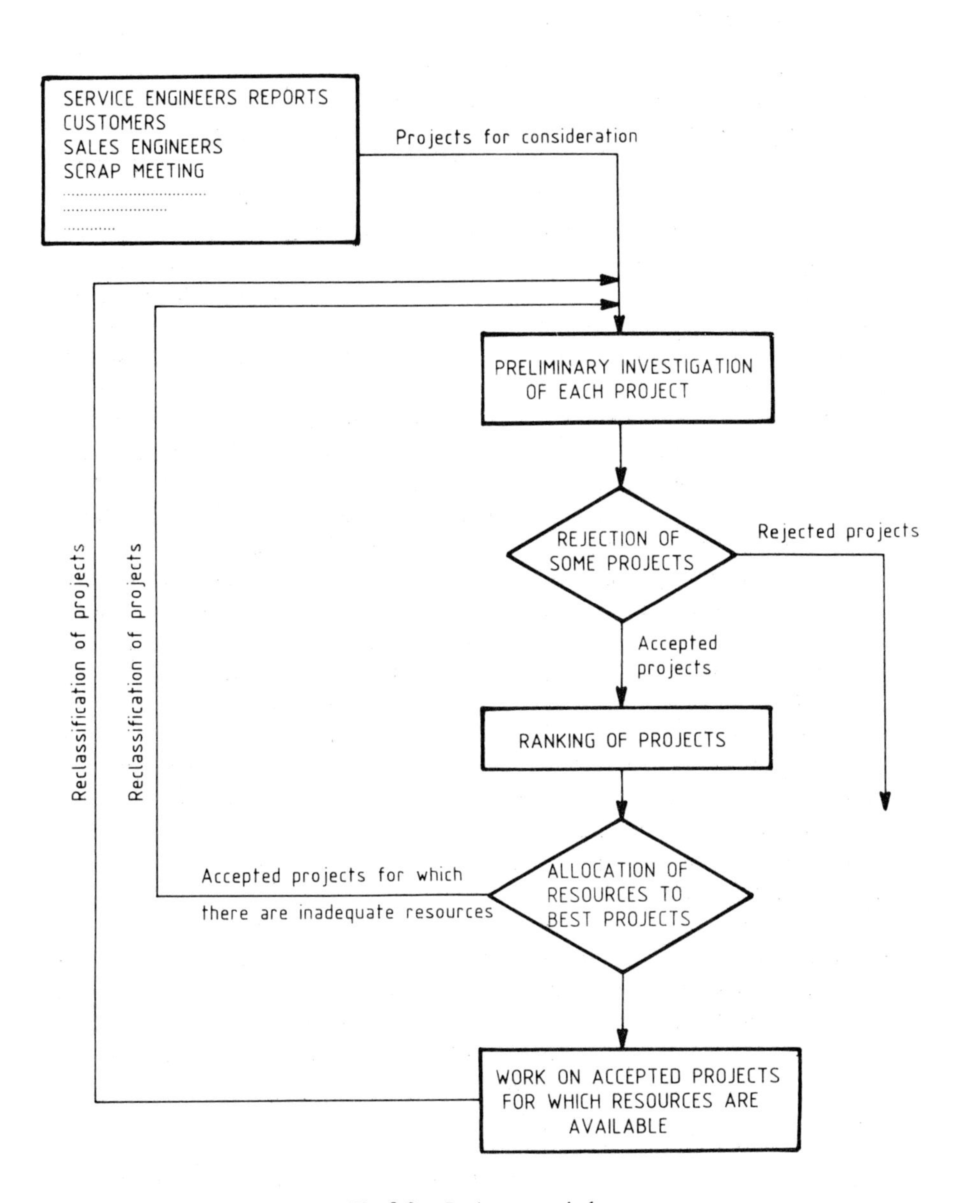

Fig. 9.2 – Project appraisal.

9.3 REGULAR PROJECT REVIEW

In deriving a decision tree to define the stages at which a project may be reviewed and accepted or rejected we have considered the project in isolation from any of the company's other business. In some cases this is fair because that project is the overriding interest of the managers. The concorde, the Severn Bridge, a new atomic power station, will be of such significance to the manufacturer that the project must generate its own management procedures. Most companies, however, have many design projects alive at any one time, and none of them may be considered in isolation from the others. The jobs going through the design process will be competing for designers and other resources. Many companies set up a procedure for reviewing all the jobs in the system, together, at regular intervals.

Typically, a committee will meet monthly to review possible new work and progress on existing work. The procedures will approximate to that described on Fig. 9.2. Possible new projects and projects on which some work has been done will be listed, discussed, and ranked by the committee. Some projects will be rejected and those that are not rejected will be allocated resources in some order of priority. This is a dynamic process; jobs which have been dormant may suddenly become desirable and worthy of resources; important jobs may be running late and have to be allowed to take resources away from other desirable projects; the cost of solving a technical problem may seem to be greater than expected, so that a project on which resources have been spent will be abandoned.

This procedure does not mean that discounted cash flow methods need not be applied or that the decision tree of Fig. 9.1 is not relevant. When projects compete for resources they must be ranked, and D.C.F. methods are important in establishing the ranking. Decisions to abandon a project or to allocate resources to the next stage of the decision tree are precisely those which a review committee will take at its regular meetings.

Normally, the committee which presides over project review is a senior one. Commonly it is chaired by the Technical Director, and senior representatives of the design, marketing, and manufacturing organizations will be members. Often men will be co-opted for their special skills or for their knowledge of a particular project. Surprisingly, most committees of this nature ask the accounting department to do any calculations involving money, although only the engineers can predict cash flows (however badly they may do it).

One feature of regular project review is that the decision points may not coincide with the ends of the feasibility study, project definition, etc. In fact, this is not significant, because the precise points at which any phases of design end are rather arbitarily defined. Provided that project reviews are held sufficiently frequently, timing should not introduce any difficulties into decision making.

It is necessary to know how much money has been spent on a project between reviews, and this means that there must be a system for recording work done. In some cases this will mean that designers must record their work, probably with a time sheet, an example of which is shown in Fig. 9.3. This particular time sheet asks the man to record his hours used on a job and also the hours that he expects to use to complete it. Information of this nature can be processed to give the type of print-out shown on Fig. 9.4. The decision maker at a review, will then have some idea of the proportion of the allowed budget that has been spent on the job and also the proportion of the work that has been completed.

NAME:	CLOCK NO.	LABOUR CODE.	WEEK ENDING:

	DIRECT				INDIRECT		DAILY TOTAL
JOB NO.							
MONDAY							
TUESDAY							
WEDNESDAY							
THURSDAY							
FRIDAY							
SATURDAY							
SUNDAY							
JOB TOTAL							
HOURS TO COMPLETE							

Time sheet.

Fig. 9.3 – An example of a timesheet.

The decision maker is thus doing more than allocate resources to a project; he is also monitoring expenditure and progress against the company's plan. This ensures that the design process is being properly managed.

JOB NO. 4321.

	Lab Code	Est Hrs	Bkd Hrs	Plus Hrs	Curr Total
Cals	01	38	15	10	25
Des	02	174	266	3	269
Lab	03	372	356	126	482
Total		584	637	139	776

Interpretation of this extract is that the original labour predicted was 38 hours of calculation, 174 hours of design and 372 hours of laboratory work. At the time of the printout, 15, 266 and 356 hours respectively had been spent with current estimates of 10, 3 and 126 hours work remaining.

The printout thus shows a current prediction of 776 hours spent against an original estimate of 584, a predicted overspend of 192 hours. The printout from which this extract is made also converts hours into cost, shows expenditure on materials and shows overspend or underspend as a variation from the previous accounting period. When a programme has been developed, time-sheet information given on a Friday can be analysed and be available to the managers on the following Monday.

Fig. 9.4 – Extract from a computer analysis of the expenditure of man hours.

9.4 CHAPTER SUMMARY

(i) If a company knows what business it is in, many projects will generate themselves because either the customer will ask for them or the need will be apparent.

(ii) Accepting or rejecting a project is a dynamic process, undertaken in stages. Each stage costs money but provides more information on which decisions may be made.

(iii) The stagewise decision process should be formal. Correctly designed, the process monitors project progress as well as aiding the decision makers.

TOPICS FOR DISCUSSION

(1) Can the future cash flows of a project be predicted

(i) at the beginning of the project's life, or
(ii) after some study of the project?

(2) How much of a design must be done before reasonably accurate predictions of the cost of making the product can be made if

(i) the product is a modification of an existing one, or
(ii) the product is a totally new conception?

(3) How common is it for the costs of advanced products (aircraft, weapon systems, nuclear power stations, hospitals, bridges, etc.) to be underestimated?

How can estimates be improved?

(4) Can a management procedure be set up to carry out the implications of Fig. 9.1?

REFERENCES

[1] Hart, A., 'A chart for evaluating product research and development projects', *Operational Research Quarterly,* 1966, **17** (4), 346–358.

[2] Smith, G., *Engineering economy,* Iowa State University Press, Ames, Iowa, (1979).

[3] Leech, D. J. & Smart, P. M., Investing in ideas in a R & D department, *Journal Mechanical Engineering Science,* I.Mech.E., London (1979).

[4] Weingartner, H. M., *Mathematical programming and the analysis of capital budgeting problems,* 1967 (Kershaw Pub. Co. Ltd London).

[5] Downey, W. G., *Development Cost Estimating,* Report of the Steering Group for the Ministry of Aviation, HMSO (1969).

CHAPTER 10

Design for reliability

Reliability is of obvious economic significance. If a product has a reputation for being unreliable the customer will not want to buy it. If a product gets a reputation for reliability it will increase its share of the market, and the customer may be prepared to pay more for it.

The customer wants a reliable product because breakdowns cost money: money for the labour of repair, money for the replacement parts, and money lost by the loss of the product while it is waiting to be repaired and while it is being repaired. If, when a breakdown occurs, there is danger to life, then the customer must face the considerable cost of recompensing the injured or of insuring against such an event.

10.1 SOME GENERAL POINTS ON SAFETY

In the last decade there have been many changes in product liabilities as has been witnessed by the number of accident injury cases taken to the Courts. Social and consumer pressures and special legislation have demanded safer products. This is seen in its most potent form in road accidents which have become a major problem in every industralized country.

Headed by the aerospace and atomic reactor fields, where a policy of 'First time safe' has had to be pursued, new concepts in safety using improved technical methods have come into being. A total system fail safe policy has to be fostered by careful specification drafting and special procedures to provide a high degree of protection to both systems and operating personnel.

Again, the role of standards and independent testing laboratories are undergoing change, and corporate and government policies and organizations for product safety are being re-examined and revised. The Health and Safety Act (1), (2) has demanded careful product surveillance to ensure that all conditions are met.

Research effort has been focused into two main areas which have been called 'primary' and 'secondary'. Primary safety aims at avoiding an accident in the first place, whereas secondary accepts that some accidents are inevitable and provides protection for the people involved.

The absence of data on the first main area presents a problem, since it is impossible to evaluate accidents which might have occurred in the absence of safety devices.

For mechanical/electrical products the interaction between operator and product is always of great importance, not only from an ergonomic point of view but because improved product performance may make it more dangerous for the average operator. In this respect it is significant that the old 1939 Special test for Drivers of Heavy Vehicles has been reintroduced so that they have to demonstrate competence of handling sophisticated controls and quickly assimilate information from warning devices.

The whole realm of product safety is bound up not only with handling qualities and correct information reading but also with mechanical and electrical reliability. But also in structural and civil engineering safety during construction as well as the completed permanent work has been called in question after the Cleddau and Yarra bridge disasters (3).

Here the need to streamline the site control of false work has been recognized by a member of the contractors' team being made solely responsible for ensuring that all the procedures are carried out. For very large jobs this temporary works coordinator now has to be a Chartered Engineer and be the signatory on the permit to load false work. Indeed the I.C.E. conditions of contract, in the interests of safety, make the contractor responsible for erection safety and the design engineer responsible for checking the adequacy of the false work (4).

In the process industries too the horrors of the Flixborough disaster have called inevitably for the need to certify critical designs from a safety angle in much the same way that civil airworthiness has been certified by the Air Registration Board (Now C.A.A.).

10.2 SOME LEGAL ASPECTS OF DESIGN

There are a number of Acts which concern the designer today. In particular, the Health and Safety at Work Act of 1974 and the Consumer Protection Act of 1961 which was amended in 1971. The latter is the only law on general product safety.

Section 6 of the Health and Safety Act provides for the first time that manufacturers, designers, suppliers, and importers of plant and components designed for use at work or substances intended for use at work will be criminally liable if they fail to take proper steps – testing, research, etc. – to see that the articles or substances concerned are safe.

Both these Acts are aspects of our legal system that deal with the interaction of society with technology. Product liability provides a model for viewing the importance of the technologist's role in understanding and providing expert guidance to the further development of technology [5].

Product liability (PL) may be defined as:

> The claim and litigation alleging that a product has been defectively or negligently designed, manufactured, or serviced, and that such defect or negligence is the proximate cause of injury to persons or property.

The trend therefore has been to move from *caveat emptor* – let the buyer take care to caveat venditor – *to let the seller beware.*

Naturally, the response of contractors and manufacturers – partly in their own self interest – is increasingly to subscribe to testing and certification programmes and to promote safety and reliability procedures, codes, and guides for all their products.

More and more the onus will fall upon design and development engineers to provide satisfactory designs which are thoroughly proven before use. Increased production and construction costs will be reflected in a higher selling price, but product safety and reliability must be justified by the economics of manufacture wherever possible. This does pose a considerable management problem.

In the United States of America, product liability has been a distinct category of law for some years. This is not, at present, the case in the United Kingdom, where the Sales of Goods Act 1979 and the Unfair Contract Terms Act 1977 have largely been the consumer's protection. Although this legislation is not as far reaching as product liability legislation in the USA it can make the manufacturer responsible for a customer's financial losses caused by faulty workmanship or design. The Consumer Safety Act 1978 will also place a considerable duty on the designer to understand his responsibility for the safe operation of the product he has designed.

Current EEC proposals on liability for defective products are likely to be adopted at some time in the future, and these will move the situation towards that which obtains in the USA. Although the proposals are more likely to affect, directly, the producer than the designer, the designer will be expected to demonstrate that he has given thought to the proving of a design.

10.2.1 The Health and Safety Act at work

Many designers have not yet realized that this new Act places a heavy responsibility directly upon them.

The Act states that 'as far as is reasonably practicable' the plant, machine, or other working article must be designed for use 'in such a way as to be safe and without risks to health when properly used'.

What the words 'as far as in reasonably practicable' mean is not defined, and one suspects will not be until a few civil and criminal actions have been decided.

The designer is thus faced with complying with a law which has no specific mechanical, electrical, or other physical provisions.

It is as though we had been told to 'drive our cars in a safe manner' without at least being warned that it is an offence to exceed a stated speed limit or to fail to stop at traffic lights, etc.

In all fairness, if the designer is required to comply with a law, then that law must be defined in a physical sense.

There are further responsibilities placed by the Act upon the designer. These are a requirement to carry out adequate testing of the product to ensure safe operation, and an obligation to supply adequate working instructions so that the product can be used in a proper manner. Normally, the designer has no responsibility for testing beyond the production prototype, and, in many cases, he is not qualified to decide whether working instructions are clear and unambiguous or not.

It is beginning to be believed by many that this Act is so general in its provisions that proper compliance will be extremely hard if not impossible.

What is required is a compulsory Code of Design and Manufacturing Practice to accompany this Act, and failing this, at least a legal requirement to comply with the relevant BSI Standards.

Where Codes of Practice do exist, the Act makes it very clear that these should be complied with, since non-observance of any Code will prove to be the deciding factor in any civil or criminal action under the Act.

There are a number of informative booklets available from The Health and Safety Commission, and designers would be well advised to obtain these.

All designers will now have to consider the safety aspect of their design very carefully indeed.

10.2.2 The Consumer Protection Act

The Consumer Protection Act 1961, which was amended in 1971, was produced to prevent the sale of dangerous goods before they cause an accident.

Before 1961, the laws about product safety each concerned one specific type of product. The Consumer Protection Act replaced most of these. The Act does not itself lay down any particular requirements, but it enables the Secretary of State to impose regulations for any goods if he or she considers that this will reduce the risk of death or personal injury. These regulations may relate to the way goods are made or the way they are labelled or packed. The Act makes it a criminal offence to sell (or possess for the purpose of selling) goods which do not meet the regulations.

The penalty is a fine of up to £100 for a first offence and a fine of up to £250 and/or a prison sentence of up to three months for a subsequent offence. No one has yet been imprisoned under the Act, but the small fines can sometimes be bumped up by bringing a series of separate charges for a number of identical items, each of which offends in the same way.

You can, yourself, take out a civil action for damages against someone who you think has broken this law. You might do this if the offence caused death, injury, or damage.

The responsibility for most aspects of consumer safety was taken over from the Home Office in October 1974 and placed in the hands of the Department of Prices and Consumer Protection (DPCP).

The DPCP does not make new regulations whenever a dangerous group of products is discovered. It may, instead, ask the manufacturers voluntarily to modify or withdraw the goods from sale altogether. Regulations are made when it appears that voluntary cooperation may not be sufficient to ensure that dangerous goods are not sold.

10.3 NEGLIGENCE

Today, because of the above Acts, the manufacturer is more vulnerable in lawsuits than ever before. So in fact are all his management personnel, including the engineers in charge of product design, development, testing, and related functions. However, not only industrial engineers, but also engineers in private practice, serving industry in these areas, share the common problem – increased liability exposure *ad infinitum.*

The engineer employee of a corporation enjoys some security since he works under the shelter of the corporate entity, which protects personal assets from attachment by lawsuit. A statute of limitations for consulting engineers is not, however, in the immediate offing. In the meantime, the small consulting engineer who cannot justify the high cost of professional liability insurance, 'walks around on marbles' waiting for that first summons.

In any case, it behoves those responsible for decision making, both manufacturers and engineers, to be informed as to what constitutes negligence. The old platitude still holds true: 'To be forewarned is to be forearmed'. Negligence problems can arise at many steps in the product design and manufacturing process. Some liability pitfalls are:

- True design errors, such as errors in calculation or those on the drawing board.
- Insufficient safety features: failure to install adequate safety devices such as guards, fail-safe switches, interlocks, monitoring devices, and safety valves.
- Deficient safety devices: employment of safety devices which fail in use.
- Deficient materials: construction from unsafe or unsuitable materials.
- Deficient processing: plan for a manufacturing process which lead to a defective product. For example, the proper heat treatment which is destroyed later in manufacture by a welding operation.
- Inadequate applications engineering: failure to plan for foreseeable uses unintended by the manufacturer.

- Insufficient research and development: failure to foresee the consequences of ordinary wear and tear and improper maintenance on the part of the user.
- Inadequate performance testing: failure to make a safety check after manufacture.
- Noncompliance with existing codes and safety standards: failure to measure up to industry standards; failure to keep abreast of scientific knowledge where the whole industry is negligent.
- Lack of liaison between engineering and marketing: failure to warn where there is superior knowledge of inherent hazards.

10.4 FACTORS OF SAFETY

To many designers a 'factor of safety' is called a 'factor of ignorance', which is really nearer the truth. The main trouble is variability of properties, and what designers usually work to are average values, with all that that word implies. There are many cases where the inadequacies of design (or production) specifications have occurred because they have been based on average values.

In Fig. 10.1 it can be seen that the material 'A' is superior in strength to material 'B', but its variability is such that the minimum strength of 'A' is less than the minimum strength of 'B'. This explains some 'inexplicable' failures; and also the uselessness of moving up a table of strengths or similar property to which there is no measure of variability.

The most important data requirements for a designer are the minimum (**not** average) strengths of his materials; for otherwise, failures will occur. Without these data he has to apply factors of safety which are crude artifices to, he hopes, blot out this area of ignorance. The end result is a waste of material; at best only a rough design. The design procedure itself has led to overdesign, a heavy and costly product, a lower performance.

10.4.1 Margin of safety

Besides using average values for the strength of his material, the designer often relates these to another largely notional factor – the load or stress imposed in the duties. As often as not these are also averages, and the fact that there may be severe overloads is again thought to be taken care of by the factor of safety.

Fig. 10.2 is a diagrammatic representation of the situation as often assumed by the designer with the sense of comfort. He has calculated the stress which will arise in the various components on the basis of an average loading, whether he realizes this is not. Then, using (average) values of strength and factors of safety, he arrives at a capability apparently satisfactorily greater than the load stress by a margin of safety which appeases any slight subsconscious feeling that his assumptions on both sides may not be exactly true.

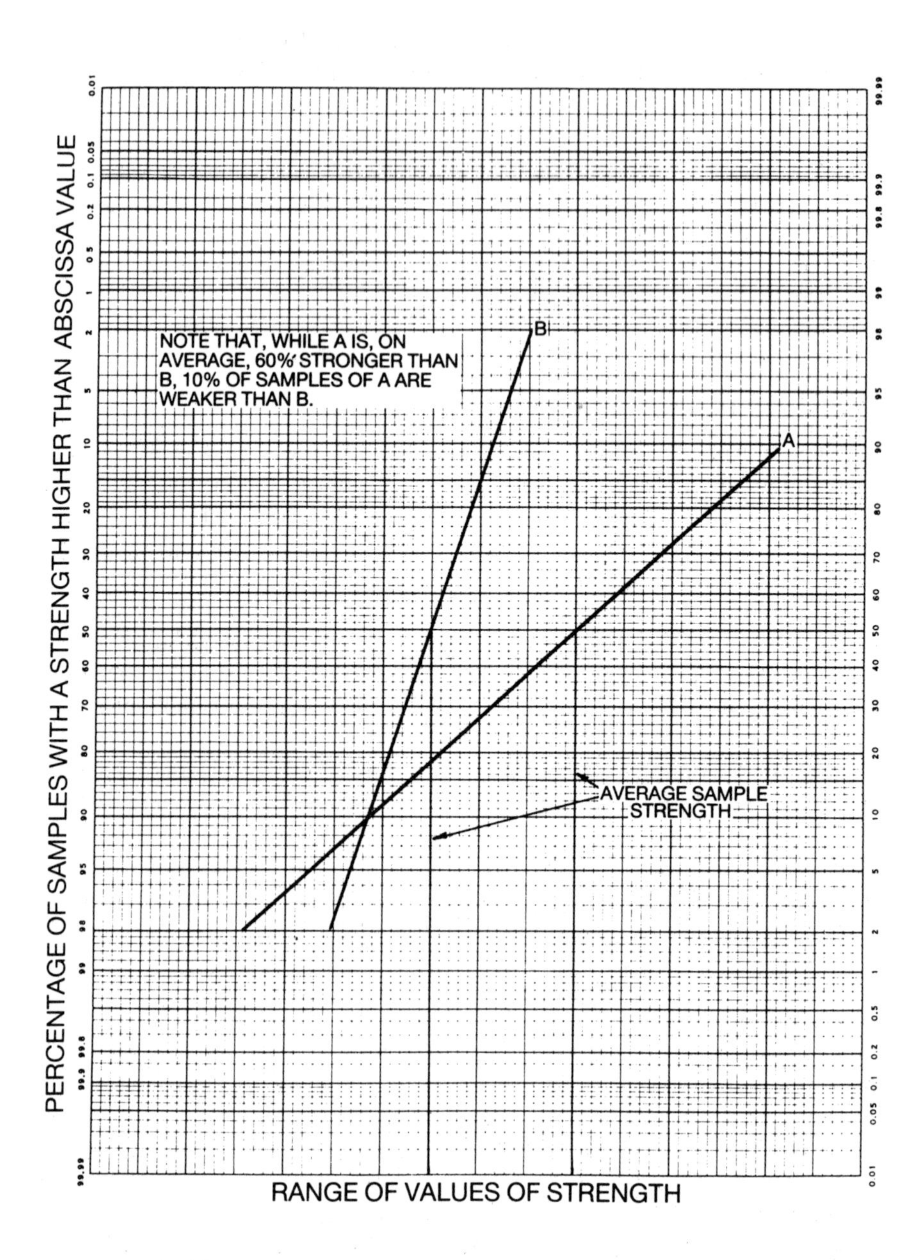

Fig. 10.1 – The pitfalls of designing on average values of strength.

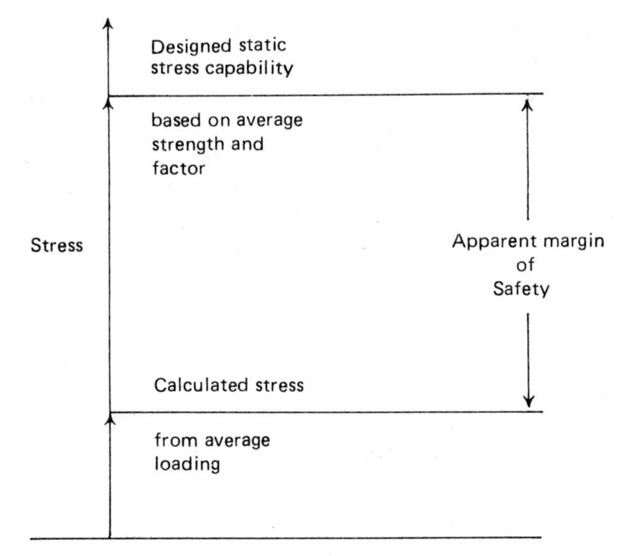

Fig. 10.2 – Misleading safety margins.

10.4.1.1 The true load

To see to what extent he is justified we must examine the bases of his two principal starting points. First let us consider the loading. To the average (or even 'average maximum') load we might have to take account of the following possibilities:

A (i) – (vi) see Fig. 10.3

(i) Variability in the load itself.
(ii) Increase in load due to misuse or abuse.
(iii) Effect of concentrations of stress, for example at corners, fillets, changes of section.
(iv) The correctness of the assumption that an assembly has been made properly, aligned truly, etc.
(v) Size variations in manufacture of components (manufacturing tolerances).
(vi) Load increases in time due to wear, corrosion, etc.

10.4.1.2 The true strength

Secondly, we may be oversimplifying the assumption about the capability of the materials to bear the load. Among the factors affecting this are the possibilities of:

B (i) – (v) see Fig. 10.3

(i) Variability in the material itself.
(ii) Weaknesses due to undetected (and undetectable) defects in the material.
(iii) Residual stresses remaining internally as a result of production method.
(iv) Optimistic presentation of the material properties by the seller (advertiser).
(v) Fatigue affects reducing the capability in time.

The true picture of the situation is now more like Fig. 10.3 which (a) shows a much reduced margin of safety and (b) shows a condition in which there will be 'inexplicable' failures, however much slide rule or computer checking is done.

It is clear from the above that there are areas for study by designers who wish not only to refine the arts and sciences of design, but also to arrive at more satisfactory, economical, and reliable designs. Although there will never be an ultimate answer to all the above features, the mere recognition of their existence will allow designers to ask the right questions and to seek answers. This will be followed by the application of the correct pressures on those whose responsibilities lie in these fields.

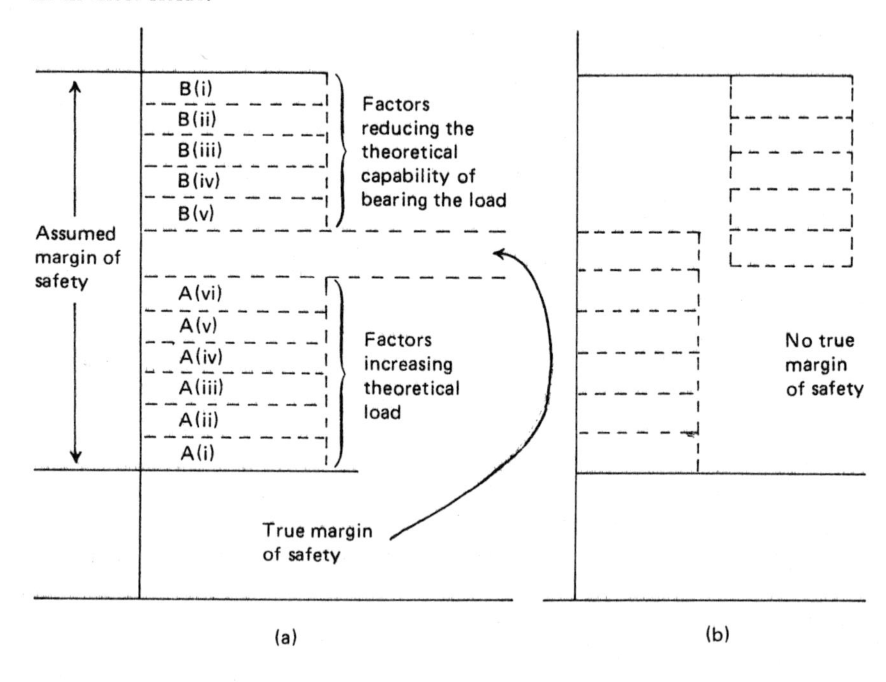

Fig. 10.3 – True margin of safety.

10.5 RELIABILITY AND FAILURE

The reliability specification for a given product can arise from a number of considerations. In addition to the statutory obligations having a reliability connotation given by the above Acts there may be contractual reliability requirements such as those set in the defence field by such specifications as DEF 05–21 *'Quality control system requirements for industry'* which includes 'control of reliability' (6).

10.5.1 Acceptable performance

A designer is often faced with the problems of designing for equipment whose life is of economic significance (rather than initial properties). This is particularly true of the growing legion of hired and rented equipments – television and radio, cars, etc. Trouble-free life here might mean the difference between profit and loss, the continuance of his job or not!

If he can take positive steps towards the design of a product which requires the ideal of no attention whatever during a reasonable (stated?) life, the rewards are large. Experience with such products will confidently enable meaningful and valuable guarantees to be offered. Satisfied customers will provide prestige and a greater share of the market. Price (within limits) will be a secondary consideration – especially in these times of publicity about 'value for money' and 'best buy'.

Where the above ideal is not possible, the next considerations must be in respect of how quickly, cheaply, and conveniently reliability restoration can take place, either the full and originally specified reliability or, if not, a level which is acceptable to the user.

Definitions of reliability usually refer to specified performance, but most people accept the reasonable view that things become old and wear, and that if this process is prolonged the situation is tolerable. Furthermore, few things for which 'a life of x years' is required, are really in use continuously for x years. Most are out of use at well-defined intervals – Sundays, holidays, nights, lunch-times, etc. In that case restoration during an interval might be perfectly acceptable as providing the sort of reliability required by the user.

Reliability is defined as the probability that the product will perform its required function under operational conditions for a specified time. Although this definition is fundamental to the determination of the costs of owning a product, it does not allow for repair. If we wish to take repair into our calculations we need to consider the Mean Time to Failure (MTTF) rather than life and the Mean Time to Repair (MTTR). We can then define *Availability* as MTTF/(MTTF + MTTR). Clearly, the designer must supply the prospective customer with information from which the cost of ownership may be calculated and this must include guarantees of life or time to failure and on the cost and time of repair.

The fact that the designer is required to commit himself to a statement of the maximum cost of failures or of the maximum downtime over a stated life means that he must be able to predict the reliability of the product before it is made, while after it is made he must be able to confirm that the promised reliability has been achieved. Even if the reliability is not stated explicitly, it must be known with some degree of accuracy before any spares-holding policy or any service support policy can be worked out. Designing so that repairable items are accessible requires the designer to be able to rank components in order of their need to be removed, and this again requires some knowledge of component reliability.

In addition to a general knowledge of reliability the designer must know how he can demonstrate the reliability that his components have achieved. In some cases the need to prove reliability involves the manufacturer in the fixed cost of the purchase of special test equipment. In all cases the variable costs of labour and consumed materials must be taken into account in the testing (usually to destruction) necessary to demonstrate reliability.

Equipment that will be used to check the reliability of a product that you have designed will not be cheap. We may have to show that the product works and continues to work over its advertized life, in a severe environment, and this will probably mean that we have to manufacture that environment. The test equipment must be capable of vibrating the product and of subjecting it to the humidity, sand, dust, or chemically or biologically corrosive contamination that it can meet in service. There will be specialist staff to operate and maintain such equipment, and there will be supporting control equipment which will generate and control the environmental cycles required by the tests.

The more quantitative approach to reliability is discussed at greater length in an appendix to this chapter.

10.5.2 More than one purpose

One of the greatest enemies of reliability is the requirement of more than one purpose to be fulfilled. In some cases, one purpose may be in conflict with another – or at least overlapping to some extent. When a designer has a single, well- defined function to embody, the way in which he can achieve the desirable result is clearly marked whatever the difficulties. But when he has to consider more than one function, he may have to decide on a compromise.

Such a compromise could mean the only partial fulfilment of each function, as well as increased complexity which always militates against reliability.

When presented with a number of functions, some of which may be incompatible to a degree, much consideration must be given to the alternatives of satisfying each separately with simple, single-purposed, reliable designs, or all together with a complex, multi-purposed, less reliable design. The questions of both quality of design and economies are paramount.

10.6 FAILURE PATTERNS

The failure regime concept can best be understood by reference to Fig. 2.3 where failure rate is plotted against time. It will be noted that this produces the familiar bath-tube curve giving three basic regimes: the infant mortality (decreasing failure rate), prime of life (constant failure rate), wear out (increasing failure rate). Such a curve applies to human life as well as man-made artefacts as evidenced in Fig. 10.4.

Probabilistic distributions of equipment life are discussed in more detail in Appendix 10.1 to this chapter.

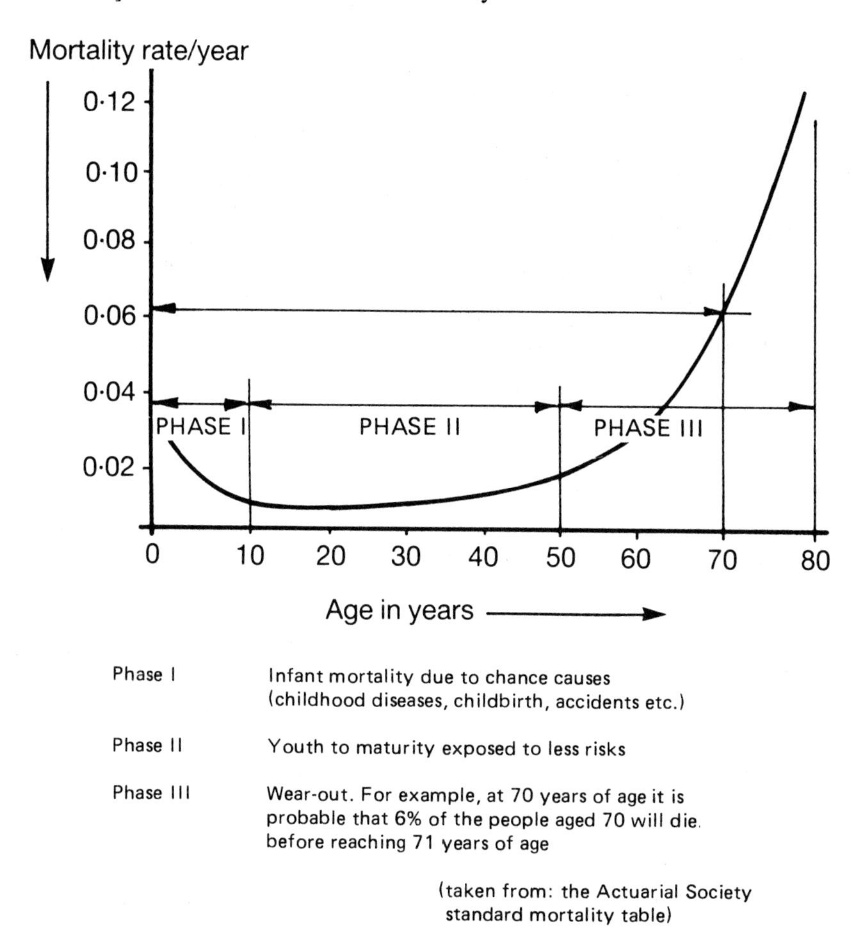
Mortality rate/year
0·12
0·10
0·08
0·06
0·04
0·02
PHASE I
PHASE II
PHASE III
0
10
20
30
40
50
60
70
80
Age in years
Phase I
Infant mortality due to chance causes (childhood diseases, childbirth, accidents etc.)
Phase II
Youth to maturity exposed to less risks
Phase III
Wear-out. For example, at 70 years of age it is probable that 6% of the people aged 70 will die before reaching 71 years of age
(taken from: the Actuarial Society standard mortality table)

Fig. 10.4 – Human failure rate.

10.7 FAILURE ANALYSIS

To design a reliable and safe product it is imperative to analyse how that product can fail, and then to design in features to avoid such failures, and in the event of any failure occurring make the product fail safe.

There has to be a correct design attitude, where a reliability philosophy is applied by management. Desirable reliability and safety growth result from good planning, designing, testing, manufacturing, and ultimately good using of the product according to a set of effectiveness oriented procedures.

10.7.1 Some techniques for P.L. prevention

Product liability prevention (PLP) is too recent in the field to have developed a comprehensive repertoire of techniques but some are now appearing. Some of the principal analytic techniques are:

Gross-hazard analysis
Performed early in design. Considers overall system as well as individual components. Called 'gross' because it is the initial safety study undertaken.

Classification of hazards
Identifies types of hazards disclosed in step 1 and classifies them according to potential severity (would defect or failure be catastrophic?). Indicates actions and/or precuations necessary to reduce hazards. May involve preparation of manuals and training procedures.

Failure modes and effects
Considers kinds of failures that might occur and their effect on the overall product or system. Example effect on system that will result from failure of a single component (a resistor or hydraulic valve, for example).

Hazard-criticality ranking
Determines statistical, or quantitative, probability of hazard occurrence. Ranking of hazards in the order of 'most critical' to 'least critical'.

Fault-tree analysis
Traces probable hazard progression. Example: if failure occurs in one component or part of the system, will fire result?

Energy-transfer analysis
Determines interchange of energy that occurs during a catastrophic accident or failure. Analysis is based on the various energy inputs to the product or system, and how these inputs will react in event of failure or catastrophic accident.

Catastrophe analysis
Identifies failure modes that would create a catastrophic accident.

System/subsystem integration
Involves detailed analysis of interfaces, primarily between systems.

Maintenance-hazard analysis
Evaluates performance of the system from a maintenance standpoint. Will it be hazardous to service and maintain? Will maintenance procedures be apt to create new hazards in the system?

Human-error analysis
Defines skills required for operation and maintenance. Considers failure modes

initiated by human error and how they would affect the system. Should be a major consideration in each step.

Transportation-hazard analysis
Determines hazards to shippers, handlers, and bystanders. Also considers what hazards may be 'created' in the system during shipping and handling.

One major technique which has come to the fore recently really combines some of the above, and is now called the 'design review'. This will be thoroughly covered in Chapter 12. It only remains to say that safety must be considered as a vital element in conducting such reviews. Any design review must become a critical analysis of a given design with special regard to the safety aspects at key points during the design process.

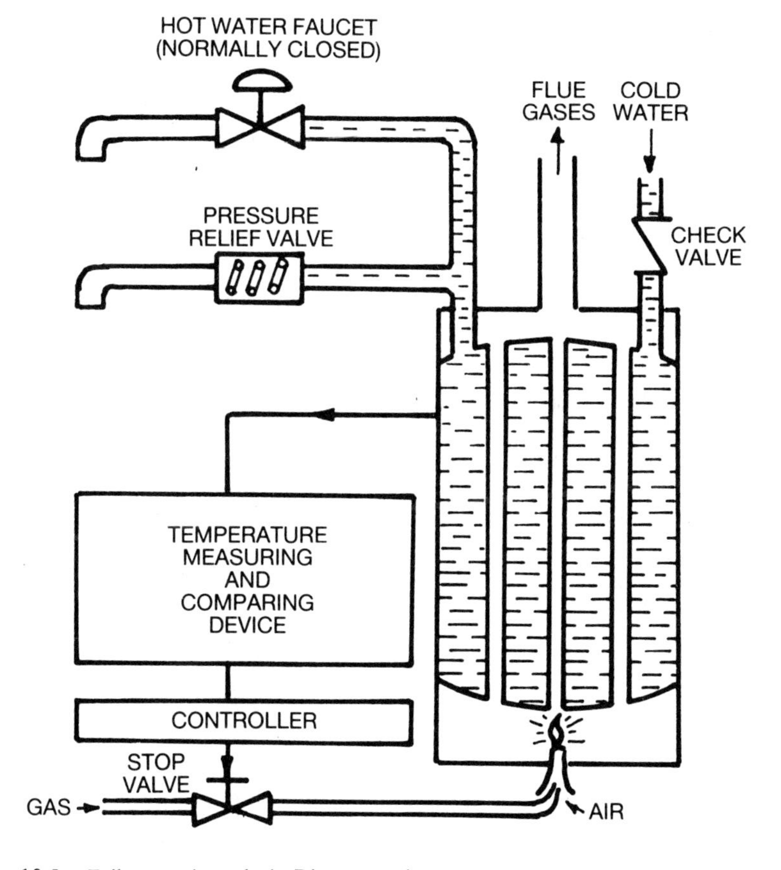

Fig. 10.5 – Failure mode analysis. Diagrammatic representation of a hot-water system.

Table 10.1

Component	Failure or error mode	Effects on		Failure frequency	Detection methods	Compensating provision & remarks
		Other components	Whole system			

Table 10.2
Failure mode analysis

COMPONENT	FAILURE OR ERROR MODE	EFFECTS ON		HAZARD CLASS				FAILURE FREQUENCY	DETECTION METHODS	COMPENSATING PROVISIONS AND REMARKS
		OTHER COMPONENTS	WHOLE SYSTEM	1	2	3	4			
Pressure relief valve	Jammed open	Increased operation of temperature sensing, controller, and gas flow due to hot water loss	Loss of hot water, greater cold water input and greater gas consumption	X				Reasonably probable	Observe at pressure-relief valve	Shut off water supply, reseal or replace relief valve
	Jammed closed	None	None	X				Probable	Manual testing	Unless combined with/other component failure, this failure has no consequence

Gas valve	Jammed open	Burner continues to operate. Pressure-relief valve opens	Water temperature and pressure increase. Water→steam		X	Reasonably probable	Water at faucet too hot. Pressure-relief valve open (observation)	Open hot water faucet to relieve pressure. Shut off gas supply. Pressure-relief valve compensates
	Jammed closed	Burner ceases to operate	System fails to produce hot water	X		Remote	Observe at output (water temperature too low)	
Temperature measuring and comparing device	Fails to react to temperature rise above preset level	Controller, gas valve, burner continue to function 'on'. Presure-relief valve opens	Water temperature to high. Water→steam		X	Remote	Observe at output (faucet)	Pressure-relief valve compensates. Open hot water faucet to relieve pressure. Shut off gas supply
	Fails to react to temperature drop below preset level	Controller, gas valve, burner continue to function 'off'	Water temperature too low	X		Remote	Observe at output (faucet)	
Flue	Blocked	Incomplete combustion at burner	Inefficiency. Production of toxic gasses		X	Remote	Possibly smell products of incomplete combustion	No compensation built in. Shut down system
Pressure-relief valve & gas valve	Jammed closed Jammed open	Burner continues to operate, pressure increases	Increased pressure cannot bleed at relief valve. Water→steam. If pressure cannot back up cold water inlet, system may rupture violently		X	Probable + reasonably probable reasonably probable	Manual testing of relief valve. Observe water output (temperature too high)	Open hot water faucet. Shut off gas supply. Pressure might be able to back up into cold water supply, providing pressure in supply is not greater than failure pressure of system

10.7.2 Failure mode analysis

This method of analysis is a technique of predicting all possible means by which a component can fail to perform its required function. FMA, when properly used, will identify all failure modes, and is also capable of assessing the probability of occurrence of such occurrences. These probabilities serve as quantitative entries into the product faults which could happen: 'How could this component fail?' and 'What would happen if the component did fail?'

It starts with a functional block diagram to determine the importance of each function. A formal interrogation is then conducted by filling in each column of the table shown in Table 10.1.

Such an analysis may range from subjective engineering judgement to an elaborate finite element stress analysis. The degree of analysis sophistication and/or test validation will determine the confidence one can place in the estimated probability of mechanism failure.

A typical example of a failure mode analysis for a domestic hot water system depicted in Fig. 10.5 is given in Table 10.2.

APPENDIX 10.1

A quantitative approach to reliability

In this Appendix to Chapter 10, some indication is given of quantitative methods in reliability analysis. Simple graphical and computer aided methods of determining parameters in the cases of Negative Exponential and Weibull distributed lives are shown. The significance of these parameters is shown in the determination of a spares holding policy, the use of redundancy, capital recovery, and consideration of optimum replacement lives.

Examples show the low accuracy with which information can be derived from limited data.

If the bath-tub curve (Fig. 2.3), showing the rate of failure as a function of time, is reasonably flat during the normal operating period, then the product will have a probabilistic life which is distributed according to the negative exponential distribution. That is, the probability of the product's life exceeding t is

$$R(t) = e^{-\lambda t}$$

where λ is the (constant) rate of failure. $1/\lambda$ is the average product life, and is usually called the mean time between failures (MTBF).

Example

Records kept on 1000 engines of the same type show an average life to failure of 14 000 flying hours. What is the probability that one such engine will survive a transatlantic flight of 7 hours?

$$\text{MTBF} = 1/\lambda \quad = 14\,000 \text{ hours}$$
$$\text{therefore} \quad \lambda \quad = 1/14\,000$$
$$R(7) \quad = \text{Probability of surviving 7 or more hours}$$
$$= e^{\frac{-7}{14\,000}} = e^{-0.0005} \quad = 0.9995$$

Note that the reliability must be near to 1, because we would not wish to own a product that carried more than one chance in several thousand of failing dangerously in a reasonable working period.

The above example considered reliability from the point of view of safety, but safety can often be related to cost. If an aeroplane crashed, killing fare paying passengers, the airline would have to pay out large sums of money to the relatives of the dead (or would have to be insured against having to do so).

Even where danger to life is not involved, considerable costs could be generated by the failure of a product. Suppose the product being considered were a pump used in, say, the Dorr process of making phosphoric acid. The Dorr process is continuous, and failure of a pump in the system could mean shutting down the whole plant until the pump is replaced or repaired. The company operating the pump will thus be faced with the cost of a spares holding policy, the cost of the maintenance when the pump is replaced or repaired, and the cost of the plant's down time. It is probable that the greatest of these costs will be the cost of down time because, while the plant is idle, people and satisfactory machinery will be unable to work, and wages and capital are being wasted.

Example

The following data were obtained from maintenance records of a pump used in the Dorr process:

13 pumps failed in less than 3 shifts
4 pumps failed in less than 5 shifts but in 3 or more shifts
1 pump failed in less than 8 shifts but in 5 or more shifts
1 pump failed in less than 11 shifts but in 8 or more shifts
3 pumps failed in less than 14 shifts but in 11 or more shifts
3 pumps failed in less than 17 shifts but in 14 or more shifts
2 pumps failed in less than 20 shifts but in 17 or more shifts
2 pumps failed in less than 27 shifts but in 20 or more shifts
1 pump failed in less than 32 shifts but in 27 or more shifts
1 pump failed in less than 35 shifts but in 32 or more shifts

31 failures were recorded

Although this information is crude, it is typical of what is available in industry. We can calculate an approximate MTBF for the pump by assuming that failures occurred in the middle of the above periods.

The accommulated lives of the 31 pumps is 282.5 shifts, so that the mean time between failures is approximately 9.1 shifts.

$$1/\lambda = 9.1 \text{ shifts}$$
$$\lambda = 0.11 \text{ failures per shift.}$$

Using this to calculate the reliability of the pump,

$$R = e^{-(0.11)s} \quad \text{where } s \text{ is the number of shifts.}$$

The mean time between failures tells us something about the spares back-up that we need, because we will require to repair or replace the pump, on average, every 9 shifts. The failures will not be regular, however, because, as the original data show, 42% (13/31) of pumps failed within 3 shifts.

The calculated reliabilities for the periods of the data may be compared with the observed failures:

Life (shifts)	Calculated reliability (R)	Calculated failures (1–R)	Period (shifts)	Calculated failures	Cumulative calculated failures	Observed failures	Cumulative observed failures
3	0.72	0.28	0–3	.28 x 31≏9	9	13	13
5	.58	.42	3–5	(0.42–0.28)x31≏4	13	4	17
8	.41	.59	5–8	(.59–.42) x 31≏5	18	1	18
11	.30	.70	8–11	(.70–.59) x 31≏3	21	1	19
14	.22	.78	11–14	(.78–.70) x 31≏2	23	3	22
17	.15	.85	14–17	(.85–.78) x 31≏2	25	3	25
20	.11	.89	17–20	(.89–.85) x 31≏1	26	2	27
27	.05	.95	20–27	(.95–.89) x 31≏2	28	2	29
32	.03	.97	27–32	(.97–.95) x 31≏1	29	1	30
35	0.02	0.98	32–35	(0.99–0.97) x 31≏1	30	1	31

There are many possible reasons for discrepancy. The sample of lives is small, and because failures are probabilistic one would not expect the actual events to reflect, precisely, the calculated events; the failure rate, λ, may not in fact be constant through the life of the pump, although this has been assumed to be the case in calculating the reliability; the failures listed may be of different component parts of the pump.

Nevertheless, the problems associated with the lack of reliability begin to emerge, and a pattern of failures is apparent.

From the way it is defined, the reliability of a product is a probability. We can design the product so that this probability (of lasting a stated time) is improved, and we can reduce the probability of failure. We cannot be sure that the time to failure of a product will exceed any given period nor be sure that it will not fail in any given time.

THE USE OF WEIBULL GRAPH PAPER

It is probably easier to make use of special graph papers to calculate the life properties of components. One special graph paper is Weibull paper†, and the negative exponential distribution is a particular case of the Weibull distribution (7), (8) where $\beta = 1$. As an example of the use of this paper, re-work the above data of the Dorr process pump.

Fig. 10.A1 shows the data plotted, and some explanation is necessary.

The vertical scale is cumulative per cent failure, and this is calculated in the following way.

(i) Arrange the n cumulated observed failures in numerical order.

(ii) Calculate q for the ith failure, where $q = \dfrac{i-\frac{1}{2}}{n}$.

(iii) Plot 100 q versus life at failure.

In the above example, we could calculate the following table.

Life at failure (shifts)	Observed failures	Cumulative failures (i)	$q = \dfrac{i-\frac{1}{2}}{n}$
3	13	13	.403 = 40.3%
5	4	17	.532 = 53.2%
8	1	18	.565 = 56.5%
11	1	19	.597 = 59.7%
14	3	22	.694 = 69.4%
17	3	25	.790 = 79.0%
20	2	27	.855 = 85.5%
27	2	29	.919 = 91.9%
32	1	30	.952 = 95.2%
35	1	31	.984 = 98.4%
	$n = 31$		

†The paper used in this example was supplied by Chartwell.

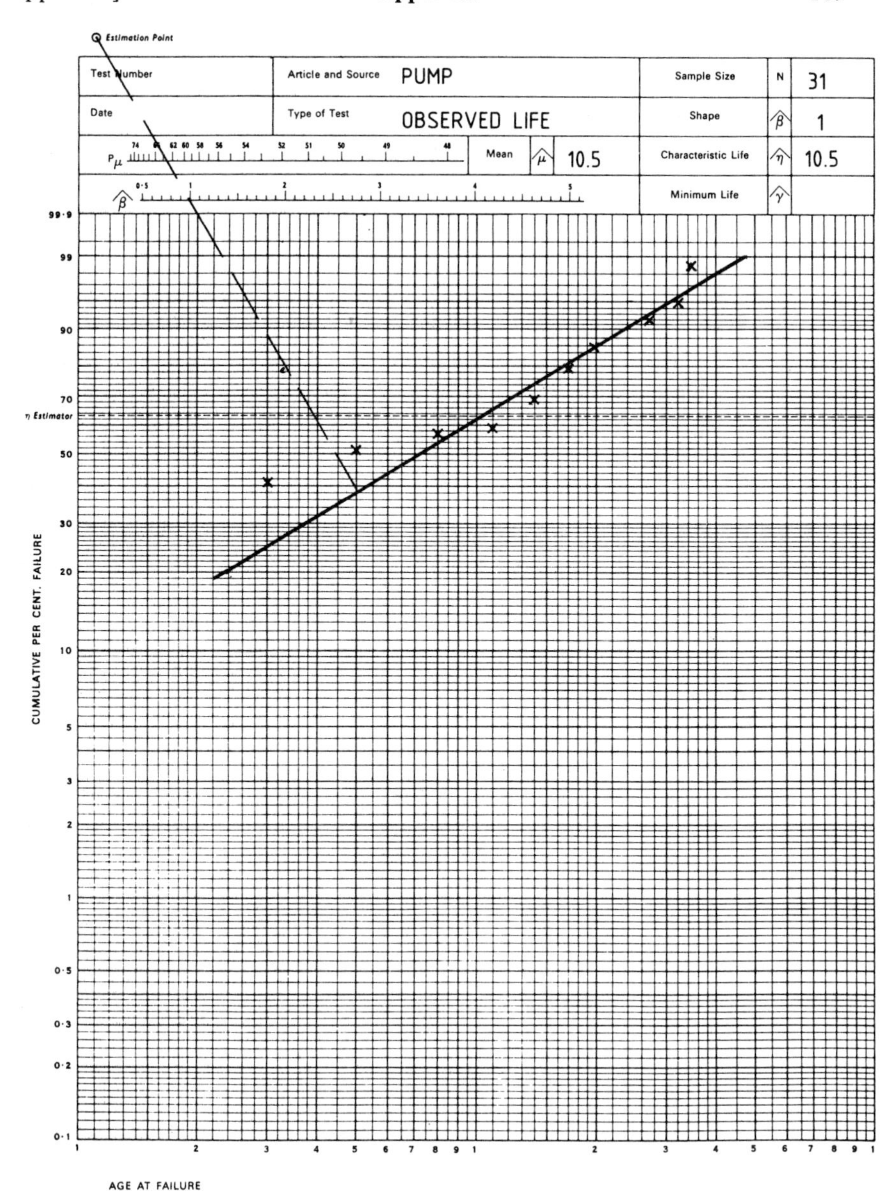

Fig. 10.A1 – Cumulative per cent failure, plotted on Weibull paper.

Having calculated and plotted the points on Weibull graph paper, we need to draw a straight line, as nearly as possible, through those points.

Because the lives plotted are assumed to be from a negative exponential distribution we would want our line to be one for which $\beta = 1$. In the proprietary paper illustrated this means that our line of best fit must be perpendicular to the construction line from the Estimation Point through the β scale where $\beta = 1$.

Again, if we assume that the lives are from a negative exponential distribution, the mean life may be read where the Estimator line crosses the line that we have plotted. In this case we read the mean life as 10.5 shifts, which does not agree with the previously determined value of 9.1. The discrepancy is caused by the inaccuracy with which the straight line is fitted to the points, but this need not concern us very much because there will always be inaccuracies in calculating parameters from probabilistic data.

It is also fairly straightforward to plot these lives using a microcomputer. Fig. 10.A2 is a Weibull plot of the data of the product lives, and this too shows that the distribution approximates to a negative exponential distribution ($\beta = 1$) with a mean life of 10.5 shifts.

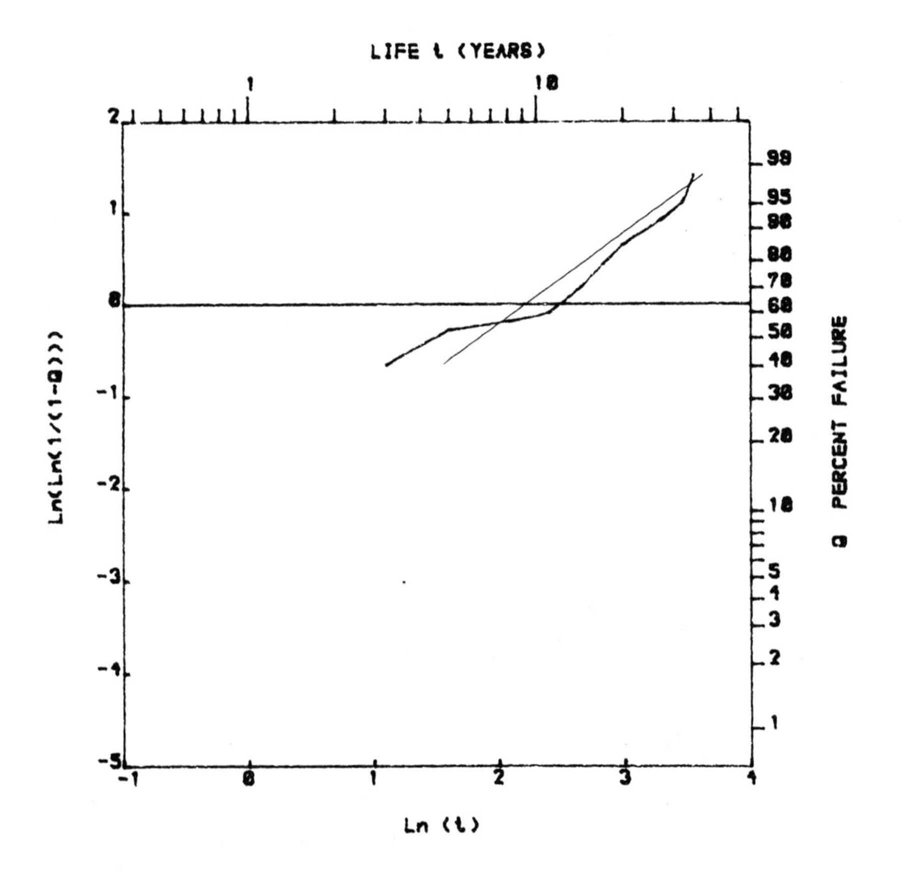

Fig. 10.A2 – Lives replotted, using a microcomputer.

Pictorial representation of the data gives some feel for whether the negative exponential distribution reasonably describes the life of the product.

The program used to plot Weibull lives has been written in MBASIC for use with a BBC Micro computer, and can be made available.

COMPONENTS IN SERIES

Unfortunately, a system often contains many parts and is such that, if any one part fails, the system cannot function (if a link fails, the chain will fail, if a tyre fails, the car cannot be used, etc.). In this case the parts of the system are said to be in series. Suppose a system has n components in series and the reliability of the ith component is $R(i)$. Then the reliability of the whole system is

$$R = R(1) \times R(2) \times R(3) \times \ldots \times R(i) \times \ldots \times R(n).$$

Example

Consider a chain of 1000 links, supporting the pylon of a suspension bridge. The chain will break if any link breaks. If each link has a reliability of 0.99999, what is the reliability of the chain?

$$\begin{aligned} R_{\text{system}} &= R_1 \times R_2 \times \ldots \times R_{1000} \\ &= (0.99999)^{1000} \\ &= 0.99005 \end{aligned}$$

Clearly these figures which imply one chance in a hundred that the chain would fail, are unacceptable. The figure for the reliability of the links is absurdly low, but notice how, with many components, reliability falls from one chance of failure in a hundred thousand to one chance of failure in a hundred.

Example

Consider the pump in the Dorr process that was the subject of an earlier example. Suppose there were four such pumps in the process and the failure of any one would stop the process. Considering only the four pumps, what is the probability that the process will work through one shift without failure?

We have already calculated that the reliability of a pump is $R = e^{-(0.11)s}$ so that the probability of the pump lasting one shift without failure is $R = e^{-(0.11)}$ $= 0.895$.

The probability that all four pumps will work for a shift, without failure, is $(0.895)^4 = 0.642$.

Increasing the number of components in a system is thus going to reduce the reliability of the system – often to an unacceptable level. Our hypothetical system with four pumps has only little more than a sixty per cent chance of getting through a shift without failure.

RELIABILITY OF COMPONENTS IN PARALLEL

If a system consists of two components which we installed in such a way that the system fails only if both components fail, we have what is called redundancy or a system with its components in parallel. Such a system is often drawn thus:

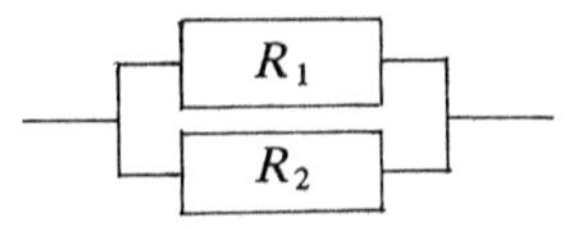

The system will fail only if both components fail. If $Q_1 = 1 - R_1$ = probability that component one will fail, and $Q_2 = 1 - R_2$, then the probability that both components will fail is $Q = Q_1 \times Q_2$. The probability that the system will not fail is then

$$\begin{aligned} R &= 1-Q = 1 - Q_1 Q_2 \\ &= 1-(1-R_1)(1-R_2) \\ &= R_1 + R_2 - R_1 R_2 . \end{aligned}$$

Example

Consider a twin-engined aeroplane which can continue its mission if only one engine fails. What is the reliability of the aeroplane in a 4-hour journey (considering only the engines) if each engine has a MTBF of 10 000 hrs?

R_E = reliability of engine = $e^{-4\lambda}$

but $1/\lambda$ = 10 000

therefore $R_E = e^{-4/10\,000} = e^{-0.0004}$.

$$\begin{aligned} R \text{ system} &= R_1 + R_2 - R_1 R_2 \\ &= 2R_E - R_E^2 \\ &= 2e^{-0.0004} - (e^{-0.0004})^2 \\ &= 2(0.99960008) - (0.99960008)^2 \\ &= 0.99999984, \end{aligned}$$

and we observe how redundancy increases reliability considerably.

Example

Suppose in our Dorr process, we regard the reliability of the pump as inadequate. We could use two pumps in parallel (assuming this to be technically feasible) to increase the reliability of the system. The reliability of each pump over a shift is 0.895, but the reliability of the two pumps in parallel is

$$\begin{aligned} R \text{ system} &= 2(0.895) - (0.895)^2 \\ &= 0.988975. \end{aligned}$$

This has not only raised the reliability of the pumps to what may be an acceptable level, but it permits repairs to be carried out on a failed pump without shutting down the system.

REPLACEMENT AND NEGATIVE EXPONENTIAL LIVES

If the the life of a component is distributed according to the negative exponential distribution, the probability of failure does not change with the life of the component. There is, therefore, no best time at which the component should be replaced; failures occur randomly, and a new unit is as likely to fail as an old one.

Even though the failures are random, we may find it desirable to carry spares, and it is useful to know how many spares we should carry.

We are all familiar with the idea of carrying a spare wheel so that a puncture will not stop us from completing a journey. Most motorists would be unhappy to start a long journey without a spare wheel, but few would find it worth spending the money to travel with two spares. The problem arises in industry because the user of a plant must have a policy for spares holding. It would be absurd to bring a factory to a standstill for lack of a cheap replacement for a failed component, but some thought would have to be given to any policy for holding very expensive, rarely used, spare parts.

If the time between failures of a component is drawn from a negative exponential distribution, it can be shown that the rate of failures is drawn from a Poisson distribution. That is, if component life is drawn from the distribution $\lambda e^{-\lambda t}$ where the mean life is $1/\lambda$, the probability that there will be n failures in time T is

$$\frac{(\lambda T)^n e^{-\lambda T}}{n!}.$$

If T is the time that it takes to obtain spares after they have been ordered, and if we are holding n spares at the time the order is placed, we need to know the probability that there will be more than n failures in time T. We thus need to know the value of

$$1 - \sum_{k=0}^{n} \frac{(\lambda T)^k e^{-\lambda T}}{k!}.$$

It is not necessary to calculate the value of this function from first principles because we can use proprietary Poisson† graph paper. Fig. 10.A3 is a sheet of such paper. Consider any value of λT. Since $1/\lambda$ is the mean time between failures and T is the period for which we are holding spares, λT is the number of average product lives for which we are holding spares.

If, for example, we were considering holding spares for 1000 hours of operation, and if the mean rate of failure (λ) were one every 500 hours, then λT would be 2 (or T would be two average lives). We can read directly from Fig. 10.A3 that, if $\lambda T = 2$, the probability

†The paper used in this example was supplied by Chartwell.

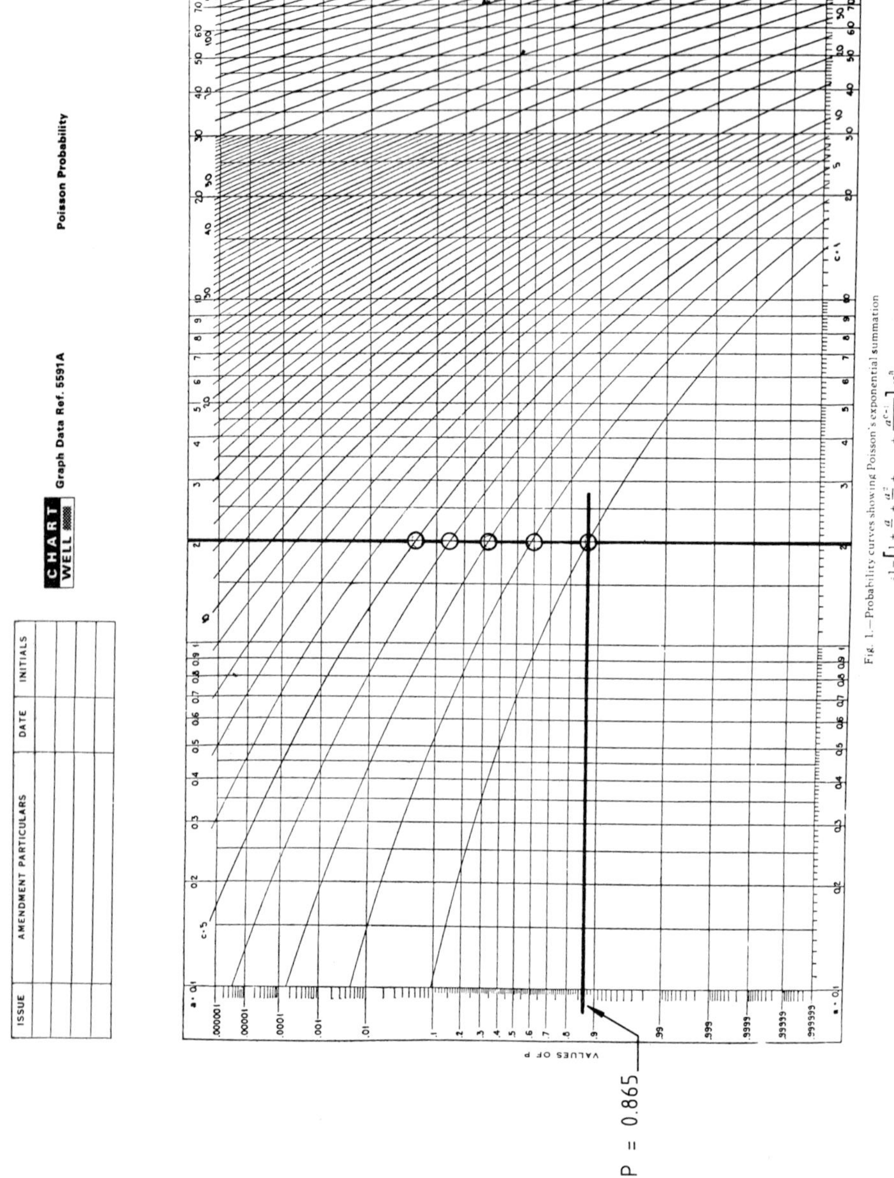

Fig. 10.A3 – Use of proprietary Poisson graph paper.

of one or more failures is 0.865,
of two or more failures is 0.594,
of three or more failures is 0.323, etc.

(Note that these figures were calculated; the graph cannot be read with such accuracy.)

These figures can be restated as the probability that we will run out of parts if $\lambda T = 2$, and we have

0 spares at the beginning of the period, is 0.865,
1 spare at the beginning of the period, is 0.594,
2 spares at the beginning of the period is 0.323, etc.

It is also simple to use a microcomputer to determine the probability of running out of spares. Fig. 10.A4 is a printout from a MBASIC program which calculates the probability of running out of spares when the MTBF of the component is known.

```
5 PRINT#-2,"     LAMBDA        N            T             P"
6 PRINT#-2
10 CLS:INPUT"LAMBDA=";LAM
20 INPUT"N=";N
30 INPUT"T=";T
40 Y=1:S=0
50 FOR K=0 TO N
60 Y=Y*K:IF Y=0 THEN Y=1
70 X=(((LAM*T)^K)*(EXP(-LAM*T)))/Y
80 S=S+X
90 NEXT K
100 F=1-S
110 PRINT:PRINT "F=";F
112 PRINT#-2, USING"    ##.####    ###.###    ####.###    ###.####";LAM,N,T,F
115 PRINT:INPUT"AGAIN Y/N ";A$
117 IF A$="Y" THEN GOTO 10
120 END
```

LAMBDA	N	T	P
0.0020	0.000	1000.000	0.8647
0.0020	1.000	1000.000	0.5940
0.0020	2.000	1000.000	0.3233
0.0020	3.000	1000.000	0.1429
0.0020	4.000	1000.000	0.0527

Fig. 10.A4.

THE WEIBULL DISTRIBUTION

Unfortunately, not all components or assemblies have lives that can be described by the negative exponential distribution. Weibull suggested that [7]

$$1 - Q = \exp[-(t/\eta)^{\beta}]$$

could be used to describe the (probabilistic) lives of many products, and experience has shown him to be correct.

Here Q is the probability of failure before time, t, and β and η are constants to be determined in the case of any particular product. η is called the characteristic life, and β is called the shape function.

We see, immediately, that in the cases where $\beta = 1$, the Weibull distribution is identical to the negative exponential distribution where $\lambda = 1/\eta$. The negative exponential distribution reflects the situation in which the failure rate, λ, is independent of time. That is, the probability of a failure in any small time interval, Δt, is $\lambda\Delta t$. This is shown in the bath-tub curve of Fig. 2.3 and Fig. 10.4. The Weibull distribution, in general, reflects a situation in which the probability of failure in interval, Δt, is $h(t) \times \Delta t$

where $$h(t) = \frac{\beta}{\eta}\left(\frac{t}{\eta}\right)^{\beta-1}$$: $h(t)$ is called the hazard function.

This hazard function is plotted on Fig. 10.A5 for some values of β. We notice that if $\beta > 1$, the rate of failure increases with time, and the greater β,

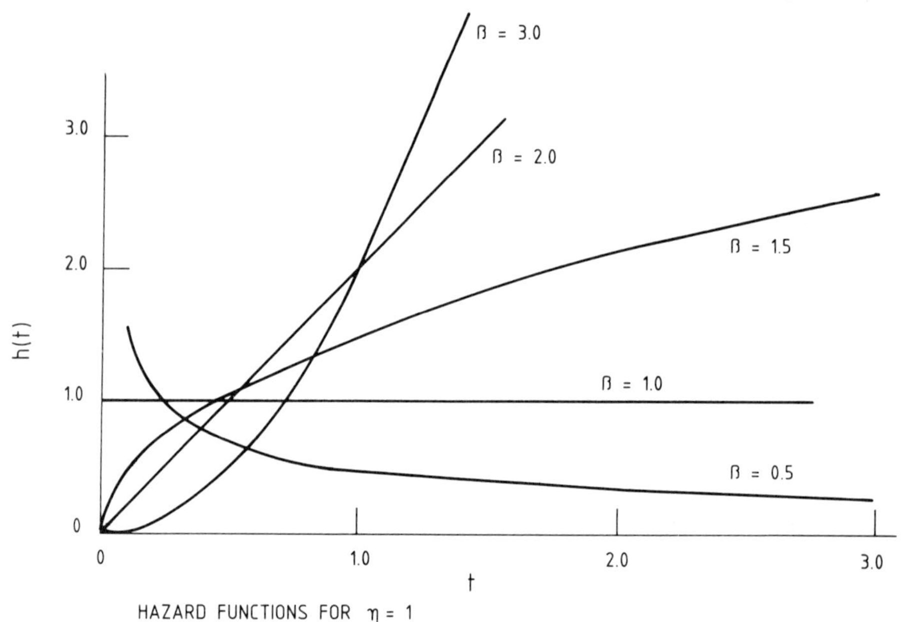

Fig. 10.A5 – The Hazard functions for $\eta = 1$.

the faster is the aging process. If a product gets worse with time, a Weibull distribution with $\beta > 1$ might be an appropriate function to describe its reliability.

As with the negative exponential distribution, it is easier to determine the parameters of the Weibull distribution by plotting lives at failure than by analysis.

Example

Suppose we have tested ten switches to failure with the following results.

Switch no.	Failed at (cycles)
1	950
2	1700
3	2250
4	2700
5	3200
6	3600
7	4200
8	4800
9	5600
10	7000

Following the practice of Hahn & Shapiro, we would argue that, for the ith failure of n, $Q = (i - \frac{1}{2})/n$. If these data are plotted we can fit a line to the points by eye, from this line we can read the estimated values of β and η directly. This is shown in Fig. 10.A6.

Alternatively we may use a computer to help us to plot the graph. Fig. 10.A7 is a plot of life data on Weibull axes, to which a line has been fitted by eye and the values of β and η printed. This has been done by a simple MBASIC program on a BBC Microcomputer.

SPARES HOLDING AND WEIBULL LIVES

As with components with exponentially distributed lives, we need to know the cost of spares-holding policy. The basis of any policy must be a knowledge of the probability that the number of spares we hold will keep the system running for a given time. Unfortunately there is no simple formula which will tell us the probability of a given number of failures in a given period.

One way of examining the effectiveness of a spares-holding policy is simulation and Fig.10.A8 is the output of 10 000 occasions of holding spare components. The program was written in MBASIC and will consider any number of occasions (although 10 000 is probably enough) of holding any small number (say 2, 3, 4, or 5) of spares. In this case, we see that on 90% of occasions, the original component and the two spares give an accumulated life of more than 1.09.

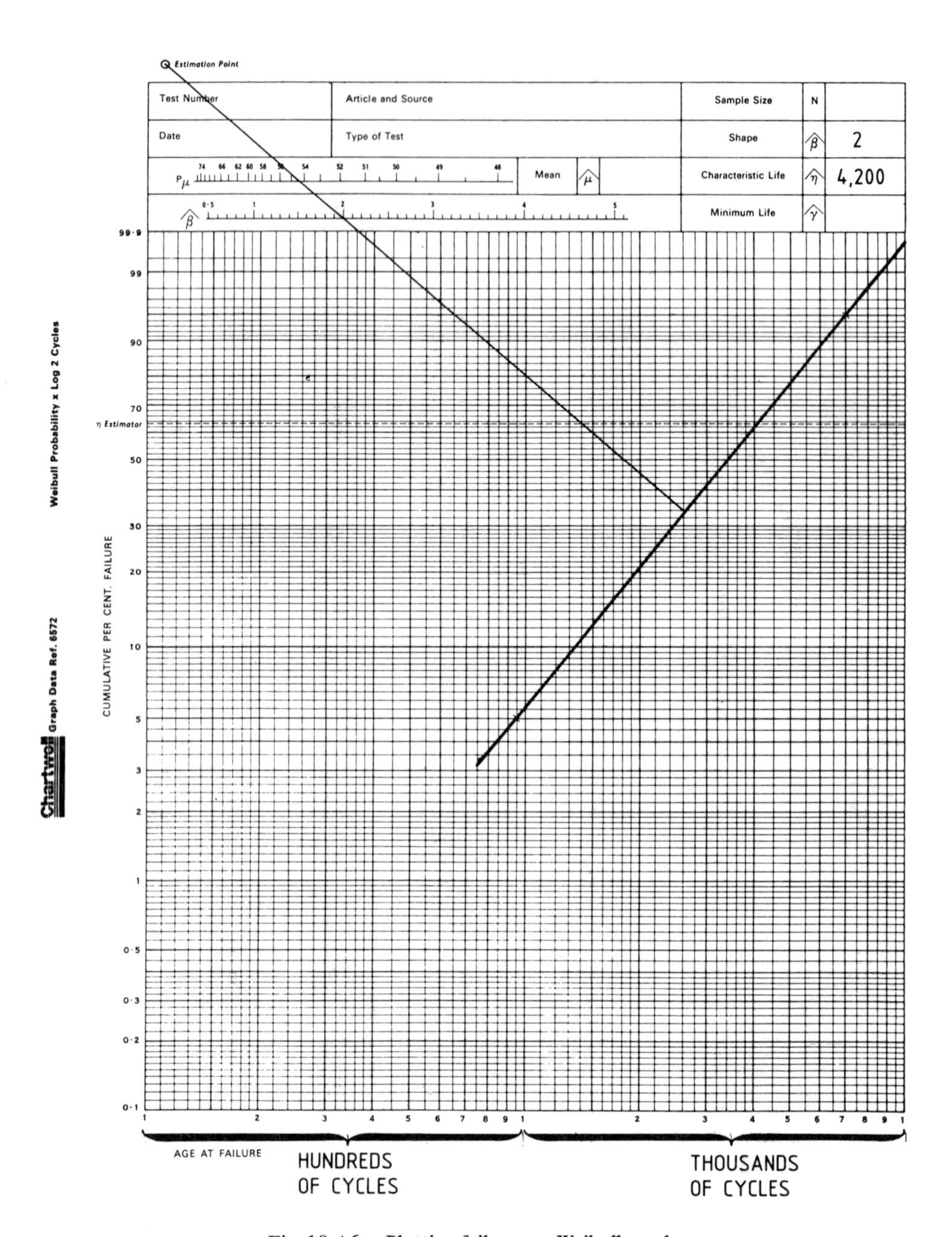

Fig. 10.A6 – Plotting failures on Weibull graph paper.

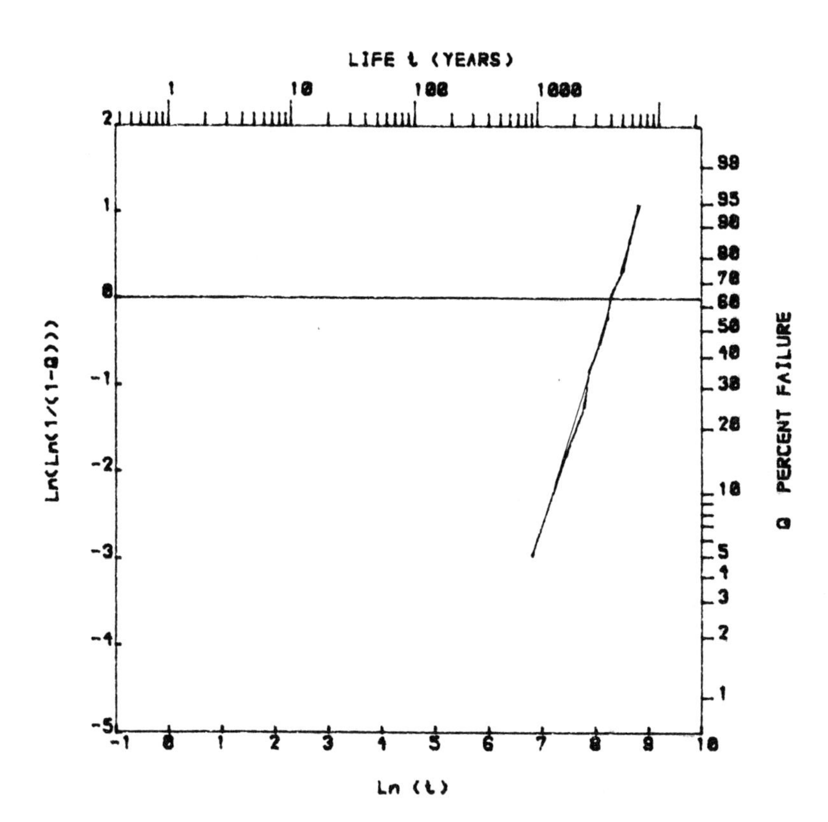
LIFE t (YEARS)
1
10
100
1000
2
1
0
-1
-2
-3
-4
-5
-1
0
1
2
3
4
5
6
7
8
9
10
Ln (t)
Ln(Ln(1/(1-Q)))
99
95
90
80
70
60
50
40
30
20
10
5
4
3
2
1
Q PERCENT FAILURE

Fig. 10.A7.

```
X(  100)=    0.5213707685
X(  200)=    0.5922966003
X(  300)=    0.6831803322
X(  400)=    0.7236173153
X(  500)=    0.8324508667
X(  600)=    0.8948119879
X(  700)=    0.9196193218
X(  800)=    0.9789986610
X(  900)=    1.0236847401
X( 1000)=    1.0579495430
X( 1100)=    1.0880138874
X( 1200)=    1.1257996559
X( 1300)=    1.1745648384
X( 1400)=    1.2140796185
X( 1500)=    1.2236077785
X( 1600)=    1.2368361950
X( 1700)=    1.2527966499
X( 1800)=    1.2782628536
X( 1900)=    1.2838032246
X( 2000)=    1.3032202721
X( 2100)=    1.3204565048
X( 2200)=    1.3548634052
X( 2300)=    1.3730409145
X( 2400)=    1.3784677982
X( 2500)=    1.3979408741
X( 2600)=    1.4061045647
X( 2700)=    1.4345340729
X( 2800)=    1.4469239712
X( 2900)=    1.4703562260
X( 3000)=    1.4853711128
X( 3100)=    1.4934949875
X( 3200)=    1.5145568848
X( 3300)=    1.5233521461
X( 3400)=    1.5411221981
X( 3500)=    1.5473415852
X( 3600)=    1.5573549271
X( 3700)=    1.5702924728
X( 3800)=    1.5836176872
X( 3900)=    1.6168859005
X( 4000)=    1.6407654285
X( 4100)=    1.6571042538
X( 4200)=    1.6631095409
X( 4300)=    1.6890625954
X( 4400)=    1.7012701035
X( 4500)=    1.7173562050
X( 4600)=    1.7303860188
X( 4700)=    1.7396216393
X( 4800)=    1.7756068707
X( 4900)=    1.7864463329
X( 5000)=    1.8245751858
X( 5100)=    1.8377664089
X( 5200)=    1.8637514114
X( 5300)=    1.8717706203
X( 5400)=    1.8890082836
X( 5500)=    1.9066731930
X( 5600)=    1.9142887592
X( 5700)=    1.9230778217
X( 5800)=    1.9390335083
X( 5900)=    1.9664990902
X( 6000)=    1.9724230766
X( 6100)=    1.9926207066
X( 6200)=    2.0111126900
X( 6300)=    2.0387535095
X( 6400)=    2.0448961258
X( 6500)=    2.0826745033
X( 6600)=    2.1074824333
X( 6700)=    2.1172666550
X( 6800)=    2.1276798248
X( 6900)=    2.1630725861
X( 7000)=    2.1768784523
X( 7100)=    2.2179317474
X( 7200)=    2.2290067673
X( 7300)=    2.2565717697
X( 7400)=    2.2746415138
X( 7500)=    2.3036580086
X( 7600)=    2.3227801323
X( 7700)=    2.3607087135
X( 7800)=    2.4000878334
X( 7900)=    2.4080219269
X( 8000)=    2.4171671867
X( 8100)=    2.4428758621
X( 8200)=    2.4698786736
X( 8300)=    2.5069646835
X( 8400)=    2.5525217056
X( 8500)=    2.5903306007
X( 8600)=    2.6537919044
X( 8700)=    2.6874475479
X( 8800)=    2.7373008728
X( 8900)=    2.7574877739
X( 9000)=    2.7624478340
X( 9100)=    2.7944493294
X( 9200)=    2.8527612686
X( 9300)=    2.8985366821
X( 9400)=    2.9323134422
X( 9500)=    2.9920706749
X( 9600)=    3.0367121696
X( 9700)=    3.1273083687
X( 9800)=    3.2339258194
X( 9900)=    3.7710123062
X(10000)=    4.4934558868

**** STOP
  END OF MACRO
```

10,000 summations of the sum of three lives (the original component and two spares)

Fig. 10.A8.

SCHEDULED REPLACEMENT

If a component has a reliability $R(t)$, that is, the probability that its life will be greater than t is $R(t)$, it may be economical to replace it before it fails. The reason for replacing a component before it fails is that we can choose a convenient, cheap time for the work – say during a maintenance shift. If the component has to be replaced during a working shift we must pay for down time as well as for replacement. We thus assume that the cost of an enforced replacement is much greater than the cost of a scheduled replacement, although the smaller we make the replacement life, the less likely this is to happen. However, if we make the replacement life short so as to have few unscheduled replacements, we get very little use out of each component.

One way of determining the optimum replacement life for a component with a Weibull distributed life is to simulate a large number of policies and to find the optimum replacement life by trial and error. Fig. 10.A.9 is the output of a MBASIC simulation program which calculated the average cost of a replacement policy.

Note that, with exponentially distributed lives, there can be no optimum replacement policy, because failure is random.

CAPITAL RECOVERY AND PROBABILISTIC LIVES

We have seen in Chapter 8 that if we know the life of a capital investment, we can determine the annual cost which is equivalent to its first cost. The formula was

$$P/A = [(1+i)^n - 1]/[i(1+i)^n].$$

When the life of the product is probabilistic, some products will last longer than others. Because the present value of a cash flow declines with its lateness, those products which live long will not compensate for those which die young.

When the life of the product is exponentially distributed, it can be shown that if the pay-off generated by the product is £A/year, then the expected net present value calculated of a large number of the product is $A/(\lambda + i)$ where $1/\lambda$ is the mean life of the product and i is the annual rate of interest.

This means that the annual cash flow that investment in a product must generate is greater than would be suggested by simple calculations based on a deterministic life. As an example, consider a component with a fixed life of 10 years when the cost of capital is 10%. Tables tell us that the product must generate 0.163 of its cost every year to pay for itself. If, however, the life of the product were from a negative exponential distribution, with a mean life of 10 years, it would have to generate 0.2 of its cost every year for us to expect it to pay for itself.

No simple formula is available for Weibull distributed lives, although, generally, the effect of the probabilistic life is less marked.

```
10 INPUT"N,BETA,ETA";N,B,E
20 INPUT"COSTS,COSTSCH";C,CS
30 PRINT#-2,"N=";N;"  BETA=";B;"  ETA=";E;
40 PRINT#-2,"  COSTS=";C;"  COSTSCH=";CS:PRINT#-2
50 INPUT"SEED  0>SE>1 ";SE
70 INPUT"T";T
75 ACL=0.0:CO=0.0
80 P2=B*LOG(E)
90 FOR I=1 TO N
100 RQ=(3.141593+SE)^2.4
110 RQ=RQ-INT(RQ)
120 SE=RQ
130 P1=LOG(LOG(1/(1-RQ)))
140 WL=EXP((P1+P2)/B)
150 IF WL<T THEN ACL=ACL+WL:CO=CO+C ELSE ACL=ACL+T:CO=CO+CS
160 NEXT I
170 AVC=CO/ACL
180 PRINT" T=´ ";T;"     AVC= ";AVC
190 PRINT#-2," T= ";T;"     AVC= ";AVC
200 INPUT"NEXT T   Y/N";A$
210 IF A$="Y" THEN GOTO 70
```

```
N= 100   BETA= 2.5   ETA= 10   COSTS= 10   COSTSCH= 1

 T=    2      AVC=    0.63886
 T=    2      AVC=    0.54510
 T=    2      AVC=    0.54547
 T=    2      AVC=    0.54539
 T=    4      AVC=    0.52395 ┐
 T=    4      AVC=    0.50755 │
 T=    4      AVC=    0.41583 │
 T=    4      AVC=    0.34376 ┘ * MINIMUM COST
 T=    6      AVC=    0.62107
 T=    6      AVC=    0.57346
 T=    6      AVC=    0.69339
 T=    6      AVC=    0.67749
 T=    8      AVC=    0.61487
 T=    8      AVC=    0.80774
 T=    8      AVC=    0.55581
 T=    8      AVC=    0.77855
 T=   10      AVC=    0.99021
 T=   10      AVC=    0.79174
 T=   10      AVC=    0.84001
 T=   10      AVC=    0.92733
 T=   12      AVC=    1.00413
 T=   12      AVC=    1.02655
 T=   12      AVC=    0.98365
 T=   12      AVC=    0.92081
 T=   14      AVC=    1.08970
 T=   14      AVC=    1.01372
 T=   14      AVC=    1.10206
 T=   14      AVC=    1.05175
```

Fig. 10.A9.

Notes

(i)It is clear that with the data chosen, the average costs are not very sensitive to the chosen replacement life for a fairly wide range of values.

(ii)Because of the probabilistic nature of the problem,the average cost at any given replacement life will vary from one simulation to another.

N = the number of replacements in each simulation
BETA and ETA are the Weibull parameters used in the simulations
COSTS = the cost of an unscheduled replacement
COSTSCH = the cost of a scheduled replaced
T = the scheduled replacement life

Where we have only one product, there is only about a fifty per cent chance of obtaining the expected return, but we can resort to confidence levels. If, for example, we have a product with the life distribution described on Fig. 10.A7 we can be 90% sure that its life exceed 1,350 cycles. If we calculate the net present value generated by a product with this life we will be 90% confident of obtaining it.

CONFIDENCE LEVELS

When we conduct tests or collect data from the field to determine the life of a product, we will accumulate only a few pieces of data. We may have information on 31 pump failures or, if testing is expensive, on only three or four equipment lives. The fewer lives we have on which to base our estimate of the mean life, the less confidence we will have in the calculation. The calculation of the mean life, for a negative exponential distribution, at any specified confidence level is a function of the chi-squared distribution [9], but some values are given in the following table:

Table 10.A1

Life is assumed to be negative, exponentially distributed.

Let θ = MTFB

Let $\hat{\theta}$ = estimated MTBF after r failures

Then for the given confidence level, $k_L \hat{\theta} < \theta < k_u \hat{\theta}$

No of failures r	Confidence level							
	99%		95%		90%		80%	
	k_L	k_u	k_L	k_u	k_L	k_u	k_L	k_u
5	0.40	4.63	0.49	3.08	0.55	2.54	0.63	2.05
10	.50	2.69	.59	2.09	.64	1.84	.70	1.61
15	.56	2.18	.64	1.79	.69	1.62	.74	1.46
20	.60	1.93	.67	1.64	.72	1.51	.77	1.37
25	.63	1.79	.70	1.55	.74	1.44	.79	1.33
30	.64	1.69	.72	1.49	.77	1.39	.80	1.30
40	.69	1.56	.75	1.40	.79	1.32	.82	1.24
50	0.71	1.49	0.77	1.35	0.80	1.28	0.84	1.21

For the Weibull distribution, the determination of the parameters β and η at any given confidence level, from limited data, is much more difficult. However, Fig. 10.A10 shows the effect of plotting limited data from known distributions, and we can see how much over-pessimism or over-optimism we could be led to, by chance.

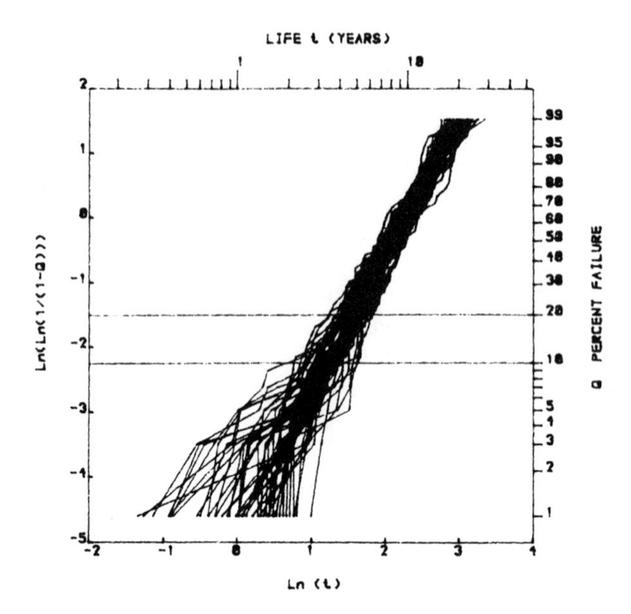

(a) 50 failed components in each sample
50 samples plotted

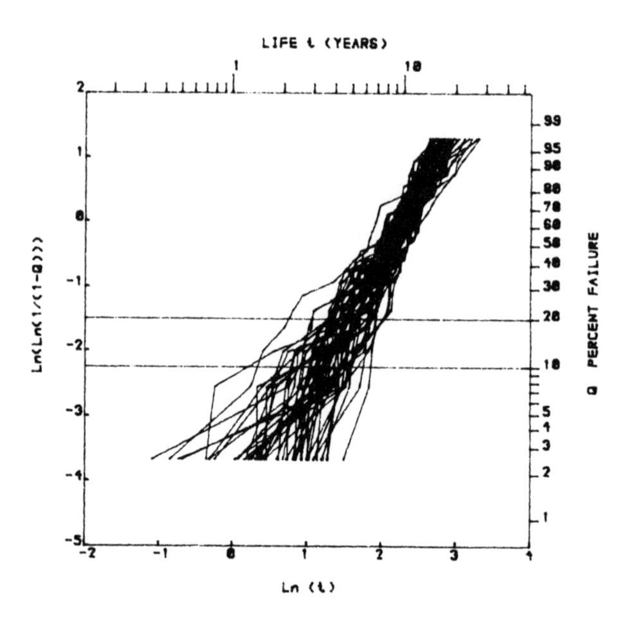

(b) 20 failed components in each sample
50 samples plotted

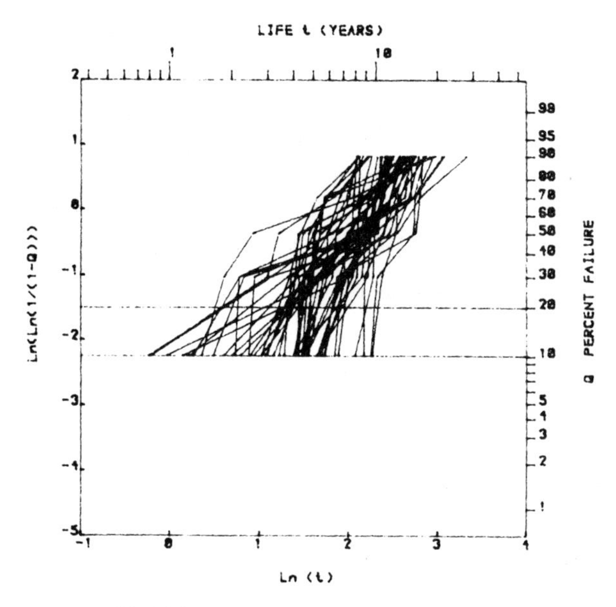

(c) 5 failed components in each sample
50 samples plotted

SUBJECTS FOR DISCUSSION

(1) Give an example of a component that is likely to have a life distributed as the negative exponential distribution.

- (i) A glass window (an old one is no more likely to be broken than a new one).
- (ii) A motor car wing on a car owned by a careless driver (the wing is likely to fail because of an accident, and this is no less likely with a new wing than an old one).
- (iii) A well-designed engine, before parts start wearing out.
- (iv) ?
- (v) ?

(2) Give an example of a system of many components, such that, if one component fails, the system will fail.

- (i) A motor car (not all components are in series but, if an engine fails, if a water pump fails, if the battery fails, the car will stop working).
- (ii) A television set (if a transistor fails, if the aerial fails, etc.).
- (iii) ?
- (iv) ?

How do you make a complicated system reliable enough to run at all?

- (v) ?
- (vi) ?

(3) Plot the following sets of data on Weibull graph paper and estimate the value of β and η in each case.

   (i) Lives of a sample of 10 products tested to failure were:
   0.75; 3.18; 3.60; 3.78; 7.07; 7.75; 9.98; 11.05; 15.26; 17.12 years.

   (ii) Cycles to failure of 20 pressure joints were:
   125; 298; 298; 428; 428; 512; 513; 709; 813; 870; 992; 1028; 1040; 1105; 1111; 1119; 1186; 1263; 1765; 1793 cycles.

(4) Give an example of component multiplication which makes a system safer.

   (i) Three automatic landing systems on theTrident airline.
   (ii) Two braking systems on a car.
   (iii) ?
   (iv) ?

(5) Give an example of component duplication which makes a system cheaper to run.

   (i) Parallel pumps in a chemical process.
   (ii) ?

(6) Discuss the merits and demerits of spares holding in a given situation.

   (i) Consider the problem of keeping combat aeroplanes available during the Falklands war.
   (ii) Consider having to shut down the operating of a tunnelling machine, in a coal mine, because of a fault in the belting system which removes the spoil.

(7) (i) What would British manufacturers lose and what would customers gain if USA style product liability laws were enforced in the EEC?
   (ii) How do the present product liability requirements in the USA affect a British manufacturer exporting to the USA?

(8) (i) What are the duties of a customer in ensuring that lack of reliability in the product he buys does not add, unnecessarily, to the cost of ownership?
   (ii) What are the duties of a designer in helping the customer to specify the reliability of a product?
   (iii) What are the duties of a designer in attempting to achieve a specified reliability, and how does he demonstrate that he has succeeded?

(9) (i) What do we mean by 'expected life' and when will this concept be useful? (The expected life will be the mean of a large number of lives, and will be useful if we are using a large number of products. It is like making a thousand bets.)

(ii) What do we mean by 'confidence' that the life of a product will exceed some stated value , and when is this concept necessary in assessing the value of the product?
(If we have only one or two products, basing returns on expected life would give a losing policy about as often as a winning one. A policy based on a 90% confidence value of product life would be far more likely to succeed but much more expensive.)

(10) How vulnerable is the designer if a product that he has designed is involved in a disaster which kills people?

Consider:
(i) a situation in which there are codes of practice;
(ii) a situation in which design standards have been legally defined; and
(iii) a situation which is new and for which no standards have yet been suggested.

REFERENCES AND FURTHER READING

[1] The Health and Safety at Work etc. Act 1974, HMSO, London, 1975.
[2] The Health and Safety at Work Act and its effect on industry, *I. Mech. E. Conference,* University of Birmingham, March 1977, Proceedings published by MEP, London, 1977.
[3] Wearne, S. H. 'A review of reports of failures', *I. Mech. E. proceedings 1979,* Vol. 193, No. 2o. I. Mech. E., London.
[4] Bragg, S. L. (Chairman), 'Interim report of the advisory committee on false work', HMSO, London, 1974.
[5] Janner, G. Product liability, Business Books, London, 1979.
[6] Quality control system requirements for industry, DEF 05–21, HMSO, London.
[7] Weibull, W., 'A statistical distribution function of wide application', *J. Appl. Mech.* Vol. 18, p. 293, 1951.
[8] Leech, D. J., *Management of engineering design,* Wiley, London, 1972.
[9] Green, A. E. and Bourne, A. J., *Reliability technology,* Wiley Interscience, London, 1977.
[10] Davies, L. M., *The Three Mile Island incident,* U.K.A.E.A., Harwell, 1979.
[11] Consumer Safety – A consulative document, Cmnd. 6398, HMSO, London, 1976.

CHAPTER 11

Documenting and costing design

Chapter 11 discusses the whole process of design from conception to the final enhancement of a successful product. The stages of design are shown in the form of a simple network; the cost of each stage is discussed; examples are given of the documentation which initiates a design activity and of that which marks its completion.

The stages of design are illustrated in Fig. 11.1.

11.1 PROJECT GENERATION

11.1.1 Invention (activity 0–1 on Fig. 11.1)

One method of generating projects is to invent new devices, but most companies allocate little or no funds to pure invention. Some will allocate a small sum of money to the maintenance of a small team conducting 'open sky' research in the hope of generating projects which will ultimately be profitable, but the probability that an invention will be profitable is very small – one only has to consider the number of patents taken out for inventions which do not sell – so that money will be allocated to undirected research or invention in the expectation that it will be written off. When the company's profit is, say, 7% of its turnover†, an expenditure of ½% or less on 'open sky' research could probably be justified because the company could afford complete failure, while only a small pay-off from an invention is required to cover the cost.

Few companies would rely on invention to generate design projects. Such ideas as the linear motor, the Wankel engine, the gas turbine, the hovercraft, or Xerox reprography required considerable investment before the clever idea could be translated into a convincing, working product, and many years of development before commercially viable products were produced. Some of these examples would have been beyond the capacity of most companies without the investment of public money, and even now are not clearly profitable. With

†7% of the cost of doing work could be a typical, before tax profit margin in a defence contract.

many inventions, even some much admired by the general public, the investment required for development is great, the period of development is long, the likelihood of success is small, and if there is a pay-off, it is usually small and always late. The invention of new products is not generally a field that interests the investor, although there are notable exceptions. Float glass, for example, represented a considerable investment in what was seen as a very risky project. The pay-off for success was thought to be great, but the probability of attaining success with the resources which could be made available, was not universally assumed to be high.

It is probable that many inventions owe their successes to the enthusiasm of the inventors rather than to thoughts of reward. It is difficult to believe that the Watts, Whittles, and Wallises of the world are moved as much by the thought of monetary rewards as by the obsessive desire to solve problems that they have set themselves.

Almost all companies which employ designers ignore, and can afford to ignore, invention as a method of generating projects, and look for markets rather than attempt to exploit inventions. This does not, of course, mean that designers do not invent or should not be encouraged to invent. It means rather that when a market or a need is defined, the designer will be expected to invent a product to meet the need instead of inventing a product and then seeking a way to exploit it.

Many firms are, in any case, in the systems business, and often put together systems from components which would not normally be regarded as technical innovations. A refrigerator system which has to be modified to work in a new environment may be a considerable technical advance over earlier systems but this advance may be achieved by the addition of valves and control mechanisms which individually do not require much mechanical engineering innovation. A racking system may be enhanced by the addition of a plastic bin which, in itself, provides no technical problems for the designer familiar with plastic moulding methods.

Particularly when the firm is selling systems, the design of the system will generate problems in the design of components. Analysing the performance of a system may well suggest that a considerable improvement in system performance will be achieved by increasing the temperature at which a turbine will work, reducing the leakage of a valve, reducing the time of a man's operation, reducing the weight or cost of a replacement part, etc., and this will lead to modifications of the specifications of components; but although the improvement may be difficult to achieve, the designer would not see his task as one of invention so much as problem solving, however ingenious his method of cooling a turbine blade or reducing the leakage of a ball valve, or however, interesting the shape of his plastic storage unit. This does not mean that innovation is not required by the designer but that his employers are far more likely to want to see innovation used to solve defined problems than for its own sake.

Fig. 11.1 – Stages of design.

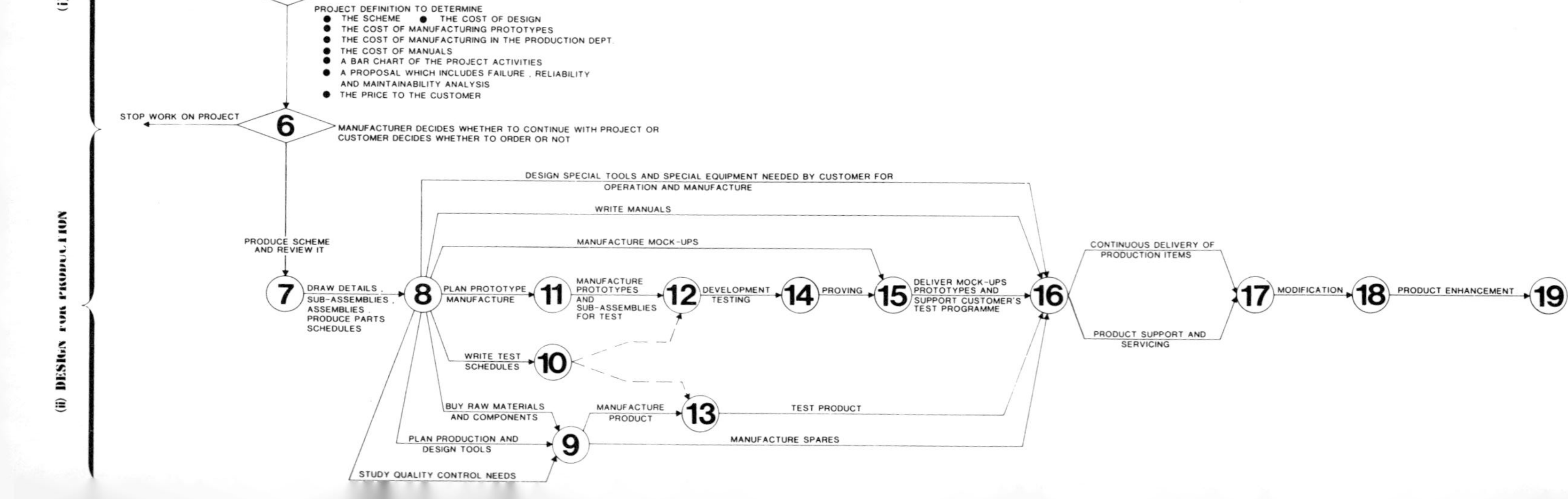

ACTIVITY	MEN INVOLVED	COST CONSIDERATION	OUTCOME AT MILESTONE
0–1	Possibly a small research team	No budget or small budget for 'open sky' research. Budget not expected to be recovered directly	Research Report
1–2	Customer Salesman Service Engineers	Time not usually debited to design budget. Cost usually treated as indirect labour cost. (May be debited to sales or service budget)	Salesman's Visit Report Preliminary Statement of Design Requirements (Fig. 11.2)
2–3	Customer Designer Salesman Managers	Salesman's time not usually debited to design budget. Designer's time sometimes allocated by Sales Dept and may not be debited to the project. Some accounting systems allow the time to be debited to the project if and when an order is received Decision is usually made at a regular meeting of a Management Committee whose members are treated as indirect labour	Design Specification (Fig. 11.3) Invitation to Tender (Fig. 11.4) Authority may be given to book time to the project (Fig. 11.5)
3–4	Designer Managers	Designer's time debited to a general number controlled by the Sales Dept or to the project. Time allowed is limited by a small arbitrary ceiling	Cost of Feasibility Study
4–5	Designers (including Development Engineers, Experimental Shop Manager) Managers	Designer's time will be debited to the project. The ceiling will have been fixed by Activity 3–4	Predicted cost of Project Definition Probably a scheme and, sometimes, a form (Fig. 11.6) will be used to transmit information on project definition tasks and their costs

ACTIVITY	MEN INVOLVED	COST CONSIDERATION	OUTCOME AT MILESTONE
5–6	Designers Experimental Shop Manager Production Engineers Reliability Engineers Technical Publications Dept Service Engineers Engineering Manager Customer Managers	Technical Staff's time will be debited to the project Ceiling times will have been fixed by Activity 4–5	Proposal for customer Scheme Cost of Prototype Manufacture (Fig. 11.7) Cost of Production (Fig. 11.8) Failure Analysis (Fig. 11.9) Reliability Analysis (Fig. 11.9) Maintainability Analysis (Figs. 11.10 & 11.11) Technical Publications Estimate (Fig. 11.12) Bar Chart of Project (Fig. 11.13) Cost of Design, Development, and Proving (Fig. 11.12)
6–7	Designers Design Review Team	Scheming time will have been determined by Activity 5–6. Modifications may be generated by the Review	Scheme Drawings
7–8	Designers Draughtsmen	Drawing time will have been determined by Activity 5–6	Detail drawings, sub-assemblies, parts lists, N.C. machine tapes, descriptions of special processes, Schedule of Parts (Fig. 11.4) Prototype Manufacturing Instructions (Fig. 11.18) Instructions to Manufacture (Fig. 11.15)
8–9	Buyers	Buyer's time is part of predicted production costs although buyer may be an indirect charge	Orders for raw materials and bought out sub-assemblies and components
8–9	Planners	Planners' time is part of predicted production costs although planner may be an indirect charge	Process Schedules (Fig. 11.16), Assembly Schedules, tool drawings, specification of special capital equipment

ACTIVITY	MEN INVOLVED	COST CONSIDERATION	OUTCOME AT MILESTONE
8–9	Quality Control Dept	Quality control engineer is probably an indirect charge on production	Production test and inspection requirements
8–10	Designer	Time for writing the test schedule will have been determined by Activity 5–6	Development Test Requirements (Fig. 11.17) Production Test Requirements
8–11	Prototype shop planner	Planners' time is part of predicted design cost although planner may be an indirect charge on the prototype shop. Note: in some companies the prototype shop is part of the Production Dept while in others it is part of the Technical (and Design) Dept	Process layouts and assembly schedules for prototype manufacture (although in many cases the prototype shop operative will work directly from the drawings)
8–16	Technical authors with some assistance from the Designer	The cost of writing manuals will have been determined by Activity 5–6 The writer's time will probably be a direct charge to the job Note that manuals may be a legitimate charge on the customer	Operating Manuals, Service Manuals (Fig. 11.19), lists of special tools (Fig. 11.20)
8–16	Designer or sometimes a specialist, test equipment, design section	The cost of designing special test equipment will have been determined by Activity 5–6. The Designers' time will probably be a direct charge to the job. The cost of special test equipment may be a legitimate charge on the customer and the subject of a separate contract	Drawings of special test equipment Orders for the manufacture of special test equipment
9–16	Production Shop	This will be part of the direct cost of production determined in Activity 5–6 but the designer must have specified the spares requirement	Spare parts to stores

ACTIVITY	MEN INVOLVED	COST CONSIDERATION	OUTCOME AT MILESTONE
9–13	Production Shop	This will be part of the direct cost of production determined in Activity 5–6	Product for customer or to stores
11–12 8–16	Prototype Shop	This will be part of the direct cost of prototype manufacture determined in Activity 5–6	Prototype for development Prototypes for customer's development programme Space models for customer's development programme
13–16	Production Shop Inspectors	Production shop personnel will be part of the direct production cost determined in Activity 5–6. Inspectors will be an indirect charge on production	Product to customer
12–14 14–15 15–16	Development Dept with assistance from designers	These will be part of the direct cost of design/development determined by Activity 5–6	Development report confirming that performance and endurance testing demonstrate compliance with the design specification Certificate of compliance with design requirements (Fig. 11.21)
16–17	Service Engineers	Service engineers are usually an indirect charge (sometimes to the Sales Dept)	Service Reports (Fig. 11.22) Modification Proposals (Fig. 11.23)
17–18	Service Engineers Development Engineers Planners The Customer The Designer Draughtsmen	If the modificaton is required to meet the specification, the costs will be charges to the departments concerned (direct charges in the case of designers, draughtsmen, and development engineers)	Forms justifying modification, specifying costs, and allocating those costs to the company or the customer
18–19	Designers	Time will be an indirect labour cost and may be debited to a fixed sales budget. Eventually, a job number for the project and a time will be allocated	

11.1.2 Identifying a market (activity 1–2 in Fig. 11.1)
If the manufacturing company is intending to exploit an invention, it is still necessary to identify a market for that invention. More commonly, however, a design project starts because some one has identified a market or a need which can be turned into a market. Anyone may identify a market, but typically either:

(i) the manufacturing company's salesmen discover a market,
(ii) a serviceman identifies a possible improvement in the company's products, or
(iii) the customer makes known his need for a new product.

Where the customer, himself, identifies his need, it is the salesman's job to make sure that his own company is invited to offer proposals and also to give his designer colleagues advance notice of the invitation.

(iv) Projects are also generated within the manufacturer's factory.

In some companies there is a suggestion scheme which invites operatives to propose improvements to the products they manufacture. Frequently such suggestions will be to reduce manufacturing difficulties observed by the operative. The manufacturing engineers, also, may well suggest design changes to ease manufacture and, indeed, should be expected to do so. Other internal sources of projects are the need to improve reliability, the need to improve the ergonomics of operation, the desire to reduce weight, the desire to take advantage of a new means of transport† and so on, virtually indefinitely.

A statement of the market or need must be made, and money must be allocated to the work required to study the requirements and their possible solutions. In some cases the necessary information is conveyed in a prescribed format, and Fig. 11.2 is an example of a form used in one company to initiate the earliest work on a project. Often the salesman's or serviceman's report is not a formal document, and, in any case, at so early a stage of project formulation it is unlikely that complete details of the requirements can be listed. Nevertheless, whoever originates the project will be expected to answer the following questions:

'Who is the customer?'
'What does the customer want, in terms of hardware?'
'What product support will the customer expect?'
'How will the customer pay for design?'
'What is the customer's programme (that is, what delivery dates will be required)?'
'What is the target price per item?'
'What is the competition?'
'What other potential sales exist?', and
'What chance has the manufacturer of getting the work?'

---

†The example of p. 26 discusses the possible redesign of a washing machine to cope with the introduction of the cheaper pick-a-back transport system.

INITIAL SALES INFORMATION

CIRCULATION NO.

ORIGIN: ISS.

CUSTOMER: DATE

Contact and position.

APPLICATION:

System
Type of unit
Qty per system
Related documents

PROPOSAL CLOSING DATE:

PROPOSAL TO INCLUDE:

WORK REQUIRED:

Technical proposal by: –
Technical estimate by: –
Program.................by: –
Quotation...............by: –

TARGET COSTS:

SALES POTENTIAL:

Estimated sales
Probability factor
Known competition

DESIGN & DEVELOPMENT COST RECOVERY:
PRODUCTION TOOLING COST RECOVERY:
ADDITIONAL FACILITIES REQUIRED:
ADDITIONAL COMMENTS:

Note that other forms include such questions as,
'What technical publications will be required?'
'Will maintenance tooling and test equipment be required?'

Fig. 11.2 – Initial sales information.

It is necessary to consider both the immediate cost of generating projects and the eventual cost if the project goes further. The time of a salesman or of a serviceman is not usually charged directly to any project that he may initiate. The serviceman would initiate a project only as a duty that is extra to those for which he is largely paid (this would be true too of an operative or manufacturing engineer who initiates a project). In many cases, however, the designer will also be involved in this project generation, and we have to consider how the designer's time is paid for. The designer would certainly be usefully employed in discussions with the customer at this time when the designer's expert knowledge can help the customer to formulate the design specification. Only a small part of the designer's time is spent at this stage of the project – probably less than 1% of the time that he will eventually require to spend if the project survives to completion – but if we take into account the number of projects which must be started for one to survive to a profitable solution, we may be dealing with 10% of the total design effort available (see, for example, the expenditure on feasibiliy studies discussed on p. 160). Because of the high probability that the project will not lead to profitable sales, and because of the difficulty of defining the project so early in its life, it may be convenient to book the designer's time to some general number (called 'Aid to Sales' in one company), although this expenditure must be monitored and will be budgeted as written-off direct labour.

As well as the immediate commitment of money to project generation the designer or manager must consider the ultimate commitment if the project eventually reaches a successful conclusion. Will the customer pay directly for all design work at the successful conclusion of project definition, or will the manufacturer be committed to spend more of his own money to see the final design stage through? Is the manufacture selling the product or his ability to design the project? The ultimate commitment may be not only to design work but to tooling, and, particularly in the case of a product designed to be produced in quantity, the tooling costs may be much more than the design costs. The bin for the racking system discussed on p. 15 will cost about £3000 to design and about £50 000 to tool for. All this will have to be spent by the manufacturer before he can get any return at all.

Even though the manufacturer is apparently spending a very small amount of design effort at this stage of the project, where a market is being identified, we see that, in addition to studying the technical problems posed, he must budget for the design effort he is using, and he must determine his expected future costs and the nature of the risks that he will be taking as the project progresses. However small the design effort to generate projects, and however successfully problems will be solved, there is no point in generating a project if the manufacturer cannot afford to contemplate the expenditure that will be demanded by a successful design.

There may then be reasons for killing a project even before it has been adequately defined. Some marketing directors will believe that there should be a formal decision at this point, to go ahead with the project or stop all work on it, but it is more likely a company will know the business it is in, and projects in that line of business will rarely be rejected at this stage since they will have been generated by need and with knowledge. If the manufacturer decides to enter a new field, such a decision is likely to be taken at board level, and a sum of money budgeted for the attempt to break into that field. In either case, the decision to proceed beyond project generation is unlikely to be taken on a job by job basis. Any sort of decision will be difficult to make, however, unless it is within the framework of a corporate plan, established by the board, which specifies the direction the company is to go and which determines the budget that will be allowed to cover work in each area.

11.1.3 Specifying the project (activity 2–3 on figure 11.1)

Identifying the market will have resulted in some description of the market, some description of the costs and risks involved, and some description of the required product. This description (at least if in the form of Fig. 11.2) may go so far as to merit being called a preliminary statement of the design requirements. So far, designers have not been involved except in a consultative capacity.

If the project originates with a customer then he will have to define his requirements precisely. Any competitive supplier, however, would expect to be aware of the requirement before the customer has written a formal design specification, and it would be to the advantage of both sides if the designer were involved in discussions with the customer while the specification is being formulated. A common procedure is for both the customer and the designer to produce formal design specifications. If the customer does not do so, the designer will want the customer to agree that his specification reflects the customer's requirements accurately. If the customer does produce a formal design specification, the designer will need to ensure that his own version omits nothing that the customer requires (the simplest way of doing this is for the designer to refer to the customer's specification by name and number and attach a copy of it to his own).

Too much importance cannot be attached to the design specification and to agreement between designer and customer that the specification correctly describes what is required. Later, the specification will provide the means of determining whether the design has been satisfactorily completed, and it will probably be the basis of any contractual agreement between the manufacturer and the customer. Fig. 11.3 is a typical design specification pro forma. This pro forma was written by the author for the Department of Industry's Industrial Technologies Education and Training Committee, with whose permission it is published. The pro forma has been annotated for educational purposes.

Design Specification Pro Forma

1 **Identifying Number**

2 **Issue Number**

3 **Function** (In basic terms, what function is the article to perform when we have designed it?) ______

______

______

______

4 **Application** (Of what system is this requirement a part?)

______

______

______

5 **Origin** (By what means, when and by whom was the requirement first made known? Usually we give here a reference to a letter, visit, telephone conversation or other discussion.)

______

______

6 **Customer's Specification** (If the customer has already written a specification, its identifying number should be quoted; if the customer has not written a specification we should say so.)

______

______

7 **General Related Specifications** (If we are required to work within the framework of existing general specifications, standards or definitions, or if existing documents are likely to be useful, their numbers should be quoted.)

______

______

______

8 **Safety** (Are any special safety precautions to be taken?)

______

______

______

Fig. 11.3 – Design specification pro forma.

9 Environment

9.1 Ambient temperatures ______________________

____

9.2 Ambient pressures ______________________

____

9.3 Vibration ______________________

____

9.4 Acceleration ______________________

____

9.5 Contaminants ______________________

____

9.6 Climate ______________________

____

9.7 Installation limitations ______________________

____

9.8 Affect on other parts of the parent system (eg compass safe distance, radio interference). ______________________

____

9.9 Other environmental factors ______________________

____

10 Number-Off and Delivery Programme

____

____

____

____

11 Price (Note that this may require a complex statement if prices reduce from prototypes through increasing batch sizes.)

____

____

____

Fig. 11.3 – continued

12 Functional Requirements

12.1 Performance and acceptable tolerances. (This will generally be a complex statement of the permissible range of many variables to be obtained in the presence of stated ranges of other variables.)

____

____

12.2 Life ____________________________________

____

12.3 Unacceptable modes of failure ____________________

____

12.4 Reliability ________________________________

____

12.5 Servicing restrictions ____________________________

____

12.6 Other functional requirements ____________________

____

13 Any Other Relevant Information

13.1 Limitations of manufacturing facilities ________________

____

13.2 Special procedural requirements ____________________

____

13.3 Other relevant information _______________________

____

14 Action Required (Preparation of proposal, preparation of detail drawings, manufacture of prototypes or manufacture of full production quality.)

____

____

____

Fig. 11.3 – continued

When the project originates within the manufacturing company, clearly the whole responsibility for producing a detailed design specification lies with the designer. Under these circumstances, the design specification is unlikely to form the basis of any contractual agreement with the customer, but its writing should still be regarded as an essential, early part of the designer's work. That is to say, however tempted, the designer should not be allowed to start scheming until he has first decided what he is trying to do. This may seem to be no more than imposing a discipline on the designer (not necessarily a bad thing), but, in fact, the decision to proceed with the project cannot be taken by the managers unless they too know what is to be designed, what are the likely costs, what are the likely rewards, and what are the likely risks, and none of this can be assessed if the design specification cannot be written.

Where the customer is inviting the designer to submit proposals, the completed design specification will be accompanied by an invitation to tender. This invitation to tender may well be much more than a short, polite letter inviting a submission, since it will contain information about tendering procedure and criteria for judgement between submissions which would not be appropriate in the specification. Fig. 11.4 shows abstracts from a formal invitation to tender which was sent with a design specification to a possible supplier.

Usually the designer's contribution to the design specification is his time. That is, the specification can usually be written without prior experimentation in either the workshop or the laboratory. The commitment will be small at this stage, and the designer may book his time to a general number which uses the money that has been budgeted for written-off, direct labour, or he may book his time to a number defining this particular project. In the latter case, the money spent will have to be identified, because it may have to be re-allocated to the written-off, direct labour budget if the project is later cancelled.

With the early tasks of a design project – invention, market identification, and specification writing – there is some difficulty in deciding what effort should be spent, because the project is insufficiently defined for us to know how long any of these tasks is likely to take. Where every project is reviewed every month, there will probably not be too much expenditure on any project between reviews, for monitoring to be possible. Where reviews are not held at regular short intervals, a small fixed number of man-hours may be allowed for each task, regardless of the project, and a review is forced when these man-hours have been spent. As an example, one company will allow the expenditure of 80 man-hours on any job before it is defined. The implication here is that the directors of the company know how many projects must be started to provide enough successful projects to keep the company in business, and how much it is necessary to spend, on average, to define each of the projects it starts.

When the design specification is available and agreed and, where appropriate, an invitation to tender has been received, a decision may be taken to stop or continue with the project. Once again, however, the decision is likely to have been made already by the manufacturing company's corporate strategy.

XYZ Aircraft Co. Ltd.

4001 Albert Street / London / WC4

1st August 19xx

Subject: Letter of Invitation to Tender.

Dear Sirs,

Variable Gxxxxxxx Aircraft

The XYZ Aircraft Co. Ltd. of London has been charged with the development and production of a Variable Gxxxxxxxx Aircraft for introduction into service during 198x.

A formal Request for Proposal pack is forwarded inviting you, as a potential supplier, to submit proposals against the requirements defined by the particular issue of the enclosed spedification.

The specification will define the requirements as we know them at the present time and any significant changes will be effected only by an official raise in issue. However you are at liberty to suggest alternative approaches or modifications to the specification if you can provide evidence that this will achieve a better overall result. All such proposals must be referred to the authority named in the specification.

In addition to drawing up and presenting your technical proposals you are requested to provide a separately bound detailed commercial proposal covering all aspects of the Standard Requirements of the Request for Proposal.

The date by which your proposals are to be received together with the quantities required is shown on the attached form.

ANY PROPOSAL OR PART PROPOSAL DELIVERED AFTER THAT DATE WILL BE CONSIDERED INVALID.

When your proposals have been received, the content will be examined formally. If any technical proposal you make falls significantly short of the specification it is unlikely that it will be given further consideration unless it has some special merit. If your proposal meets the specification the factors which are most likely to affect its success in competition with other proposals are PRICE and your ADJUDGED ABILITY TO COMPLETE THE TASK WITHIN YOUR CAPABILITIES AND ON TIME.

for XYZ Aircraft Ltd.
Charlie Brown
Procurement Manager.

Fig. 11.4 – Letter of invitation to tender.

11.2 THE FEASIBILITY STUDY

11.2.1 Determining the cost of the feasibility study (activity 3–4 in Fig. 11.1)

The feasibility study is usually regarded as a study that is undertaken to confirm the technical and commercial feasibility of the project before any more money is spent. As has been suggested, however, it is more logical to regard the feasibility study as a means of defining the requirements and the cost of the project definition. If it does this, the feasibility study will not have demonstrated any technical or commercial reasons for stopping the project.

In asking how much the feasibility study will cost, we have the difficulty that the only information available is the design specification – it is not yet known how the problems will be solved. It is likely that some approximate solution to the design problem must be sketched so that an approximate cost based on experience of feasibility studies of similar projects can be ventured. Alternatively, regular project review or a small fixed allowance of man-hours which will force a review before expenditure becomes excessive, may provide adequate control. One company believes that its expenditure on feasibility studies is controlled quite simply by the time available to make the studies (that is, before market conditions change or before the need has been met by a competitor).

It is likely, however, that once expenditure is authorized on a feasibility study, the project has an identity and work will be charged to it even though subsequent failure may require any expenditure to be re-allocated to the budget allowance for written-off direct labour.

This may be signalled by the issue of a form which requests and authorizes the expenditure of a fixed sum of money on a number of tasks thought to be essential in the feasibility study: Fig. 11.5 is an example of such an authority to work on a project which also asks for an estimate of the cost of project definition.

11.2.2 The feasibility study (activity 4–5 in Fig. 11.1)

The feasibility study is to determine how much must be spent on project definition, and to do this, project definition must be planned. If the project has been generated internally the work of project definition may be, to some extent, arbitrary; but if project definition actually consists of making a proposal to a customer, there is no great difficulty, in any given area of technology, in deciding what work must be done in order to make such a proposal. What the customer will want to know will depend on the type of product that he is buying, but, except in the simplest of cases, he will want:

(a) to know the cost of each of the products he buys,
(b) a failure analysis,
(c) a reliability and life analysis,
(d) an analysis of the product support requirements,

FROM: Technical Sales

ENQUIRY NO.

ISSUE NO.

TO:

DATE:

SALES ENQUIRY Please carry out a preliminary examination of the specification attached and assess the number of hours necessary to put forward a scheme to the customer by:			
CUSTOMER:			
APPLICATION:			
SYSTEM:			
TYPE OF UNIT:			
QUANTITY PER APPLICATION:			
PRELIMINARY SPECIFICATION:			
HOURS AUTHORISED TO TECHNICAL DEPARTMENT	01 02	Calcs. Design.	
HOURS BOOKED TO ISSUE			
FURTHER HOURS REQUIRED BY TECHNICAL DEPT. (State what is required, e.g. proposal, scheme etc.)	01	Calcs.	
	02	Design	
	03	Laboratory	
		Total	

Fig. 11.5 – Authority to work pro forma.

(e) an analysis of the cost of repair and maintenance,
(f) an analysis of the spares holding requirements,
(g) an analysis of the performance of the product in all relevant environments,
(h) an analysis of the cost of operating the product,
(i) to know the total cost of design.

The feasibility study is not intended to provide this information but merely to decide what will have to be done and how much it will cost to provide the information. To do this a scheme drawing of the proposed product will be required, although it will be a matter of some judgement to decide when the scheme is of sufficient detail to determine the costs of the studies required by project definition. The completion of the feasibility study is sometimes marked by the completion of a form of the type shown in Fig. 11.6. In whatever form the information resulting from the feasibility study is transmitted, however, it must contain a scheme and a list of the tasks to be undertaken during project definition and the resources required to do them. If the design project is of sufficient complexity, it may be desirable for the information to be given in the form of a critical path network with resource requirements.

If the project has been generated externally, the information of Fig. 11.6 and any previous assessments of the expected pay-off from the project will enable the decision to be taken to permit money to be spent on making a proposal to the customer. If the project has been generated internally, the decision will be whether to spend money on project definition although the completion of the feasibility study and the start of project definition may be more arbitrarily defined, or even avoided altogether, if regular reviews of the project are undertaken.

The work of the feasibility study would also be expected to give a better prediction of the long-term costs and benefits of the project, if it is taken to completion, than have been available before, and this will influence the decision maker. Assuming that it is considered worthwhile to continue work on the project, the decision makers will authorize the expenditure of a stated amount of money on project definition.

11.3 PROJECT DEFINITION (activity 5–6 in Fig. 11.1)

If the project is externally generated, the project definition will result in a proposal to the customer. If the project is internally generated, the project definition will provide the information on which will be based the decision to embark on the final stage of design. In either case, the decision maker will need approximately the same information, of which the following list may be regarded as typical.

FROM: Technical Sales

ENQUIRY NO.

ISSUE NO.

TO:

DATE:

<table>
<tr><td colspan="4">SALES ENQUIRY</td></tr>
<tr><td rowspan="4">This enquiry covers the necessary technical work required to provide the Sales Department with: –

(State what is required, e.g. proposal scheme, technical estimate, etc.)</td><td>01</td><td>CALCULATIONS</td><td></td></tr>
<tr><td>02</td><td>DESIGN</td><td></td></tr>
<tr><td>03</td><td>LABORATORY</td><td></td></tr>
<tr><td colspan="2">HOURS AUTHORISED : Total</td><td></td></tr>
<tr><td colspan="4">CUSTOMER:</td></tr>
<tr><td colspan="4">APPLICATION:</td></tr>
<tr><td colspan="4">SYSTEM:</td></tr>
<tr><td colspan="4">TYPE OF UNIT:</td></tr>
<tr><td colspan="4">QUANTITY PER APPLICATION:</td></tr>
<tr><td colspan="4">PROVISIONAL SPECIFICATION NO.</td></tr>
<tr><td colspan="4">PROPOSAL DATE TO CUSTOMER:</td></tr>
<tr><td rowspan="4">EXTRA TIME REQUIRED</td><td>01</td><td>CALCULATIONS</td><td></td></tr>
<tr><td>02</td><td>DESIGN</td><td></td></tr>
<tr><td>03</td><td>LABORATORY</td><td></td></tr>
<tr><td></td><td></td><td></td></tr>
<tr><td colspan="4">REASON FOR EXTRA TIME:</td></tr>
<tr><td colspan="4">NOTE: If it becomes apparent that allowed time is insufficient, this form must be returned with time columns completed. No hours beyond those authorized may be booked until a further issue of this form has been raised.</td></tr>
</table>

Fig. 11.6.

(a) *The cost of manufacturing the product*
This will be the cost of manufacturing the products by the cheapest methods available for the number which will ultimately be ordered. There may also be a number of products made in the prototype shop, and the cost of these will also be required. A design scheme of considerable detail will be required to enable costs to be predicted, and the assistance of manufacturing engineers will be essential both to review the scheme and to determine manufacturing costs from it.

The engineers concerned will probably be asked to commit themselves formally to their estimates, and Figs. 11.7 and 11.8 are examples of forms used for this.†

(b) *The failure analysis*‡
The failure analysis, which is intended to prove that any failure is safe, can, surprisingly enough, often be done when the scheme is quite rudimentary. Whether a circuit fails open or shorts, whether a component is duplicated, whether a valve fails open or closed, whether brakes fail on or off, can often be determined from a scheme which is little more than a system diagram. Sometimes the failure analysis will generate tasks for the final design stage; for example, designing a compressor wheel and case such that the fragments of the wheel will be contained in the casing if the wheel bursts will require to be proved by experiment during the final stage of design.

A formal report of the failure analysis will be required, and an example of extracts from such a report is shown in Fig. 11.9.

(c) *The reliability analysis*
The reliability analysis is intended to prove that the reliability of the product will be at least that required by the specification. Again, much of the analysis can be done with a comparatively rudimentary scheme, as much of the work is really system analysis. It is not possible to calculate or predict the reliability of a product which is yet to be designed unless the designer has access to very complete records of the lives and failures of the components which will be found in the new design or of components with sufficient similarity to those in the new design to permit 'read across' evidence to be used.

Where evidence does not exist from which the reliability of components can be assessed, it will be necessary to generate experimental work which will form a task for the final stage of design. For example, a newly designed diaphragm which will be subjected to load reversals may need to be tested to destruction in

---

†The differences between Figs. 11.7 and 11.8 are partly due to their being derived from different companies.

‡Leech, *Management of engineering design,* Wiley, London, 1972, discusses a method of failure analysis.

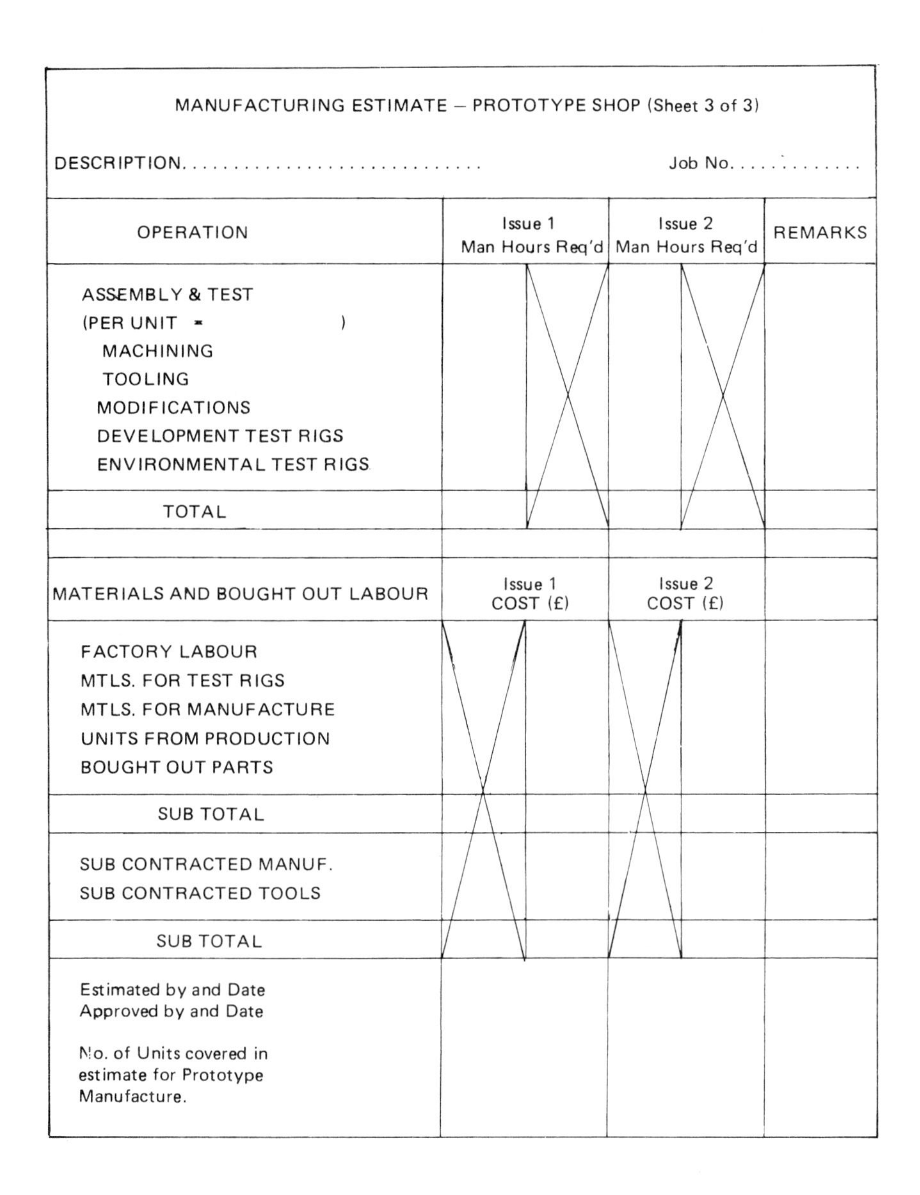

MANUFACTURING ESTIMATE – PROTOTYPE SHOP (Sheet 3 of 3)

DESCRIPTION. . . . . Job No. . . . .

OPERATION	Issue 1 Man Hours Req'd	Issue 2 Man Hours Req'd	REMARKS
ASSEMBLY & TEST (PER UNIT =) MACHINING TOOLING MODIFICATIONS DEVELOPMENT TEST RIGS ENVIRONMENTAL TEST RIGS			
TOTAL			
MATERIALS AND BOUGHT OUT LABOUR	Issue 1 COST (£)	Issue 2 COST (£)	
FACTORY LABOUR MTLS. FOR TEST RIGS MTLS. FOR MANUFACTURE UNITS FROM PRODUCTION BOUGHT OUT PARTS			
SUB TOTAL			
SUB CONTRACTED MANUF. SUB CONTRACTED TOOLS			
SUB TOTAL			
Estimated by and Date Approved by and Date No. of Units covered in estimate for Prototype Manufacture.			

Fig. 11.7 – Request for an estimate of cost of manufacturing a prototype component.

MANUFACTURING ESTIMATE			
CUSTOMER. . . . . PART No. . . . . Issue . . . . . . DESCRIPTION	Sheet No. . . . . . . .of . . . . . Compiled by. . . . . Date . . . . Job No.		
		£	p
MATERIALS			
PRODUCTION LABOUR	DIRECT LABOUR . . . . HRS AT . . . . OVERHEAD %		
ENGINEERING LABOUR	DIRECT LABOUR . . . . HRS AT . . . . OVERHEAD %		
ASSEMBLY and TEST	STAFF . . . . HRS AT . . . . DIRECT LABOUR . . . . HRS AT . . . . OVERHEAD %		
INSTALLATION & COMISSIONING	STAFF . . . . HRS AT . . . . DIRECT LABOUR . . . . HRS AT . . . . OVERHEAD % EXPENSES		
	SUB TOTAL PROFIT %		
TOTAL			
REMARKS			
SELLING PRICE:	DESIGN £ ____________ DEVELOPMENT £ ____________ MANUFACTURE £ ____________ TOTAL £ ____________ EACH		
APPROVED. . . . . SALES MANAGER . . . . DATE.			

Fig. 11.8 – Request for an estimate of cost of manufacturing a component.

PART NO.	DESCRIPTION	NO. OFF	FAULT	TROUBLE RATE	FAILURE RATE	EFFECT ON EQUIPMENT	INDIC.	EFFECT ON SYSTEM
				PER 10^6 HOURS				
To be allocated	Seat and Ball	1	Damage or Contamination	4	3	Valve allows drainage of fuel or air from the bay at differential pressures below 1.psi This is not detrimental Reverse leakage of air into the bay could occur.	None	None except in case of fire.
	Ball guide and spring register	1	Fracture	1	1	As above	None	None except in case of fire.
	Spring	1	Loss of rate	1	-	Valve allows drainage of fuel or air from bay at pressures below 0.75psi Non return function is not affected.	None	None

Fig. 11.9 – An extract from a failure analysis.

the final stage of design (and, of course, will have to be re-designed if the number of reversals to failure is too few).

A formal report of the reliability analysis will be required, and while this may be a separate document, the information may be given with the failure analysis, as is shown in Fig. 11.9.

(d) *Analysis of repair and maintenance costs*
(e) *Analysis of product support costs*
(f) *Analysis of spares holding requirements*

These three analyses are related to one another and are typical extensions of the reliability and life analysis. In many cases, a large part of each of these analyses will be common to every product made by the company, and many manufacturers produce a publication advertising their general product support facilities.

Where particular reference is made in the specification to special maintenance requirements, it will be necessary to demonstrate that these requirements will be met. An example of this is the portable test equipment which must be supplied to check the systems of a military aircraft immediately before a mission; such equipment would be regarded as a fundamental part of the system to be sold, and, indeed, the system would have to be designed with such check out procedures in mind.

Fig. 11.10 is an extract from a maintainability analysis, while Fig. 11.11 is a vignette of the fault summary of a component which shows estimated costs of fault diagnosis and rectification.

An important part of these analyses is the prediction of the cost of the manuals which must be supplied when the product is delivered to the customer.

<table>
<tr><td rowspan="2">SUPPLIER
C. Brown & Co.</td><td colspan="3">PURCHASER XYZ & Co.</td></tr>
<tr><td colspan="3">PURCHASER'S SPECIFICATION ABC 123</td></tr>
<tr><td colspan="2">EQUIPMENT DESCRIPTION Pressure Valve Relief</td><td colspan="2">Part No. 654321</td></tr>
<tr><td colspan="4">JUSTIFICATION FOR SERVICE LIFE
Similar valves in xxxxxx, both 'On Condition', give 45,000 hours subject to scheduled maintenance requirements below.</td></tr>
<tr><td>SCHEDULED MAINTENANCE OVERHAUL REQUIREMENTS
Visual external check
Security check</td><td>FREQUENCY
On condition</td><td colspan="2">JUSTIFICATION
System rarely used so that wear does not present a problem. Ingress of foreign matter is unlikely to jam the valve.</td></tr>
<tr><td>STORAGE REQUIREMENTS
Store in a xxx container
Before installation check free action of piston according to maintenance manual.</td><td>FREQUENCY
5 yearly</td><td colspan="2">JUSTIFICATION
Similar valves have 5-year shelf life.</td></tr>
<tr><td colspan="2">DETAILS OF REQUIRED TEST FACILITIES
None</td><td colspan="2">DETAILS OF ANY SPECIAL SUPPORT EQUIPMENT REQUIRED BY CUSTOMER
None</td></tr>
<tr><td colspan="2">ESTIMATED OVERHAUL COST (£)
ACTUAL</td><td colspan="2" rowspan="2">REMARKS</td></tr>
<tr><td>MATERIALS
£4.50</td><td>MAN HOURS
3.5 | TOTAL COST
£39.50</td></tr>
</table>

Fig. 11.10 – Equipment maintainability data.

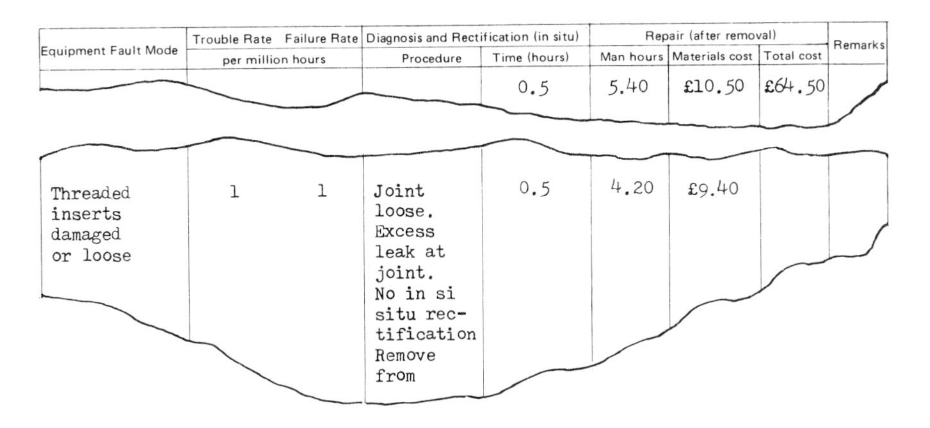

Equipment Fault Mode	Trouble Rate	Failure Rate	Diagnosis and Rectification (in situ)		Repair (after removal)			Remarks
	per million hours		Procedure	Time (hours)	Man hours	Materials cost	Total cost	
				0.5	5.40	£10.50	£64.50	
Threaded inserts damaged or loose	1	1	Joint loose. Excess leak at joint. No in si situ rectification Remove from	0.5	4.20	£9.40		

Fig. 11.11 – Abstract from an equipment fault summary.

Although the manuals will not be written until the final stage of design, the cost of the work involved must be predicted since the prices of the manuals will be quoted when the submission is made to the customer. Note that the estimating form of Fig. 11.12 allows for the writing of manuals.

(g) *The performance analysis*

The performance analysis of the product in all relevant environments is so particular to the design that it will form a major part of any submission to a customer. Generally, the analysis will take the form of calculations proving that each of the performance requirements of the specification will be met. Some experimentation may be necessary to support any calculations.

The performance analysis will also provide the basis of the development programme, for many of the calculations will have to be supported eventually in the laboratory.

(h) *The cost of design, development, and proving*

The scheme which has survived analysis will provide the information needed to determine the number and complexity of details, assembly, and sub-assembly drawings, and most companies are able to predict with reasonable accuracy, the average cost of a drawing of any given size. The cost of the design drawings can therefore be predicted with reasonable accuracy.

The performance analysis will have yielded a draft, at least, of the development testing that will be required, while the reliability and life analysis will have indicated the laboratory proving and endurance testing that will be required.

Sheet 1 of 3

TECHNICAL DEPT. ESTIMATE

DESCRIPTION . . . . Job No. . . . .

CUSTOMER. . . . .

DRG NO . . . . ISSUE. . . . . SPECIFICATION NO . . . ISSUE . .

ACTIVITY	MAN HOURS	REMARKS
Prepare Specification Reliability Analysis Performance Analysis Stress Analysis Customer Liaison and Provision of Information		
SUB TOTAL		
Scheme Amendments Detail Drawings, S/A's and M/A Installation Drg. Schedule of Parts Tolerance Study Test Schedules Work Arising from Development Maintenance Analysis Provision of information for writing Maintenance Manual Operating Manual		
SUB TOTAL		
Development Development Report Writing Approval Tests (See Sheet 2) Manufacture of Test Rigs		
SUB TOTAL		
RECORDS		
TOTAL MAN HOURS		
MATERIAL REQUIREMENTS	COST (£)	
Material for Test Rigs Sub Contracted Resting Bought Out Test Equipment		
SUB TOTAL		
Units Required for Development Units Required for Approval Tests		
SUB TOTAL		
Less Possible Reclamation of Development Units		
TOTAL COST		

Fig. 11.12 – Sheet 1

Sheet 2 of 3

TECHNICAL DEPT. ESTIMATE

DESCRIPTION . . . . JOB No. . . . .

DETAILED BREAKDOWN OF APPROVAL TESTS

TEST		COSTS	MAN HOURS @ £ /HR	TOTAL COST
Vibration	/Functioning			
Acceleration	/Functioning			
High Temperature	Functioning			
Low Temperature	Functioning			
Humidity	/Functioning			
Fungus Resistance	/Functioning			
Sand and Dust	/Functioning			
Salt Spray	/Functioning			
Radio/TV Interference				
Fire Resistance				
Fire Proofness				
Flame Proofness				
Other (State)				
TOTAL				

REMARKS (INCLUDING REFERENCE TO READ ACROSS EVIDENCE).

Fig. 11.12 – Sheet 2

Sheet 3 of 3

TECHNICAL DEPT. ESTIMATE

DESCRIPTION . . . . JOB No. . . . .

COST OF PROTOTYPE MANUFACTURE

ACTIVITY	MAN HOURS @ £ /HR		COST (£)
Assembly and Test Machining Tooling Modifications Development Test Rigs Approval Test Rigs Other			
SUB TOTAL			

DESCRIPTION		COST (£)
Factory Labour Matls for Test Rigs Matls for Manufacture Units from Production Bought Out Parts Other		
SUB TOTAL		
SUB CONTRACT MANUFACTURE SUB CONTRACT TOOL CHARGES		
SUB TOTAL		
TOTAL COST		

No. of Units to be Manufactured in Prototype Shop

No. of Units Supplied for Testing

REMARKS

Fig. 11.12 – Sheet 3

These predictions of the drawing and testing requirements will yield two further outputs: the first is a bar chart or critical path network of the work that will be required during the final design stage, together with a list of the resources required for each task; the second is an estimate of the costs of drawing, development and proving that will constitute the final stage of design.

Fig. 11.13 is an extract from a bar chart which was a prediction of the tasks required by the final stage of design while Fig. 11.12 is a form used in the estimates of costs.

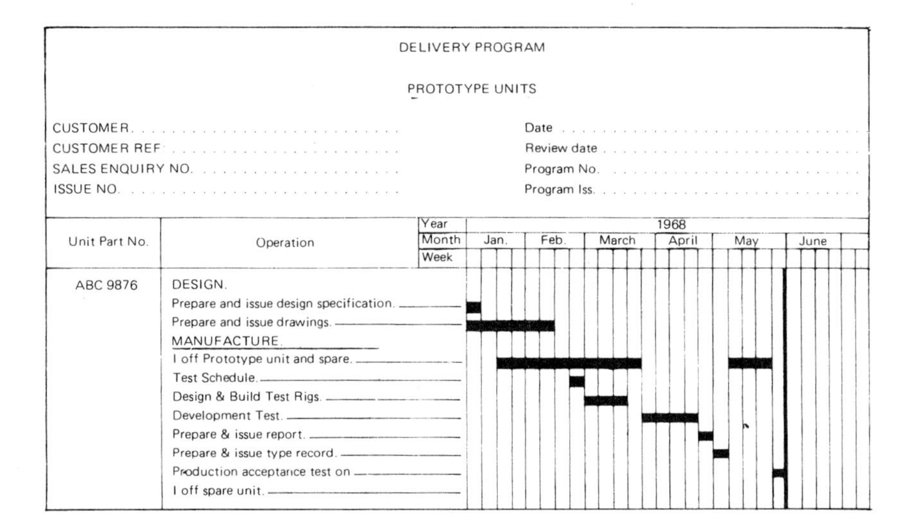

Fig. 11.13 – Extract from bar chart predicting the tasks required by the final stage of design.

One of the difficulties of estimating design costs becomes apparent from Fig. 11.12, for while most companies have the experience with which they can predict with acceptable accuracy the cost of a drawing or the cost of a test, unless the product is perfect from the drawing board, there will be failures, modifications, and repeat tests, and such events cannot be predicted. How does one predict the drawing work that will arise from development or the cost of subsequent modifications to prototypes?

(i) *The proposal*

If the project has been externally generated, project definition will result in a proposal to the customer. If it has been internally generated, it will result in a report to the directors of the manufacturing company. In either case the proposal or report is likely to be a publication that is something more than a mere compilation of the analyses and cost predictions that have been generated by project definition. Many designers would hope, at this stage, to call on the services of professional technical illustrators and professional writers in order to make a coherent and attractive report from the separate analyses and predictions.

If the project is externally generated, the customer may decide to order the product. If the product is internally generated, the directors of the manufacturing company may decide to allocate money to the final stage of design. In either of these cases, the project will enter the final stage of complete, detail design.

11.4 COMPLETE, DETAIL DESIGN

11.4.1 Scheming (activity 6–7 in Fig. 11.1)

Although the feasibility study and project definition will have yielded a scheme which has survived analysis, it is likely that when the decision is made to embark on the final stage of design, the design scheme will have to be redrawn. This is partly because, when the proposal is made to the decision maker at the end of project definition, there are likely to be changes suggested, even when the proposal is accepted. The submission itself is likely to suggest changes which could be made, with profit, to the design specification. But the main reasons for modifying the scheme arise because its purpose has changed. Originally the scheme was intended to solve design problems and to permit the analyses which will demonstrate that the methods of solving the problems will work. Once the final stage of design is started, the scheme is required to be an instruction to the detailing draughtsman and a basis of the design reviews which will be conducted before details are released for manufacture.†

In some companies, the scheme is not issued for detailing until one of the design reviews has pronounced it acceptable – possibly by the signatures on the scheme of the reviewing committee.

11.4.2 Detail drawing (activity 7–8 in Fig. 11.1)

Detail drawing is in one sense the most difficult part of design, because it is in the detail and sub-assembly design that the costs of manufacture are determined (at least within the constraints already imposed by the scheme). It is therefore important that the drawings are reviewed by manufacturing engineers and buyers as they are made.

The need to ensure that a detail is designed in such a way that it is cheap to manufacture is obvious. Less obvious is the need to review the sub-assemblies to ensure that they really are useful breakpoints during manufacture, for there is no point in creating a sub-assembly which the fitter will easily build through unless that sub-assembly must exist for ease of maintenance. Where appropriate, discussion between manufacturing engineers, service engineers, and the designer will be required to ensure a reasonable trade-off between ease of manufacture and ease of maintenance. A further case for reviewing sub-assembly design arises

† Chapter 12 discusses procedures for design review.

from the need for inspection and adjustment. There will sometimes be a trade-off between ease of manufacture, ease of inspection, and ease of adjustment which can only be found by discussion between experts; for example, should adjustments be made with shims during inspection giving a simple design but an expensive inspection procedure, or should a means of adjustment be built into the design with the reverse effect?

The detail drawings become instructions to many people, and must be issued formally to at least buyers, production planners, quality control engineers, the prototype shop, technical authors, service engineers, and spares compilers.

Normally, what will be issued is a main assembly drawing, all sub-assembly drawings, all detail drawings, and a schedule of all details and assemblies. The schedule may be in the form shown in Fig. 11.14 which relates details and sub-sub-assemblies to sub-assemblies and the main assembly. Some companies do not issue a separate schedule but ensure that the main assembly drawing calls up all sub-assemblies and details. The form in which lists of drawings are issued depends on the methods used to call up Bills of Materials.

XYZ & CO. LTD.

SCHEDULE FOR

DESCRIPTION . . . .
Part No. . . . .

Sheet No. 1 of. . sheets
Compiled by. . . . . . .
Date . . . .
Job. No. . . . .

Issue No.	Alteration No.	Modification No.	Date	Remarks

INSTALLATION DRAWING NUMBER
PRODUCTION TEST SCHEDULE

Stage	Part Number	Made From	Number/Stage	Number/Inst.	Part Description

Fig. 11.14 – Schedule of details and assemblies.

11.4.3 Buying (activity 8–9 in Fig. 11.1)

The schedule of parts and all other manufacturing drawings will be issued, as a matter of course, to the buyers, together with a manufacturing instruction listing the number of products to be made, the required delivery date, and any other information relevant to manufacture. The buyers will need this information to place orders, outside the company, for raw materials and bought out parts and components. Fig. 11.15 is an instruction to manufacture which provides some information (particularly, number off) without which the buyer cannot do his job properly.

Sheet . . . . . . of . . . . . . Sheets

MANUFACTURING INSTRUCTION

From: Design Dept.

To Planning Dept.
Estimating Dept.
Buying Dept.
Inspection Dept.

Description of Component. . . . .
Part No. . . . .
Job No . . . .

PLEASE MANUFACTURE, BUILD AND TEST: –

Description	Schedule or	Issue	Remarks/Action/No. Off

Fig. 11.15 – Manufacturing instruction.

11.4.4 Planning and tooling (activity 8–9 in Fig. 11.1)

The schedule of parts, all the manufacturing drawings, and the manufacturing instruction will be issued as a matter of course to the manufacturing engineers who will:

(a) produce the process schedules and assembly schedules which will be the manufacturing instructions used in the production department. If there has already been good liaison between the draughtsman and the planner, the details and sub-assemblies should already be suitable for the planners' purposes. Fig. 11.16 is an extract from a process schedule defining the manufacture of a machined detail.

(b) produce the requirements for tools, probably draw the tools and get them manufactured and delivered.

PROCESS SCHEDULE

DESCRIPTION				ORDER WK	PART Nº	
Piston Guide				34	123456	
TOOLS AVAILABLE	MAT'L AVAILABLE	REORDER SLIP Nº	QUANTITY	REQ. WK	BATCH Nº	SPLIT
	10-9-8x		40 + 4	42	5432	

	MATERIAL SPECIFICATION (OR MADE FROM)	QTY PER 100	PART Nº (CHECK)	ISSUE
1	50 mm diameter brass to BSS.249	10 m	123456	3

2	OPERATION NUMBER	10	20	30	40	50	60	70	80	85	90	100	105	110	120		
3	MACHINE GROUP	3	3	3	5	5	4	14	4	4	4	4	18	25	22	90	

4	OP. Nº	MC GP.	DESCRIPTION OF OPERATION	T/A EACH (MINS)	SETTING TIME (MINS)	M/C LOAD PER 100 (HRS)	TOTAL LOAD	QTY. SCRAP	QTY. GOOD	DATE COMPLETED
5	10	3	HERBERT NO.4	18.2	10/2	30.4				
6	20	3	HERBERT NO.4	16.0	13/5	27.0				
7			FIT PROTECTOR COM.5930*							
8	30	3	HERBERT NO.4	6.7	$6\frac{1}{2}/1$	9.5				
9	40	5	VERT/MILL	4.75		8.0				
10	50	5	VERT/MILL	3.00	$1\frac{3}{4}/\frac{1}{2}$	5.0				
11	60	4	C/LATHE	3.70	$2\frac{3}{4}/1$	6.2				
12	70	14	BROACH	3.20	$1\frac{3}{4}/\frac{1}{2}$	5.4				
13	80	4	C/LATHE	8.00	$2\frac{1}{2}/\frac{1}{2}$	13.4				
14	85	4	C/LATHE	1.55	$2/\frac{1}{2}$	2.6				
15	90	4	C/LATHE	5.60	4/1	9.4				
16	100	4	C/LATHE	3.60	$2\frac{3}{4}/1$	6.0				
17	105	18	HONE. DELAPENA							
18	110	25	BENCH DE-BURRS	4.00		6.7				
19	120	22	SMART AND BROWN POLISH	8.50		14.2				
20										
21		90	DELIVER TO STORES							
22										
23										
24	13		MACHINE SHOP	86.80	$49\frac{1}{4}/13\frac{1}{2}$	143.8				
25										

Fig. 11.16 – Process schedule.

11.4.5 Designing inspection procedures (activity 8–9 in Fig. 11.1)

The schedule of parts and all other manufacturing instructions will be issued as, a matter of course to the Quality and Reliability Department who will determine the inspection procedures that will be used during production. It is unlikely that inspection procedures can be determined completely, however, until development has confirmed the significance of the measurements to be taken. Some progress can be made, however, by consultation with the designer who is producing the schedule of required development tests.

Buyers, manufacturing engineers, and quality control engineers are unlikely to be members of the Design Department. They may, for example, be accountable to the production director, while designers are accountable to the technical director of the company. Good liaison across departmental boundaries is essential, and, in some companies, a project team consisting of a designer, a manufacturing engineer, and others (in one company, a designer, a manufacturing engineer, and a marketing engineer) is set up to 'walk the product through the shops'. In other companies, a multidisciplinary team is set up to manage the project by 'management by objectives' procedures.

Apart from problems caused by departmental boundaries, buying, planning, and inspection will necessarily wait on development testing, and considerable feedback is required from the laboratory to these activities.

Buyers, manufacturing engineers, and quality control engineers will not generally have their work charged as a cost of design. They will either have their work charged as production department overheads or it will be charged to the cost of manufacturing the product. Where a multidisciplinary team is set up to manage the product its cost may be regarded as either a design cost or a manufacturing cost.

11.4.6 Test schedule writing (activity 8–10 in Fig. 11.1)

The development tests must be planned by the designer responsible for the scheme, although he would expect to cooperate with the development engineer when writing the schedule of tests. Many of the test requirements will have been generated during project definition by the need to confirm that specification requirements have been met, and many will have been generated by uncertainties about the solutions that have been proposed to some of the design problems. Fig. 11.17 is an extract from a test schedule. Such a schedule provides the instructions against which the development engineers will work.

11.4.7 Prototype planning (activity 8–11 in Fig. 11.1)

Most development testing is done on a prototype because prototypes will be available before products are made in the production department, and also because some manufacturing problems will have to be solved before production methods can be fully planned. Usually the prototype shop operatives will work directly from the drawings rather than to process schedules, but even so, some

TEST SCHEDULE

FOR

5. Functional Tests (L xxx Flow Rig).

The tests are to be carried out with a nominal 24 volt DC supply. All pressure measurements are to be made using static pressure tappings.

With the valve set up having a downstream volume of 10 litres and the cock terminating this volume set to give a flow of 10 kg/m in at room temperature when the valve downstream pressure is 4 bar. the performance is to be recorded under the following conditions:-

5.1 Room Temperature Tests (Datum Valve Removed).

The tests are to be carried out by applying a false datum pressure equivalent to the valve inlet pressure.

5.1.1. Failure Test

With the solenoid de-energised, i.e. valve closed, the valve inlet pressure must be increased to 10 bar. The solenoid is to be energised and the pressure switch must operate at between 5 and 5.2 bar to close the valve head. The downstream pressure may overshoot to 10 bar pressure but must not exceed 5.2 bar for more than 1.1 secs.

5.1.2. Valve Head Tests (Minimum Operating Pressure).

Repeat tests 4.1 and 4.2.

5.2 Control Characteristic Test (with Datum Valve Re-Fitted).

5.2.1. Steady State Test.

Energise the solenoid and then increase the valve inlet pressure slowly from 0 to 10 bar and decrease slowly back to 0. Between 4.5 and 10 bar the valve downstream pressure must be stable and lie between 4 and 4.75 bar on both rising and falling inlet pressure.

Note: If the above test requirements are not met due to overshoots, hysteresis etc, then the following procedure is to be carried out:-

The unit is to be subjected to a 15 minute soak on the flow rig with an inlet pressure of 10 bar and a flow in excess of 10 kg/min at a through air temperature of 160°C. At the end of the soak period, the unit will be cycled open and closed 10 times.

After cycling the functional tests at high temperature (Para 6) may be carried out at this stage.

After the functional tests at high temperature, the unit is to be to room temperature and tests 5.2 are to be carried out.

5.2.2. Slam Acceleration Test.

The solenoid must be engergised and the valve inlet pressure from 0 to 10 bar at a rate of 3 bar/sec. The same bandwidth of 4/4.75 bar must be obtained. An over allowable provided the pressure switch is not actuated. stabilise within 5 seconds of applying maximum inlet be measured on a suitable recorder.

Prepared by	C. Brown	Issue No.	8	9	10
Checked By	P. Nuts	Alt No		8	
		Issue No			

Fig. 11.17 – Test schedule.

measure of planning and buying will be required in the management of prototype manufacture.

The schedule of parts and all detail and assembly drawings will therefore be issued to the prototype shop together with a prototype manufacturing instruction. The prototype manufacturing instruction will differ from the manufacturing instruction, because the number required will be different, the delivery dates will be different, and the standards to which the products will be made may be different. For example, some sub-assemblies may be made for testing to destruction, or some prototypes may be simple space models for use by the customer in his early design of the parent system. Fig. 11.18 is an extract from a Prototype Manufacturing Instruction.

From: CHIEF DESIGNER — Date . . . .
Sheet No . . . .
To: PROTOTYPE SHOP MANAGER — No. of Sheets . . . .
Copy to: PROTOTYPE SHOP PLANNER — CHIEF DEVELOPMENT ENGINEER
PROTOTYPE SHOP BUYER — PRODUCT MANAGER. . . . .
PROTOTYPE SHOP ESTIMATOR
CONTRACTS OFFICE

TYPE		SUBJECT				
REQ'D FOR	BENCH TESTS	CUSTOMER TESTS	FINAL STANDARD		JOB NO.	
No. OFF			PRODUCT			
ACTION REQUIRED		SCHEDULE OR DRAWING NO.	No. Off	Date Req'd	COST CODE	REMARKS

Fig. 11.18 – Protytpe manufacturing instruction.

11.4.8 Manual writing (activity 8–16 in Fig. 11.1)

The schedule of parts, details, and assembly drawings will be issued to the Technical Publications Department, who will prepare the operating manuals and maintenance and repair manuals for use by the customer. If the manuals are

DISASSEMBLY

1. Preliminary

A. Before commencing to disassemble, determine reason for removal from service so that the degree of disassembly can be assessed.

B. If removal has been due to malfunctioning, proceed as in TROUBLE SHOOTING to verify and then establish the cause of failure, after which fit replacement part or repair as necessary.

C. A unit removed for overhaul should be disassembled, its componenets inspected and renewed as necessary, re-assembled and tested in accordance with the relevant sections of this manual.

2. Components to be discarded

A. It is recommended that the following items are automatically discarded at overhaul and replaced by new prior to re-assembly:-

Seal washers. Tab washers.

3. General

A. See the CAUTION under 'data', sub-section 4 of DESCRIPTION AND OPERA

B. Disassembly should be carried out under clearn conditions. No sp tools are required.

C. The figures in parenthesis throughout the text identify items as on the Exploded View, Fig. 501, in the ASSEMBLY

4. Disassembly Procedure

A. Unscrew banjo bolt (6) and collect body (7)

B. Remove the two countersunk head (2). Ease the element clear.

C. Remove the four screws together and lift of the body coll Collect the

Fig. 11.19 – Extract from a servicing manual.

expensive to produce, as would be the case when designing any product with a more than trivial technical content, costs should be kept separate from other design costs because the supply of manuals may be the subject of a separate contract from that for the supply of the product. Even if the manuals are not paid for separately, the customer will want to know what fraction of the price he is paying is for manuals, and as with the cost for design, he may wish to amortize it over a guaranteed order for products. Generally, however, the number of manuals to be supplied will depend on the manufacturer's product support organization and the customer's maintenance organization rather than the number of products to be supplied.

Although the manuals may be produced by a Technical Publications Department which is separate from the Design Department, there must be good liaison between the designer, the technical author, the technical illustrator, and the service engineer to ensure that the design makes good operating, maintenance, and repair methods possible.

11.4.9 Service tool design (activity 8–16 in Fig. 11.1)

Servicing and repairing the product is likely to require special tools and special test equipment. Some of these will have been designed to be used during production, but some will be specially designed for maintenance and testing in the field. The designers of such equipment may be part of the design team or may be manufacturing engineers, but in either case, liaison between designers, manufacturing engineers, quality control engineers, and service engineers is essential if good maintenance procedures are to be designed. Liaison with the Technical Publications Department is also essential because the operating, maintenance, and repair manuals will necessarily refer to the special tools and test rigs required. Special tools and special test rigs may be the subject of a contract with the customer that is separate from the contract for the supply of the product. Again, not only will the customer want to know what he is paying for such items, but the quality and nature of such equipment will depend on the customer's maintenance organization and the manufacturer's product support organization rather than on the number of products being sold.

11.4.10 Manufacturing prototypes (activities 11–12 and 8–15 in Fig. 11.1)

The Prototype Shop will manufacture:

(a) products and sub-assemblies which will be used in the development and proving of the product. Some of these sub-assemblies will be tested to destruction, but some may be sold. The cost of the products and sub-assemblies which cannot later be sold to the customer are a charge on the design and development programme. It is to be expected that design faults will be found during attempts to manufacture prototypes, so that the cost of the prototype manufacturing programme will not be only the calculated

SPECIAL TOOLS, FIXTURES, AND EQUIPMENT

The following special fixtures and services are required for setting and testing the unit:

A. Test Head, Part No. ABC 1234 together with a suitable dial micrometer.
B. Setting Piece, Part No. ABC 2345.

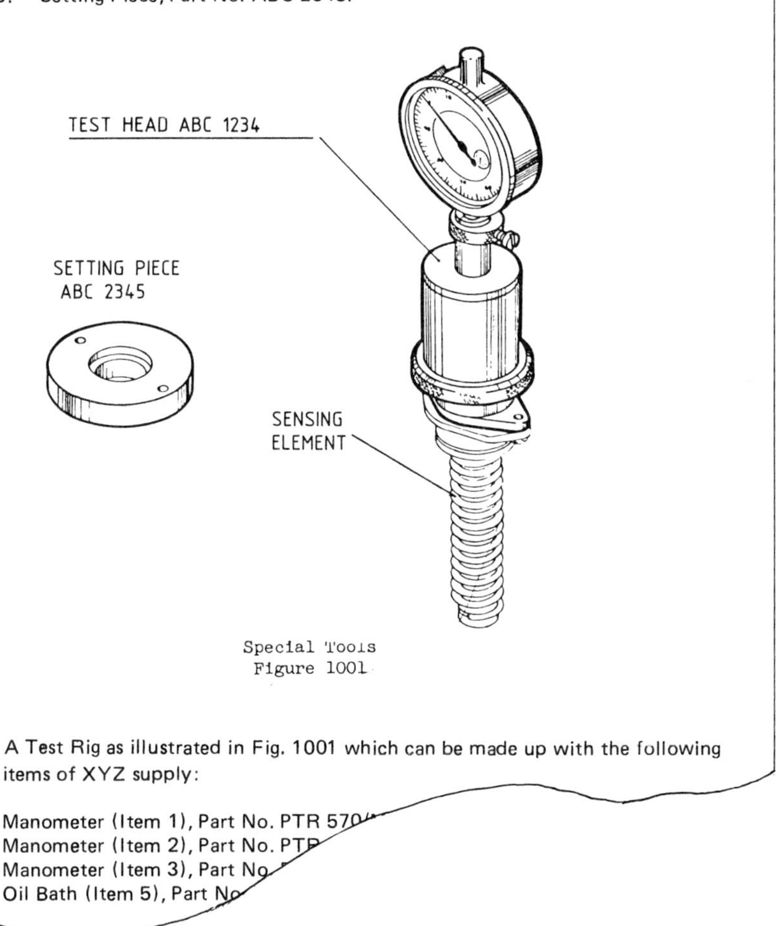

A Test Rig as illustrated in Fig. 1001 which can be made up with the following items of XYZ supply:

Manometer (Item 1), Part No. PTR 570
Manometer (Item 2), Part No. PT
Manometer (Item 3), Part No
Oil Bath (Item 5), Part No

Fig. 11.20 – Extract from a manual which refers to special test equipment.

cost of making parts to drawings; it must include the cost of the re-working and redesigning that will be necessary until manufacturing (and sometimes development) problems have been solved.

(b) products and sub-assemblies which will be supplied to the customer for use in his development programme. Some of these products may be simple space models which the customer may need to complete the design of the parent system, but some may be working components of varying degrees of sophistication for use by the customer in the development of the parent system. Generally the customer's requirement for prototype equipment will be stated in his invitation to tender. The cost of this equipment will be itemized in the tender and will not be regarded as part of the cost of design.

11.4.11 Development testing of prototypes (activity 12–14 in Fig. 11.1)

Development testing is either the most expensive part of design or it generates the greatest expenses in design. The testing, at least initially, will be to the programme dictated by the test schedule. Except in the case of very simple design problems it is almost certain that the product or its sub-assemblies will not work when first tested. This means that a test will probably be followed by modification to the design and then a re-test. This makes it difficult to predict the cost of drawing and the cost of development, because any prediction implies that probable redesign and re-test has been allowed for. This has been mentioned in the discussion on cost prediction during project definition, but it is also apparent in Fig. 11.12 that designers are aware of the need to allow for design failure in estimating the cost of development. Usually, although formal development reports are written, interim bulletins must also be written to ensure rapid feedback of information to the drawing office, so that rapid modifications may be made when testing shows design faults.

A formal development test report will be required by the customer to demonstrate that the product, when fully developed, complies with the specification.

11.4.12 Manufacture of the product with the production tooling, and manufacture of spares (activities 9–16 and 9–13 in Fig. 11.1)

Manufacturing using production tooling will not only wait on the production of process schedules but, with luck, will also wait until the manufacture of prototypes has revealed any manufacturing problems which were not detected during the design reviews. Often, however, the required delivery dates and the time taken to draw, plan, and tool are such that production processes must be proved before prototype manufacture and development has proved the design. One or more of the products made using production tooling will be required for testing in order to demonstrate that the differences between such products and the prototypes have not led to differences in performance, reliability, life, etc.

The cost of production is clearly not a charge to design except where the design department may have to buy a production item for test purposes.

The transition from supplying – to the customer – products made in the prototype shop, to supplying products made on production tooling, is often very gradual because of the difficulty of meeting the customer's delivery programme with the lead time required for tooling. Under such circumstances the prototype shop may be used as a short order shop, and in some companies, the prototype shop budget assumes that a certain percentage of the shop's time (perhaps 20%) will be spent on short order production work. One of the problems of design can be the difficulty of getting test items made in a prototype shop cluttered with short order production work.

11.4.13 Proving the prototype (activity 14–15 in Fig. 11.1)

When the prototype works, in the sense that it meets the performance requirement of the specification, it will still be necessary to prove that it meets the reliability and life requirements. Some of the reliability and life requirements will have been demonstrated by read-across evidence from the use of sub-assemblies in other products, some endurance testing needs will have been demonstrated by the reliability and life analyses, and some demonstration of reliability may not be possible except through service. An example of the need to prove life in service is when life testing cannot be accelerated; so that, for example, a life of 1000 hours can only be demonstrated by 1000 hours of operation. The product is likely to be in operation before its full life has been demonstrated, so that the guaranteed life will have to be less initially than that required by the specification, being increased as (first) testing and (later) perhaps operation, warrants. If a product is to be offered ultimately without a life limitation but on an 'on condition' basis, this status can only be achieved by proving the reliability in service.

The expense of proving will usually derive from two sources. Much of the testing will be automatic in that the product can be cycled, automatically, through various performance regimes for many hours, so that there will be little labour required by the tests; but the automatic control equipment and the apparatus for generating the environmental conditions will be very expensive.†

The other cause of high development costs that can arise is failure to meet the requirements, so that redesign and re-testing become necessary. As previously noted, this is difficult to predict.

The laboratory report of the proving will be required by the customer to confirm that the product complies with the specification. Probably, in fact, one report will be supplied to discuss both the development testing and the endurance testing. In some cases the customer will require a certificate of compliance with the design requirements, and Fig. 11.21 illustrates such a certificate.

The cost of proving the product is, of course, wholly a cost of design.

†Vibration environment requirements and atmospheric environment requirements are discussed in Leech, *Management of engineering design*.

3G.100 : Part 1 : 1973

APPENDIX A

STANDARD FORM OF DECLARATION OF DESIGN AND PERFORMANCE

NOTE. Before specifying equipment, check with *(manufacturer's name)* that this declaration is the latest issue.

D.D.P. Number: ....

Issue Number: ....

Approving Authority ....

(Name and address of manufacturer)

....

....

....

DECLARATION OF DESIGN AND PERFORMANCE

of .... *(name of equipment)* ....

Basic Number: ....

Identification code *(if different from basic number)*: ....

Description: ....

Weight: ....

Overall dimensions and position of the centre of gravity
(or reference to drawing if attached): ....

Design specification reference: ....

Drawing schedule reference: ....

Quality control procedure reference: ....

Development contract No. *(if applicable)*: ....

Modification standard reference: ....
(if affecting this declaration)

System or wiring diagram *(if appropriate)*: ....

Installation drawing number: ....

Maintenance, repair and overhaul manual reference and
issue numbers: ....

Test report references: ....

Fig. 11.21 – Extract from certificate of compilance with design requirements.

3G.100 : Part 1 : 1973

Fault analysis reports *(if appropriate)*: ....

Particulars of approvals held for the equipment *(and a brief summary of application and experience of equipment)* ....

Particulars of approvals held for similar equipment *(and a brief summary of application and experience of that equipment)* ....

Airworthiness requirements with which the equipment complies *(when applicable)* ....

Performance

(To be declared under sub-headings relevant to the particular equipment i.e. electrical and mechanical input and output characteristics peculiar to the equipment, time rating and duty cycle, etc.)

Declarations

(State here those declarations required by the relevant subsections of British Standard 3G.100 relating to environmental and operating conditions for aircraft equipment (i.e. those stated in Table 1) and also those additional declarations required by the relevant equipment specifications.)

NOTE. The limits of declared performance and those implied by the declarations are not intended to be absolute but to indicate the performance which has been proved by test.

Limitations

(The equipment designer shall state any known limitations not specifically covered by the above declarations of which the user should be aware.)

Departures from specification

(The equipment designer shall list any departures from British Standard 3G.100 and from the design specification to which the equipment is declared.)

I hereby certify that the information contained in this declaration of design and performance is accurate and the declaration is made under the authority of

.... *(Company's design approval reference number)*

.... *(Manufacturer's name)* cannot accept responsibility for satisfactory operation of equipment used outside the declared conditions set out above, without their agreement.

Signed .... *(Approved signatory)*

on behalf of .... *(Manufacturer's name)*

Date ....

Extracts from BS 3 G 100: Part 1: 1973 are reproduced by permission of the British Standards Institution. Complete copies of the document can be obtained from BSI at Linford Wood, Milton Keynes, MK14 6LE.

Fig. 11.21 – Page 2

3G.100 : Part 1 : 1973

TABLE 1. DECLARATIONS REQUIRED

	Information *(i.e. range, grading, etc.)*	Test report references or data	Effective clause in 3G.100 (or 2G.100)*	Relevant clause of individual equipment specification (if appropriate)
(1) The minimum life or overhaul period of the equipment, *(based on test evidence and/ or operational experience)*			Part 2: Section 1	
(2) Temperature pressure. *State grade letter and suffix numeral*			Part 2: Subsection 3.2	
(3) Tropical exposure			Part 2: Subsection 3.7	
(4) Performance in ice formation or accumulation conditions (when applicable). *State test employed*			Part 2: Subsection 3.9 or Part 2: Subsection 3.10	
(5) Waterproofness. *State grade* (when applicable)			Part 2: Subsection 3.11	
(6) Vibration. *State category* a. *without antivibration mounting, and/or* b. *with specified antivibration mounting*			Part 2: Subsection 3.1	
(7) Acceleration – *State Class Grade† and Category for crash conditions*			Part 2: Subsection 3.6	
(8) Explosion-proofness. *State category* (when applicable)			Part 2: Subsection 3.5	
(9) Fire resistance. *State grade* (when applicable)			Part 2: Subsection 3.13	
(10) Electro-magnetic interference: *State frequency ranges covered* (when applicable)			Part 4: Section 2	
(11) Magnetic influence. *State compass safe distance*			Part 2: Section 2	
(12) Sand and dust (when applicable)			Part 3: Subsection 3.X	
(13) Fluid contamination (when applicable). *State fluids, etc.*			Part 3: Subsection 3.12	
(14) Salt mist (when applicable)			Part 3: Subsection 3.8	
(15) Mould growth (when applicable)			Part 3: Subsection 3.3	
(16) Differential pressure (when applicable). *State test employed*			Part 3: Subsection 3.4	
(17) Bump and shock (when applicable). *State test employed*				
(18) Flammibility and smoke/toxic fumes (when applicable). *State test employed*			Part 2: Section 1	

* Refer to British Standard 3G.100 : Part 0 for latest information regarding sections published.
† The limiting value in the case of grade D shall be stated.

Fig. 11.21 – Page 3

11.4.14 Manufacturing spares (activity 9–16 in Fig. 11.1)
Manufacturing spares will not be a direct charge to design because those spares that are supplied to the customer will be paid for by the customer. The designer is involved, however, because the whole programme of spares provisioning must be thought out as a deliberate exercise involving the designer, the product support organization, the repair department, and the customer's maintenance department. The problem will be similar to that of inventory control, and will be generated by the reliability analysis. The manufacturer must carry spares that will be required often, but cannot afford to carry so large an inventory that the customer will never have to wait for a spare part. The specification may demand a maximum permissible cost of down-time from which a service level may be calculated with an acceptable approximation. The cost of the designer's time spent in discussing spares provisioning must be allowed for in determining the cost of design.

11.4.15 Supporting the customer's test programme (activity 15–16 in Fig. 11.1)
While the product is being designed and developed, the customer will be designing and developing the parent system. For example, if an aircraft navigation system is being designed, the customer may be designing the aircraft into which it will be installed; if a plastic bin is being designed, the customer may be building the stillage of which the bin is part; if a valve is being designed, the customer may be designing the submarine in which the valve will be installed. The customer will require the designer's assistance with his own test programme from time to time, and some money must be budgeted to pay for this. Such assistance cannot be refused because, apart from needing to keep the customer's goodwill, work on the parent system may generate modifications to the design specification and hence changes to the design of the product.[†]

11.4.16 Supporting the product in service (activity 16–17 in Fig. 11.1)
The need for product support is fairly obvious, and has been discussed. With almost any designed product beyond the lowest level of complexity, the manufacturer will expect to provide support. Usually the manufacturer will maintain a service department, and usually such a service department is treated as an overhead in that the service man's time is not charged to the design or manufacture of the product with which he may be concerned. Some procedure must exist, however, by which the service man will report on the work that he is doing, so that the designer becomes aware of design faults which do not come to light until the product is in service. Fig. 11.22 shows extracts from service

[†]It has been suggested, perhaps cynically, that the first competition for a design job determines who will get the job but not the price. There will be so many changes to the specification after the acceptance of the tender that the eventual price will be determined on a 'costs plus' basis since it will be negotiated after changes have been requested and made.

. Valve together ...
...ed on the engine (as specified on XYZ Rquiremen...
...52); it was agreed there was a danger of all the nuts being unscrewed when the valve is removed from the engines and ABC are asked to take some form of safeguard action to prevent this.

ACTION REQUESTED

XYZ to supply 'read-across' evidence:-

(i) of us of R.751 Silicone Rubber 'O' ring as a static joint seal at temperatures exceeding 300°C, and

(ii) of the use of S80 piston sliding brass guide without lubrication.

(i) ABC Ltd. to investigate means of preventing valve-securing nuts being accidentally unscrewed when valve is removed from bulk-hea...

(ii) ABC Ltd. to supply XYZ with copies of new format of "Accesso... Development Modification Proposal"forms.

Written by C. Brown Order/Code No. 4321

ACTION: Those present (1 copy each)

Fig. 11.22(a) – An extract from a liason engineer's report.

SERVICE QUERY NOTE — XYZ Cº LTD.

DATE QUERY NOTE RAISED	ISSUE Nº
CUSTOMER PQR	COMPONENT xxxx R Piston PART Nº 987654
REQUESTED BY SERVICE	OTHER TYPES AFFECTED NONE

REQUEST 1. Extension of Repair scheme 123 to provide oversize bushes in in +.002", +.004", +.006",+.008", +010" steps.
2. To enable existing pistons having oversize bush bores to be salvaged immediately, authority to D— Plate (copper) standard bushes is requested. The D— Plating to be not more than .003" on diameter.

REASON Salvage
N.B. There are no new pistons available from XYZ stores

LIMITATIONS The total number of pistons to be repaired under "Request 2" above to be restricted to 20 only.

SUPPORTING DOCUMENTS

APPROVAL . DESIGN DEV. STRESS

IF QUERY IS REJECTED BY : DESIGN/DEV./STRESS
REASON

QUERY ANSWERED BY
CONCESSION YES/NO Nº ISSUE Nº
MODIFICATION YES/NO Nº CLASSIFICATION

REMARKS

(b) A more formal method of reporting from the field.

ABC CO LTD
SWANSEA WEST GLAMORGAN GT BRITAIN

MODIFICATION PROPOSAL NO.	(FOR D.O. USE) MOD. NO.
EQUIP. TYPE & DESCRIPTION	MRI/DIS.
	CUST.COMP.MOD.
AIRCRAFT OR ENGINE.	DATE RAISED:
TITLE.	DATE ISSUED:
REASON.	ORIGIN:

DESCRIPTION.

EFFECT ON PRODUCTION TEST PROCEDURE

IS C.O.D. AFFECTED - YES/NO

MOD. APPROVAL TESTS REF. NO.	IDENTIFICATION:

LIABILITY. T.A.C. or CUSTOMER.	INT. CLASSIFICATION (RECOMMENDED) CLASS	WEIGHT CHANGE

INTERCHANGEABILITY AFFECTED. IS/IS NOT

MILITARY		CIVIL MAJOR/MINOR
MANDATORY	A/-	MANDATORY
FULLY RETROSPECTIVE	B/-	RECOMMENDED RETRO.
REPAIR & OVERHAUL	C/-	RECOMMENDED AT O/H
NEW PROD. ONLY	D/-	OPTIONAL
		INFORMATION.

APPROVED FOR:

DESIGN	DEVELOPMENT	STRESS & WT.	CUSTOMER'S TECH. APPROVAL	REF:

ADDITIONAL INFORMATION

WORKS ORDER/CODE NO.

Fig. 11.23 – Modification proposal form.

reports intended to acquaint the designer with faults which may need correcting by design changes. There are two reasons for this. Firstly, the designer must build a history of the faults of his designs so that errors will not be repeated. Secondly, where the customer has a right to expect the fault to be rectified at the manufacturer's expense, information about the fault must be brought to the designer. A third, incidental, reason for the feedback of information from the service engineer to the designer is that it is a means of project generation.

11.4.17 Modifying the design (activity 17–18 in Fig. 11.1)

When a fault in design is demonstrated in service and is required to be rectified, or when the experience of operation convinces the customer that a change in design must be made, some procedures must be adopted for initiating the work and for ensuring that it is properly paid for. Clearly, some changes must be made immediately the need for them is seen (for example if a design fault in the brake system of a car is found), while some changes can wait until the product is to be serviced anyway (for example, changing the material of switch contacts a to obtain a longer life). Some changes must be paid for by the customer (where he has changed his requirement).

Frequently, the changes are initiated by a service man, but they may be initiated by the customer, an operator of the product, an operative manufacturing the product, or anyone else involved in its manufacture or use. Properly, the design department has a formal system for dealing with modifications, to ensure that costs are correctly determined and allocated and priorities assessed. Fig. 11.23 illustrates one company's modification proposal form.

11.4.18 Design enhancement (activity 17–18 in Fig. 11.1)

A successful design should be only one stage in a continuous sequence of successful designs. Experience with each product suggests enhancements which can form the basis of the next generation of products. It has been suggested that some designers actually design products with potential for stretch in performance, but whether that is the case or not, every successful design generates a family of designs – a successful engine generates a family of increasingly powerful, increasingly reliable engines; a successful washing machine generates a family of increasingly reliable, cheaper washing machines; and so on. The end of one design is the generation of the next.

CHAPTER 12

Management of engineering design change

In this chapter the reasons for generating engineering changes are given, and the means of classifying such changes is suggested. The need for having a logical engineering change procedure so that agreed design changes can be properly planned and controlled for incorporation is stressed. The use of design reviews, which may also be the means of initiating changes, is discussed as a management technique. This provides a formal documented and systematic study of the design at it proceeds, and may be looked upon as a product liability prevention system. Finally, mention is made of configuration management for controlling product variations to meet different classes of customers.

Engineering change is often the focus for widespread concern, and it tends to become a very sensitive area for those undertaking design work. Many responsible people within a manufacturing concern are affected by the introduction of an engineering change, and they naturally want, and should have, their say about any changes. When a change provides vital safeguards for safety and reliability, its incorporation automatically becomes mandatory. Here a simple system is required which provides good speedy communication to give effect to the necessary change. Other types of changes require a more rigorous evaluation system. Any company that lives by selling what it manufactures must be prepared to change. It must be prepared to manufacture new products when the old are no longer required in the market place, but must also be prepared to modify a product to make it cheaper to produce or more attractive to the customer. At the heart of these changes is the designer and his team, and the management of engineering change dictates much of the organization of the company, the way it operates, and the way it spends its money.

Changes are disrupting to an orderly routine process, and no matter where the request for a change may originate and no matter what beneficial effects may result from its incorporation – reduction in cost, better saleability of the product, or more satisfied customers – the charge of obstruction is always levelled against the engineering design department as it is they who initiate the change. The design team become the scapegoats for increased tooling costs, missed delivery dates, and all other evils.

The technical definition for a change must be an engineering department responsibility resting squarely upon the design department with development engineering approval. However, its introduction into the product depends, except for a mandatory change, on many considerations, such as the procurement of material parts, or the inventory of parts already made, tooling charges, tool procurement, costs and replacement kits in the field, as well as many other factors. Rarely, if ever, does a product go onto the market without some engineering change being effected during its initial introduction phase. Indeed, company products are very often plagued by innumerable changes if their complexity is high. Fig. 12.1 indicates the flow of information required for coping with design changes, and nearly always there will be a delicate interplay between design and development. Such difficult engineering feats as the design of an aircraft gas turbine engine, for example, may undergo as many as 3000 or more changes during their initial production.

In a well engineered and properly developed product, engineering changes to drawings and specifications are disruptive to the normal flow of production work. But in competitive situations affecting performance and cost they are often unavoidable. It must be clearly understood that they will drastically upset all the elements making up the body of a manufacturing unit and their introduction into a production line, however modest, will require stringent appraisal of all pertinent factors. No engineering changes should therefore be lightly undertaken, for delays will generally ensue. However, the need to deliver on time is crucial for most enterprises today if they are to remain competitive. Consequently there is a requirement to be so organized that changes can be made quickly and easily. A typical example occurs in the case of castings. The designer may suggest a casting as the cheapest way of meeting a particular design problem, but if the lead time is six months it may be better and cheaper to design both the hardware and the procedures so that changes can be undertaken and tackled quickly with the aid of N.C. machining. In such cases the cost saving may offset the cost of the equipment. A typical example in one company was where a certain design called for casting requiring a lead time of 6 months, whereas changes to a N.C. machined fabrication could be made in two shifts. Here, although the cost of the N.C. machinery and support staff was considerable, the smaller interruption to the production programme and the shorter lead time justified the expense.

It has been shown in Chapter 6 that the average turn-round time of jobs can be extended considerably if we attempt to use more than about 80% of the available resource time. Simple queuing theory would lead us to expect that if changes added 5% to the work load, job turn-round times could be doubled.

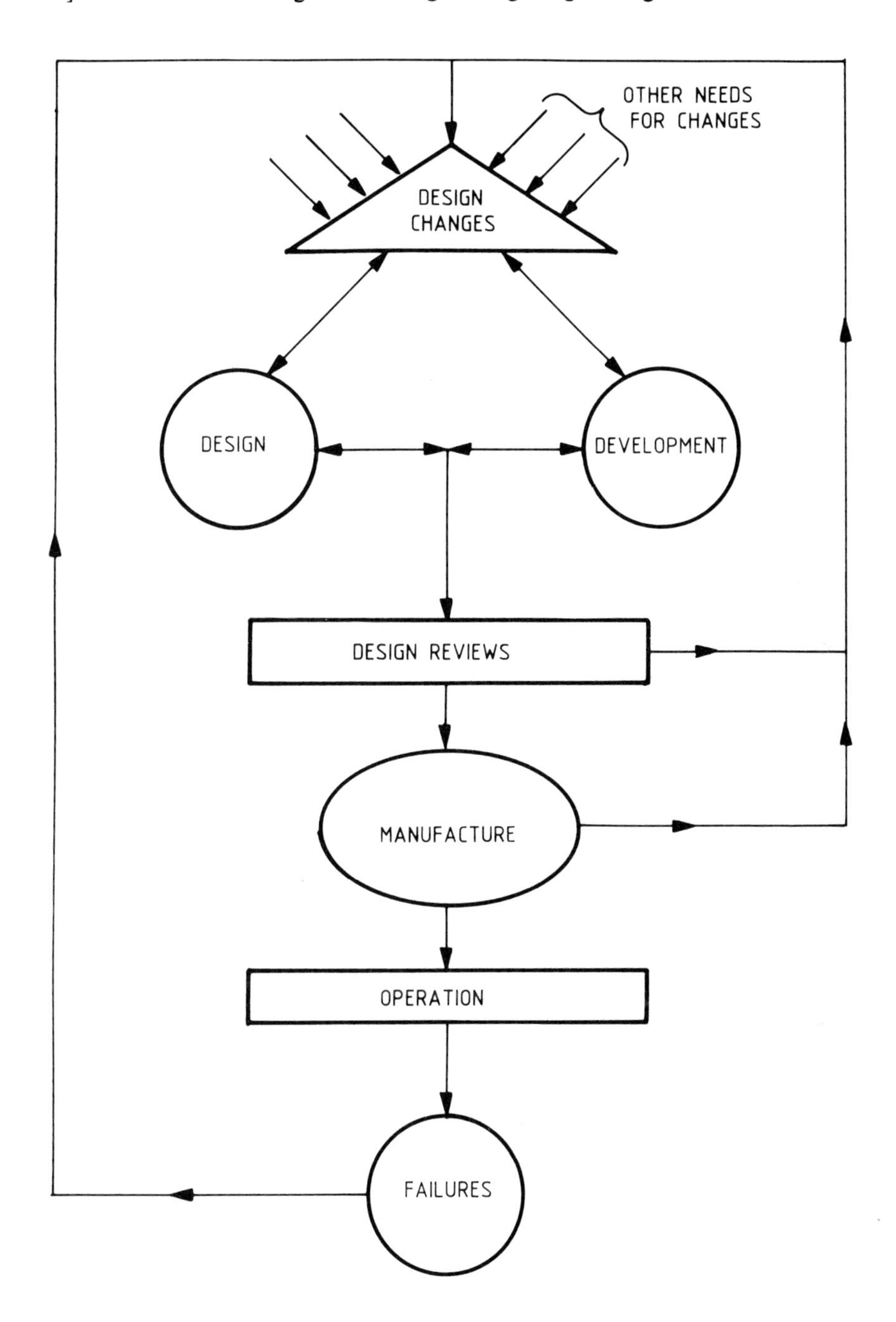

Fig. 12.1 – Flow of information required for coping with design changes.

The simulation in Chapter 6 and work by Das [1] tend to confirm that changes will extend delivery dates considerably and so they must be managed very carefully. A well-established engineering change procedure must govern all the issued and released engineering documents, such as layouts, drawings, specifications, parts lists, and assembly instructions, as well as an agreed schedule for incorporation of the change concerned.

Before considering the essentials of a change procedure it is worth mentioning some of the major reasons for changes occurring with particular reference to existing products and changes which are made to new products after the first issue of manufacturing drawings.

12.1 REASONS FOR ENGINEERING CHANGES

The most dramatic change occurs when a management decision is taken to design and manufacture an entirely new product, although this is by no means necessarily the most important change in a company. Most changes are small and are forced on the organization by the need to remain profitable. A motor car may owe its succes to a series of small changes over the life of the model. At the end of ten or more years the car being manufactured may look almost the same as the prototype, but many changes will have kept down its price, improved its reliability, reduced its cost of ownership, and generally have maintained or increased the profit it makes for its manufacturers. At least one washing machine manufacturer deliberately maintains a policy of not making major model changes in order that, by a process of continuous development, they will produce a machine of high reliability (for which they have found the customer prepared to pay a high price), a machine that costs less and less to make, with decreasing servicing and warranty charges, and a machine that costs less to transport. The washing machine they produce now is, of course, different from the one they produced ten years ago, although it looks much the same. The improvements have been brought about by a number of small changes, each clearly defined in objective, each carefully costed, and each manageable because it was limited and clearly defined.

In a totally different field, when the drawings of a new aerospace project are issued it is assumed that many will be changed as development and manufacture show modifications to be desirable. It has been argued that continuing this process of change through development and early production can considerably reduce the cost of later production aircraft and space vehicles. There will, therefore, be many circumstances where it becomes essential to incorporate changes, and some of the main reasons for generating design changes will now be set out.

12.1.1 The need to reduce manufacturing costs continually

As manufacturing technology improves over the life span of a product it is frequently desirable to take advantage of such improvements. Initially, the design team produces a set of drawings from which manufacture is either impossible or unnecessarily expensive. In the first case, change is essential, in the second, merely desirable. It is rare for an expert production engineer to examine a manufacturing process, as first suggested, without being able to suggest ways of reducing cost, even without resorting to new technology, but one difficulty is that problems are never completely solved on paper. However much the drawings are studied before issuing them, there are innumerable ways of changing the design to make the product easier to make. When eventually the hardware becomes available and it is possible to observe assembly, then desirable changes can be spotted. This is really the continuation of the design process from the drawing board to the hardware. Unfortunately, too often designers do not watch the product being made until a cost reduction exercise is forced on them by a dissatisfied customer.

12.1.2 The need to make the product work

What the designer draws may not be possible to manufacture. When changes make manufacture possible, the product will often not work. This is to be expected, for even the giants of engineering history, Stephenson, Royce, did not get their drawings right first time. This means, however, that design engineers expect to develop a product in the laboratory and build it in experimental workshops [2]. Such a practice can lead to development which will generate changes – perhaps two thirds of the drawings may be changed. The whole process here is fraught with danger unless all changes made are notified to the design office. Too often prototypes and lash-ups are titivated by development engineers and technicians to make a product work in the laboratory, but the changes are not fed back for incorporating in the manufacturing instructions.

12.1.3 The need to improve serviceability

Sometimes the manufacturer is responsible for servicing his own products. This will certainly be the case during any period of warranty, but is also the case when the manufacturer leases his product (copying machines and television sets are examples of this). Even where the manufacturer takes responsibility for servicing, the customer will have down time costs that the manufacturer does not share. If the customer feels that he is losing too much money through poor serviceability, he will take his business somewhere else next time. In other words it is not just the prime cost that the customer is interested in but the ownership costs [3].

12.1.4 The need to improve the life, reliability, and maintainability of the product

Life, reliability, and maintainability are, of course, facets of serviceability. Sometimes the manufacturer is aware of his own costs in servicing, sometimes he does something about it, sometimes he lives with it. Few manufacturers seem to be aware of the profit they can make out of selling life, reliability, and maintainability. The washing machine manufacturer, previously mentioned, believes that he can charge £35 more for a machine than his competitors because he has a deserved reputation for a highly reliable product. It is not so easy to calculate the losses inflicted by a reputation for poor reliability, but going out of business is the ultimate price that may be paid.

Reliability may be defined as the probability that a machine, device, or system will operate for a given period without failure under the specified operating conditions. It is really the extension of quality through time (see Chapter 10). Reliability is therefore concerned with failure; ultimately it can only be measured by observing failure, and this implies that an accurate statement of reliability can only be made after the product under consideration has failed. Reliability parameters can only be obtained by measuring small samples, before product launch, through simulated tests [4]. Hence the reliability of a product is influenced directly by design and development testing, quality control, commissioning and handover, and subsequent maintenance. Finally the need for adequate field defect and performance data feedback becomes imperative, as indicated in Fig. 12.1. Such data can generate further design changes.

Design reviews will also endeavour to improve reliability, safety, and maintenance factors, and will cause additional changes to occur during the actual design process. Design reviews are discussed later in the chapter.

It is becoming more usual for customers to demand contracts which state reliability objectives which have to be demonstrated as a condition of acceptance or type approval. In some cases financial penalties are imposed by the customer for failures in service. Hence maintainability as well as reliability here are influenced by both producer and customer. From the customer's point of view availability is all important; and it represents the ability of a product to carry out its presecribed function without excessive down time [5]. This in turn depends upon reliability and maintainability, and there will be trade-off decisions between these two factors which may well be taken at design review meetings. Maintainability is influenced by the design. For example, the use of alarms, test points, ease of access, built-in redundancy, etc., the human factors – operator skills, training, etc. and the environment of operation – tools, spares holding, logistics, etc. All of these factors can and do affect the saleability of a product, and some can only be ultimately determined over a long period of development which imposes successive incremental design changes. The curve of total ownership cost against availability is shown in Fig. 12.2. There will always be a value of availability corresponding to minimum cost, but this may well occur at a higher availability than the manufacturer's minimum cost [6].

12.1.5 The need to cheapen distribution

From time to time, cheaper methods of transport become available, cheaper storage systems, and new outlet markets. The washing machine case previously

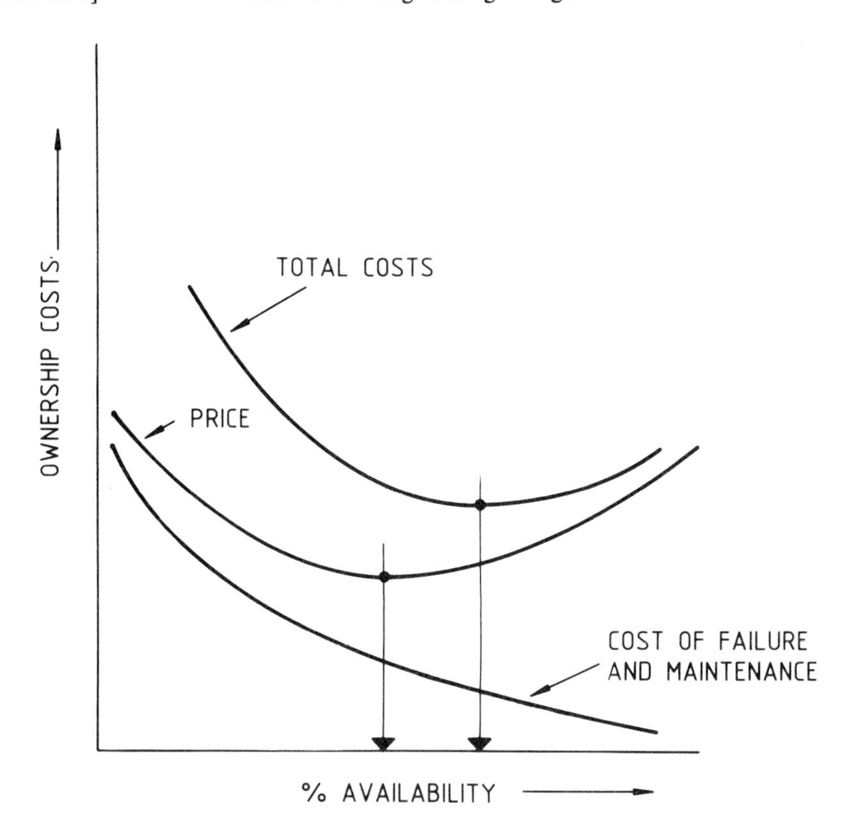

Fig. 12.2 – Total ownership cost *vs* availability.

mentioned was not redesigned to take advantage of the cheaper transport afforded by a 'pick-a-back' system, but the packaging was. This was a finely balanced decision after considerable study and assessment of the more severe accelerations that would be met in the new transport system.

12.1.6 The need for adequate production support

Even when the drawings and the method of manufacture of the product are acceptable it may well be that production support is inadequate and procedures must be changed. If the operative is waiting for a tool to be sharpened, for a component to be supplied, for a jig to be emptied, then his support must be strengthened. Perhaps a new procedure for sharpening more drills is required, or a greater reservoir of sharp drills, perhaps a better production control system to predict and avoid shortages. In some cases a job can even be held up while a payment system is being devised.

Good management will ensure that operatives are encouraged to complain if they believe something is wrong with the job. It is not easy to build feedback from the operative into the reporting system (see Fig. 12.1); probably a formal system will do less to encourage such feedback than a sympathetic environment which may take years to develop [7]. Some companies are moving away from payment by results in the hope of encouraging constructive feedback from the operative. Again, companies are endeavouring to use 'quality control circles' not merely to improve quality but to improve communications. A Q.C. circle in Japan originally consisted of a group of operatives, chaired by a foreman, who discussed the quality of the products made. In the UK and United States the quality control circle has been increased to include members of different specialities, but usually contains a preponderance of operatives.

12.1.7 The need to change the design specification

The customer may change his mind. Experience with the product or with the parent system of which it is a part may lead him to realize that he needs something different. Changing the specification will, of course, be his responsibility, but it needs good records to find out how much his changes will cost him [2]. Also a designer may change the specification because he realizes that he is designing the wrong thing, or the right thing in the wrong way.

So there are many reasons why changes become necessary, and if it is recognized that some change is inevitable, how can the proposed changes be controlled? How can it be ensured that the need for any change is seen, and then defined? How are the value and cost of the change to be calculated and if, after analysis and evaluation, the change is seen to be desirable, how is it incorporated with the least disturbance to the organization?

These aspects of the management of engineering change will now be discussed.

12.2 ORGANIZING, PLANNING, AND CONTROLLING ENGINEERING DESIGN CHANGES

Assuming that a design can be made, how is it possible to reduce the cost of manufacture when the designer and planners are concentrating on their next jobs and the operative is thinking of his bonus? Perhaps this function could be performed by the value engineer or an industrial engineer (the rate fixer will think his job done when he has timed the method, even if the method is poor), but the existence of these engineers is recognition that the designer's job is only half done. Would it not be better to recognize that the first set of drawings and the first set of process schedules will be capable of improvement? Some kind of team may be necessary to 'walk the design through the shops'; a team which will not merely deal with crises but will set itself the task of generating the changes which will reduce manufacturing costs. One American firm does this by using a

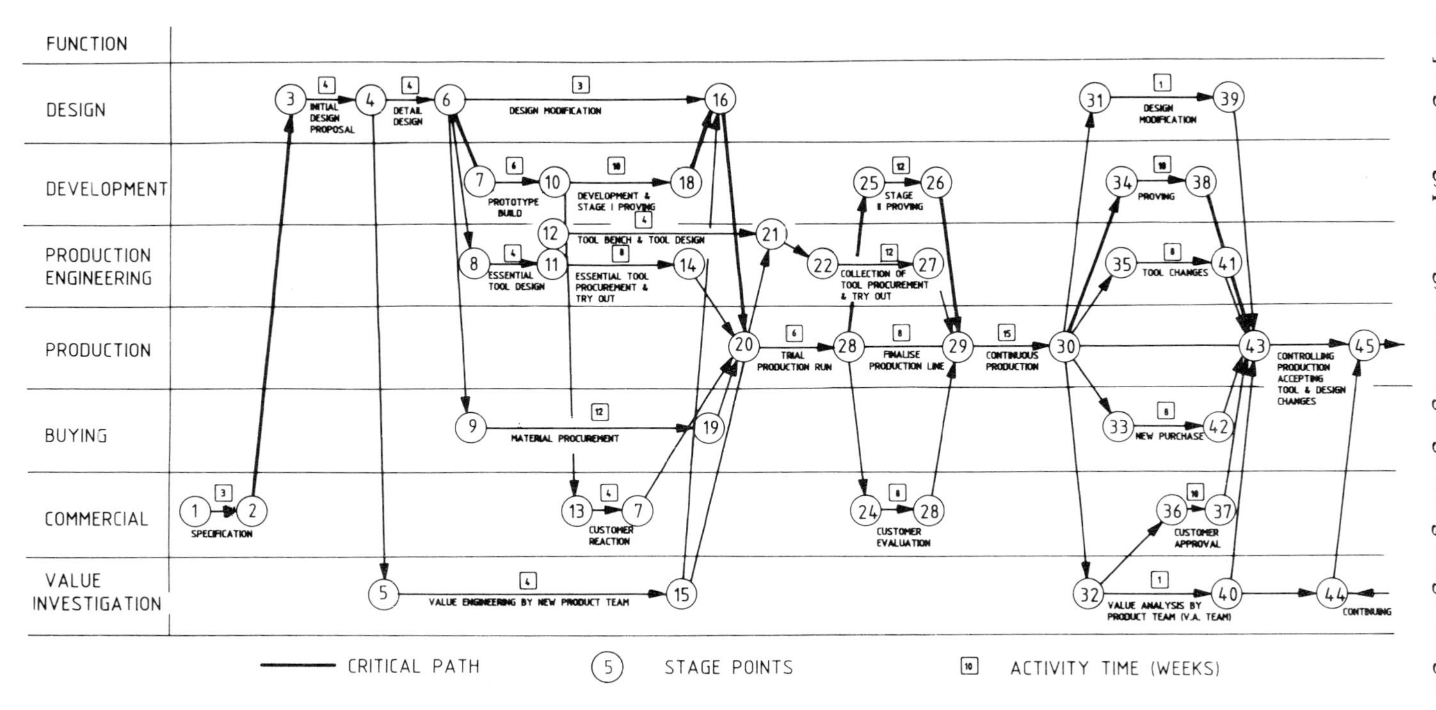

Fig. 12.3 – Critical path network for production introduction.

project team consisting of a designer, a manufacturing engineer, and a marketing engineer; another company uses a larger project team whose members' specialist skills vary with the project (although the team almost always has a designer and a manufacturing engineer).

Such teams prevent the 'compartmentalization' which seems to exist in some companies where the design office forgets the drawings when they have been issued. Nevertheless, such a team generally lives only until the delivery of early production hardware and some effort has to be made to continue to reduce the cost of the hardware as long as it is in production. As continuous production takes place an on-going value analysis approach then becomes necessary.

The whole translation process for introducing a mass-produced product can be depicted on an arrow network which is zoned by functions. Such a network can be seen in Fig. 12.3. Here it will be noted that the work commences with the compilation of a specification and moves to the design department where the initial design is completed and then the process moves to development for Stage I proving. Some modifications will inevitably be generated from development, and may be by the value engineering team or even by the customer or the buying department.

The initial pilot production run may then be undertaken when final evaluaation and complete tool procurement will take place. Here again, further modification may be necessary and could be generated by the tool or production department or the customer. Finally, as production continues modifications may still be required to reduce cost and ease production bottlenecks. These may be generated through new technology, new tooling, or value analysis work.

The question now arises of how to inject a change procedure into such a process without causing excessive disruption. Such a procedure must first evaluate the extent of any proposed change and then the cost and value of the change.

12.2.1 The extent of the change

A change to the design of a product involves much more than a change to the drawings. A change to the method of manufacture requires more than a change to the process schedules. In any case, the change must be defined, not only for the sake of the draughtsman or planner but also to provide a criterion against which we may check whether the requirements have been met [8]. If a change of drawing is required, it is possible that new process schedules will be required, and that production schedules will need to be changed. The change will need to be manufactured, proved, and then supported in the field by servicing instructions and spares. Changing the drawings may well be the least expensive of all these necessary activities. The company that makes changes must clearly have the facilities which permit these activities; a project office defining the change, a drawing office, a planning department, a prototype shop, a development laboratory, a department writing support manuals, and a department which determines spares support policies. There is little doubt that where computer aided design is used great benefits accrue in the area of changes to meet special customer requirements (see Chapter 13).

12.2.2 The cost and value of the change

A change, however small, is as much a new project as designing and building the Concorde, and like any project is only worth undertaking if its value is greater than its cost. Costs vary considerably in the ease with which they are predicted. The direct costs of specifying a change, changing the drawings, planning the change in manufacturing method, manufacturing the changed product, and making changes in the field may be estimated in terms of more or fewer man-hours. Direct costs may also be calculated in terms of the materials consumed, the cost of running test rigs, and the cost of units destroyed in tests. Easy to calculate but difficult to apportion (to each change) are the overheads created by the ability to manage change [9]. Such overheads must take account of the cost of a computer data bank or filing system, of the regular meetings of senior staff (usually classed as indirect labour), of the capital cost of development and test facilities, of the capital cost of manufacturing facilities (would we use N.C. machining instead of sand casting if changes were not expected?), and of the cost of the procedural system which controls changes.

The value of the change may be so great that there is no real need to calculate it. Where the change is made for safety reasons, the cost of not making the change may be virtually infinite and the change may be legally mandatory. Where the change is made to reduce manufacturing or distribution costs, its value is entirely predictable (at least its value per product – the total value may depend on future sales which can only be predicted inaccurately).

Where the change is made to reduce the customer's cost of ownership by improving reliability, improving maintainability, or increasing life, it will be difficult to predict the value of the change unless it has been forced by threats to cancel an order. In less dramatic cases, much experience is required to estimate the value of changes – the company that believed that its washing machine's reputation for reliability was worth $100 in the selling price had twenty years' experience of marketing the machine. Often a change is not one which the accountant will value as readily as he will cost. The accountant will know the interest charges generated by stock holding but be unaware of the losses caused by running out of spares, and similarly, he will calculate the cost of improving quality but be unaware of the profit which will be generated by selling a product with a good reputation for quality.

In some cases it is, of course, possible to calculate the savings that the customer will derive from the change over the life of the product and share the present value of those savings between the supplier and the customer. Often, however, such predictions can only be very approximate.

12.3 ENGINEERING CHANGE PROCEDURE

Having determined the extent, cost, and value of a proposed change, an engineering change proposal (ECP) must be prepared. This will require information from a number of departments, so that the planning and control of the change may be undertaken. The procedure will follow along these lines:

Originating source	– Provide description and reason for the projected change.
Engineering records	– Assign number, clarify description, add necessary drawings, etc.
Design/Development Engineers	– Plan technical evaluation and establish technical processing.
Design	– Define and detail change.
Development	– Review, clarify, and recommend.
Sales	– Provide service information.
Production	– Review and provide effectivity (point at which a change can be inserted and becomes effective).
Change committee	– Review – the change committee acts as a 'gate' accepting, rejecting, or calling for more information on the E.C.P.

To aid the work of the Change Committee and make sure that all aspects of the change have been considered, two valuable techniques may be used:

(A) Categorize the change
(B) Use the critical evaluation form.

Most firms find that three broad classes of changes will cover their requirements:

Class I changes	– Those affecting interchangeability of units, spares, etc. Those affecting performance or safety. Those affecting price (or contractual particulars).
Class II changes	– Those which do not fall into the category above but are covered by drawing changes or parts list numbers. They could be tolerance amendments unrelated to interchangeable fits, material changes unrelated to physical properties, or changes in process sequence.
Class III changes	– Those not covered by I and II but cover minor alterations to documentation to correct errors or ambiguities.

Some companies may prefer to define change by categories based on the purpose of the change, and these may be superimposed on the first broad classes of changes. Examples of these second class of changes are:

Mandatory changes – For reason of safety or to permit a product to meet the performance specification.

Record change – A change to correct a drawing, specification, or parts list.

Cost reduction changes – A change to reduce the cost of the product or component.

Routine changes – All other changes not covered by the above.

The first category of change passes without waiting for a cost evaluation. Likewise with the second, but for the cost reduction category there has to be a verfication that a cost reduction potential actually exists before it is passed by the change committee.

The critical examination sheet of work study fame is shown in Fig. 12.4 and is useful for evaluating a change. The secret is to demand a written answer to each question in turn. This enables the design/development engineers or project engineers to think around the change and ensure that all factors have been adequately considered [10, 11].

Armed with detailed information about the change and its proposed categorization a comprehensive engineering change proposal can be drawn up which makes the work of the committee much easier.

The class of change will determine the programme of incorporation.

Class I changes require approval by the customer, and such approval may be ratified by a contract amendment.

Class II and III can be covered by the change committee without troubling the customer.

A typical engineering change proposal (ECP) for an automotive product can be seen in Fig. 12.5. It will be noted that the costs incurred are clearly indicated.

If the Committee after due consideration passes the ECP, the change procedure might be as follows:

Engineering records – Prepare engineering change note and parts list.

Design – Prepare final drawings, schedules, parts lists, etc.

Development & test – Where necessary carry out substantiation tests to ensure that the revised part is at least equal in quality, durability, and performance etc. to the original.

Customer – Give final approval if required.

Change committee – Release and issue the engineering change note (ECN).

Production control – Issue a change effectivity report for engineering records stating from which model, unit, product the change has been incorporated.

A typical ECN is shown in Fig. 12.6.

DESCRIPTION OF ELEMENT:

	The Present Facts		Possible Alternatives	Selected Alternative for Development
PURPOSE	WHAT is to be achieved? Consider element in isolation. (Note: What is not how or why.)	WHY do it? Put down all valid reasons.	WHAT else will be achieved? The answer here should never be nothing. There are three main alternatives: (a) Non-achievement. (b) Part-achievement. (c) Avoid the necessity for achievement. Expand these.	WHAT should be achieved? Divide into short-term and long-term. Under long-term might go suggestions for future research, customer education etc. The economics of the situation must be borne in mind throughout.
MEANS	HOW is it to be achieved? Tabulate information: (a) Materials employed. (b) Equipment employed. (c) Consider safety.	WHY do that way? All reasons should be given Tabulate if necessary under each heading.	HOW else could it be achieved? Consider *all* conceivable alternatives for each main heading.	HOW should it be achieved? Each heading should be considered first in isolation then collectively. The selected items should then be gathered together to produce the *best, safest* and *cheapest* method.
SEQUENCE	WHEN is it to be achieved? What are the previous and subsequent significant activities and what are the time factors involved?	WHY then? Give the significance of the sequence.	WHEN else could it be achieved? Again consider all alternatives.	WHEN should it be achieved? Bear in mind economics.
PLACE	WHERE is it achieved? If appropriate give location of site or factory.	WHY there? What governs the location.	WHERE else could it be achieved? Consider all alternatives.	WHERE should it be achieved? Answer may be given in relation to some other activity. Consider limitations and cost of building design etc.
PERSON	WHO achieves it? Grade of worker or workers, designation, title etc.	WHY that team or person? Give reasons for this team or group or individual.	WHO else could achieve it? Consider all alternatives.	WHO should achieve it? It may not be possible at this stage to select an individual or a group without further discussion elsewhere.

Project Engineer. . . . .

Date: . . . .

Fig. 12.4.

ENGINEERING CHANGE PROPOSAL

<table>
<tr><td>DATE</td><td colspan="2">PROJECT NO.</td><td>ECP NO.</td></tr>
<tr><td colspan="3">TITLE</td><td>CLASS OF CHANGE</td></tr>
<tr><td>ITEMS CONCERNED</td><td>AMOUNT £</td><td>RELATED EXPENSES</td><td>AMOUNT £</td></tr>
<tr><td>CAPITAL TOOLS</td><td></td><td>DEVELOPMENT</td><td></td></tr>
<tr><td>MACHINERY & EQUIPMENT</td><td></td><td>PRODUCTION</td><td></td></tr>
<tr><td>EXISTING TOOLING, JIGS, FIXTURES</td><td></td><td>PREPRODUCTION</td><td></td></tr>
<tr><td>TEST EQUIPMENT</td><td></td><td>SPARES</td><td></td></tr>
<tr><td>TOTAL</td><td>£</td><td></td><td>£</td></tr>
<tr><td colspan="4">REASONS FOR CHANGE:
1. DESCRIPTION
2. DISCUSSION</td></tr>
<tr><td colspan="4">IS STOP-WORK ORDER REQUIRED? YES NO
IS EFFECTIVITY REPORT REQUIRED YES NO
ARE PARTS COMMON TO SEVERAL MODELS YES NO</td></tr>
<tr><td colspan="3">RECOMMENDED FOR APPROVAL</td><td>APPROVAL/REJECTION</td></tr>
<tr><td colspan="3" rowspan="4">ESTIMATED ENGINEERING MAN HOURS TO MAKE CHANGE
ESTIMATED EFFECT ON PRICE
ATTACH ALL SUBSTANTIATING DATA</td><td></td></tr>
<tr><td>DATE</td></tr>
<tr><td>RELEASED</td></tr>
<tr><td>DATE</td></tr>
</table>

Fig. 12.5

ENGINEERING CHANGE NOTICE

<table>
<tr><td colspan="4">DATE</td><td colspan="3">PROJECT NO.</td><td>ECN NO.</td><td colspan="2"></td></tr>
<tr><td colspan="7">TITLE</td><td>CLASS OF CHANGE</td><td colspan="2"></td></tr>
<tr><td colspan="10">DISPOSITION OF PARTS</td></tr>
<tr><td colspan="4">PARTS IN HAND</td><td colspan="2">IN PROCESS</td><td colspan="2">ASSEMBLED</td><td colspan="2">SHIPPED</td></tr>
<tr><td>PART NO.</td><td>USE</td><td>REWORK</td><td>SCRAP</td><td>USE</td><td>SCRAP</td><td>REWORK</td><td>USE</td><td colspan="2"></td></tr>
<tr><td></td><td></td><td></td><td></td><td></td><td></td><td></td><td></td><td colspan="2"></td></tr>
<tr><td></td><td></td><td></td><td></td><td></td><td></td><td></td><td></td><td colspan="2"></td></tr>
<tr><td></td><td></td><td></td><td></td><td></td><td></td><td></td><td></td><td colspan="2"></td></tr>
<tr><td colspan="10">ESTIMATED EFFECTIVE DATE ________________
EFFECTIVITY REPORT REQUIRED YES NO
PARTS LIST AFFECTED ____
DRAWINGS AFFECTED ____</td></tr>
<tr><td colspan="10">DESCRIPTION OF CHANGE</td></tr>
<tr><td>ITEM</td><td colspan="9"></td></tr>
<tr><td></td><td colspan="9"></td></tr>
<tr><td colspan="4"></td><td colspan="6">WRITTEN BY:__________DATE:__________
APPROVED BY:____________________
PROJECT ENGINEER</td></tr>
</table>

Fig. 12.6

ORGANIZATION

In order to consider an appropriate organization to carry through such a procedure it is helpful to construct a critical task analysis so that action sequence can be charted. A typical example of such a task analysis is shown in Fig. 12.7 for an automotive part for which the failure rate has been high in service.

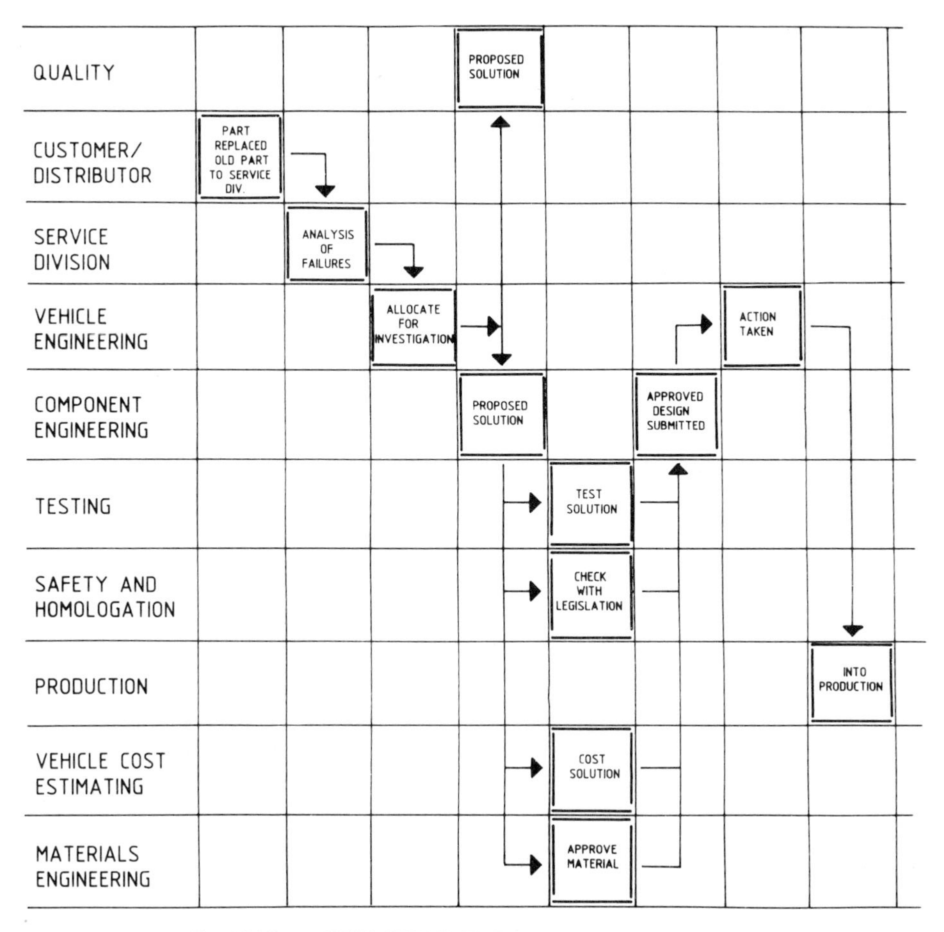

Fig. 12.7 – A critical task analysis.

The change committee is charged with the responsibility of final approval or disapproval of any change generated. The committee should be composed of personnel drawn from engineering, manufacturing, sales, quality and reliability, and safety, who must be senior people having authority to take the necessary decisions. The representatives must have a good technical understanding, and the committee should be chaired by a project manager or, in the case of a large contract, by the customer's representative.

The committee really forms an arm of general management, and as such has the authority to sanction expenditure up to a certain limit. If a change cost exceeds this threshold the appeal must be made to top management for granting additional funds, and if necessary a change to the contract. For this purpose a comprehensive case for the change incorporation must be raised which will include, in the case of a mass-produced item, a rate of return which represents the rate at which the investment is recovered by the cash proceeds to be received in future years of production. When calculating annual savings all the cost factors concerned must be included, namely piece cost, warranty effect, complexity effect, obsolescence cost. There should also be added to any investment the necessary administrative cost per change release.

Having considered the reasons for generating design changes and how these design changes may be incorporated effectively and efficiently, it is necessary to consider the use of design reviews in the management of change.

12.4 THE DESIGN REVIEW

Design is a multidisciplinary exercise, and while the designer himself will be a generalist, specialist knowledge must be available to him. Design reviews are one way of providing a communication forum for the specialist concerned with a particular design effort.

Innovation is an important part of design, but most designs evolve as a series of modifications from an existing solution to the designer's problem or from a first innovative solution. The sequence of changes which occurs from the first design scheme to the set of details which define the product which is finally sold at a profit must be carefully managed to ensure that it is directed towards profit. Again, design reviews are a contribution to such management of change. Innovation is important because the first design provides limits beyond which it will be impossible to reduce cost or improve performance or quality. Nevertheless, experience shows that a good innovation can be improved by modification (an example is the ability to reduce the cost of an economic aeroplane even after the thousandth has been made).

12.4.1 The main factors to be considered in engineering design

While there are many other factors to be considered in every engineering design, four are paramount as far as design reviews are concerned: Function, Safety, Economy, and Effect on Environment.

Function – The artefact as designed and produced must fulfil its intended function over its desired life span. Functions may be very complex. For example, a structure may have to carry a specified load but at the same time be watertight and fire resistant. Moreover, the design load may be variable over a wide spectrum.

Safety – Most engineering design concerns the public, whether or not they know it, and the lives of the public often depend on the engineering designer's work. The consequences of failure can be very grave. The designer's estimate of adequate safety must be as rigorously analytical as possible, but will often be partly a matter of judgement: he should seek to evaluate his designs by appropriate tests on prototypes and models.

Economy – Generally the product must be produced at the least possible cost, except in dire emergencies. The designer seeks to give the best value for money, so that design decisions are often based on economic considerations right down to the smallest detail. But also, today, the designer must consider the balance between prime cost and operating and maintenance costs: with ever increasing labour costs these latter may be crucial factors [12, 13].

Effect on environment – Asethetic and ergonomic factors must be considered in any engineering design. The emphasis placed on appearance will vary with the nature of the product. Thus, for a domestic appliance, appearance would be a key factor, whereas in the design of a radar station appearance does not matter so much. Since aesthetic and ergonomic quality are inseparable from functional and material decisions, it follows that they must be considered at the beginning of the design process and not added as cosmetics at the end.

12.4.2 The process of generating a design

Engineering design covers a wide field, varies from product to product, and the process depends upon the degree of innovation involved. The basic system into which any engineering design must fit involves both men and general environment. Fig. 12.8 depicts the total system life cycle for a product or project in which it will be noted there are four distinct phases, Planning, Acquisition, Use, and Retirement. In the acquisition phase, design and production takes place, as previously stated in Chapter 5.

Engineering design is an iterative problem solving process. Progress is made in a cyclical manner with constant feedback taking place as the design is defined and delineated.

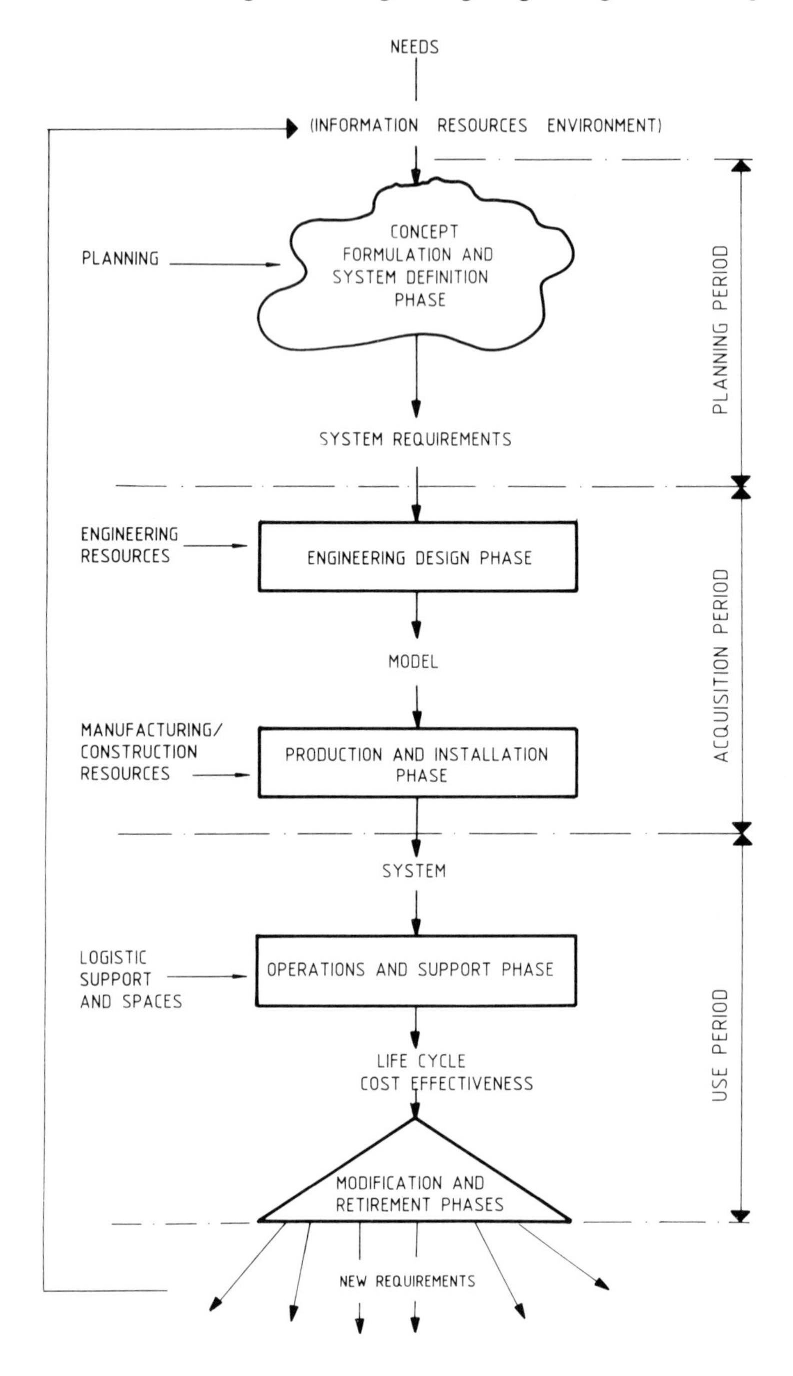

Fig. 12.8 – Total system life cycle.

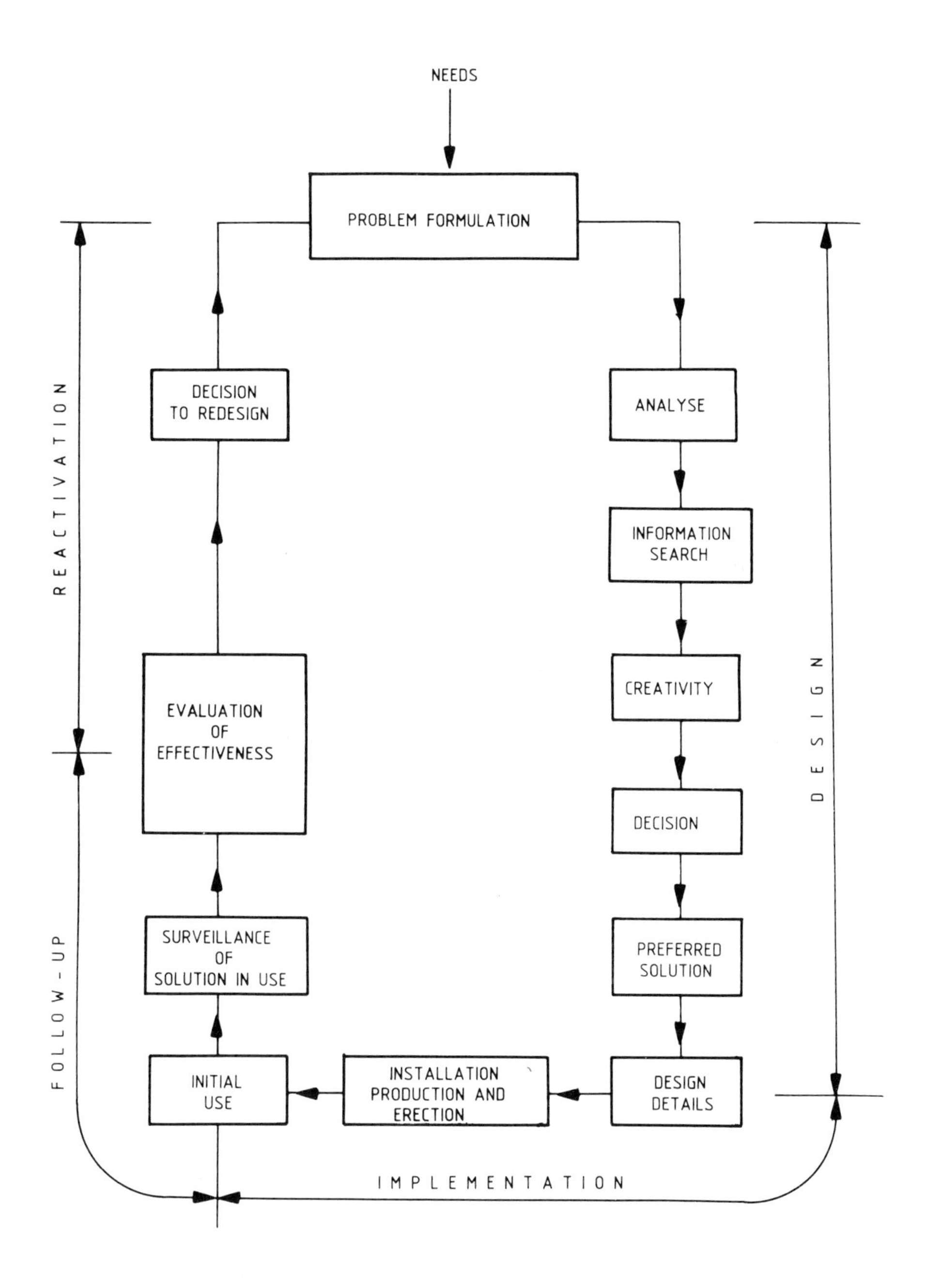

Fig. 12.9 – The design cycle.

The essential steps through design cycle have been represented in various ways, and Fig. 12.9 depicts one of the these. At all these stages iterations take place.

First, the need has to be identified and translated into engineering requirements. Possible ways and means of solving the engineering problem posed then have to be generated, and the best solution chosen must have regard to all the circumstances prevailing. The detailed design brief may then be written and work started on the realization of the preferred solution. For this purpose engineering analysis and development work will be necessary. Economics will generally dictate that a mathematical model is built in the first instance, and later on 'lash-up' full scale models and prototypes to validate that the realization is satisfactory.

The final stage is communication of design intent to another group of people to achieve production and construction of the chosen and proven solution.

Design management is concerned with managing this translation process which, it will be noted, starts off with verbal statements and is then gradually refined into specific numerical quantities and finally into a special engineering stylistic language enabling 3D objects to be depicted on a 2D surface by means of engineering drawings. The design message will often need to be augmented by the use of other media such as models, photographs, written documents, and punch tape, etc. The whole process is a team effort requiring the assistance of many specialists.

12.4.3 The role of the critic in engineering design

Critics in the realm of art are numerous; they tell us what is good and bad about new books, paintings, music, sculpture, and other objets d'art.

The critic, to be useful, has to be objective and catalytic. Critics perform four main activities when executing their work: analysis, comparison, quotation, and judgement [14]. The analysis includes an objective description of the work and defines the basic problem that is solved together with an appraisal of the strategy used. Comparison is then made between the limitations of the form and the achievement, and between this achievement and other comparable design work. Quotation is perhaps his great stock in trade; he quotes from other designers and their works. Finally, the critic judges the workmanship and the total effect.

The art critic differs from the design critic in several important ways. The artist is usually free to chose his subject, whereas the engineering designer has to satisfy real needs. A work of art is only criticized when it is finished, but the engineering design team must defend their effort at all stages of the realization process. Lastly, art criticism is an open-loop process and may not affect the originator directly, but the engineering design criticism impinges directly on those who carry out the design, and it is they who must justify all decisions taken. It is this interaction between a design team and their critics that should generate the best solution to a particular piece of design, and this can be formally achieved by design reviews.

The role of the critic can be visualized by referring to Fig. 12.10 where it will be noted that his criticism starts from criteria which stem from the design brief and possibly criteria generated during the actual design process. By drawing on their experience and knowledge (memory bank) and using their judgement and comparing with other designs, they generate informed criticism which has to be examined by the design team.

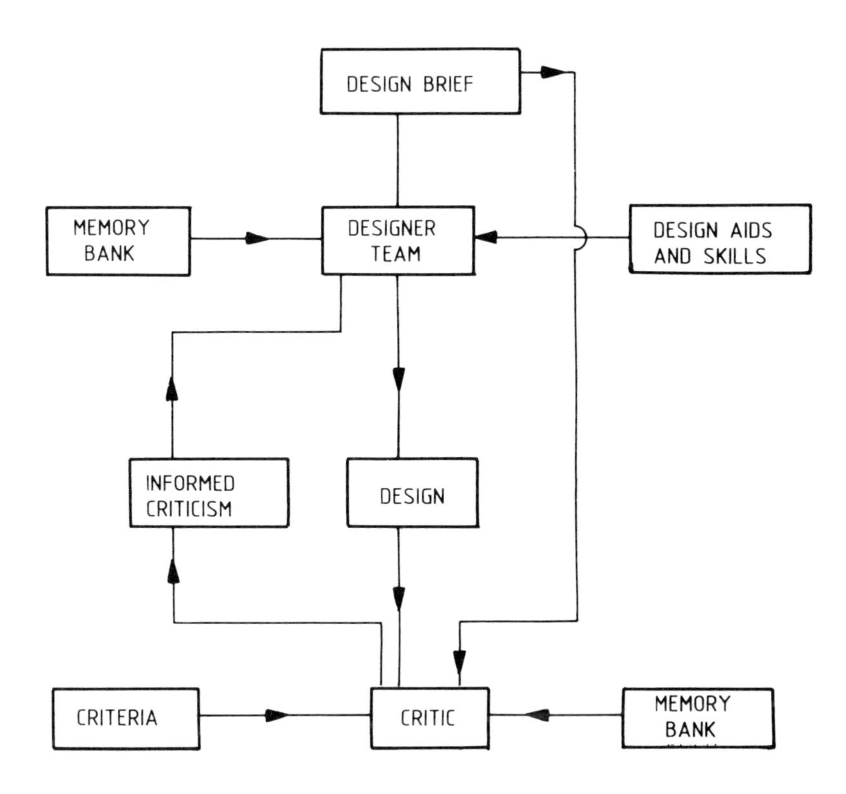

Fig. 12.10 – The role of the critic.

12.4.4 The elements of a design review

A design review is a formal synergistic, documented, and systematic study of design by users and specialists who are not directly associated with realizing the design. It is a management technique for ensuring that an independent evaluation of a product or project is achieved prior to its final handover to a customer [15, 16].

The basic objective therefore of any design review programme is to provide assurance to the customer and the supplier management, by formalized documentation of the decision logic, that the design satisfies the specification of requirements. It must also ensure that the proposed design is producible, bearing in mind the capability and capacity of the production facilities, and that it can be tested against the requirement.

It will also cover such aspects as the scaling of spares and repair policy. If conducted properly a review will, by its very questioning approach, reveal hidden snags and defects.

In view of the Health and Safety at Work Legislation the design review may be looked upon as a product liability preventer. There are, of course, other liability prevention tools such as hazard analysis, failure mode and effects analysis, as well as quality control and reliability methods. But the formal documentation and interrogation process involved in a design review ensures that companies maintain an accurate record of the design assumptions and criteria throughout the design process.

To achieve success, designs must be reviewed by engineering, manufacturing, marketing, purchasing, quality and safety people, and other specialists, as a group and not in series.

It must be stressed that a design review is not merely a 'design check' absolving the designer of all responsibility. A review is to ensure that the design is as soundly based as possible, but it must not become design by committee.

It is now becoming accepted that some assurance is required that new equipment or plant does not contain features that could lead to operational failure, and minimum life cycle costs are becoming vital for all customers [4].

12.4.5 The mechanics of conducting a design review

Companies should, at this time particularly, consciously seek to conduct design reviews for all their future products and also on any major 'face-lifts' of existing products where cost, performance, interchangeability, or appearance are concerned. Other modifications will be handled by the modification control procedure already discussed.

The exact procedure for conducting a design review will vary according to the nature of the work undertaken by various companies. However, there are some general steps which will be common to all design reviews. These concern the selection of a Chairman, identification of products or projects for review, schedule of meetings, choosing participants [16, 17].

The review Chairman

He or she is responsible for convening and scheduling the meetings, issuing an agenda, minutes, and final report and, most important of all, ensuring that the recommendations are carried out. The review chairman must have a broad understanding of the design requirements and an adequate knowledge of the technology involved, and may be selected from design, system or project management departments as appropriate to the review subjects, but should not be directly associated with the design or design team. The chairman must seek to be 'impartial', and his knowledge of a particular product or project will come from the prior circulation of specified documents.

Design review participants

The types of people who will attend any design review will vary with the nature of the product or project and with the type of review being undertaken. There will always be a core group whose responsibilities cover the administration and chairing of the meetings and the actual designers who prepare and substantiate their design decisions and methods. The preliminary design review (PDR) will survey the concept or proposed design solution. This will be looked at by reliability and safety experts, etc.

Other members will be needed for the intermediate (IDR) and final review meetings (FDR). Some typical examples are set out in Table 12.1.

Dates for the reviews

Specified times for design reviews will need to be incorporated into the product/project design and development schedule. Such a schedule will depend upon the significance of the particular product/project under consideration, the complexity and the amount of previously proven components and sub-assemblies used in the design [18].

Generally, reviews can be grouped into milestone categories which are applicable to any design/development effort such as:

- The preliminary design review – concept or proposal
- The intermediate design reviews – prior to preparation of detail drawings
- The pre-release design review – prior to release for pilot production
- The final design review – prior to the start of full production.

There may well need to be several intermediate design reviews depending upon the type of product or project under consideration. There may also need to be a final acceptance review in the case of projects prior to the delivery and handover to the customer. With products this may take the form of an improvement design review. Taken over the product life cycle, these reviews will be spaced as indicated in Fig. 12.11, and for a project can be represented on a generalized milestone flow chart of the design development effort (see Fig. 12.12).

Information and data

At all review meetings it is imperative that the necessary information and data are distributed to all participants at least ten days before the meeting is convened. Such information will differ in specifics, depending on the milestone at which

Table 12.1
Design review group responsibilities

FUNCTION OF GROUP MEMBER	RESPONSIBILITIES	TYPE OF DESIGN REVIEW		
		PDR	IDR	FDR
Chairman (often project manager)	Convenes and conducts meetings and issues reports	✓	✓	✓
Design engineer(s)	Prepares and presents design and substantiates decisions	✓	✓	✓
Reliability	Evaluates design for the specified goals, **MTBF**, etc.	✓	✓	✓
Quality control	Ensure that functions of inspection and test are adequately conducted		✓	✓
Manufacturing engineers	Ensure that design is producible with minimum cost and time		✓	✓
Operational engineer(s)	Ensure that installation, commissioning and maintenance considerations are taken into account in the design	✓	✓	✓
Product safety	Assures that design is able to meet all known regulations	✓	✓	✓
Packaging and Shipping	Assures that the product is capable of being handled without damage		✓	✓
Materials engineers	Ensure that materials selected will perform satisfactorily and review possible alternatives		✓	
Purchasing engineers	Assure that acceptable bought-out items and materials to meet cost and time target are obtained		✓	
Design engineers not associated with the design and consultant specialists	Evaluate overall design to ensure that performance, cost, time, aesthetics, ergonomics, etc., are satisfactory.	✓	✓	✓
Marketing	Assure that customer requirements are fully stated and understood	✓		

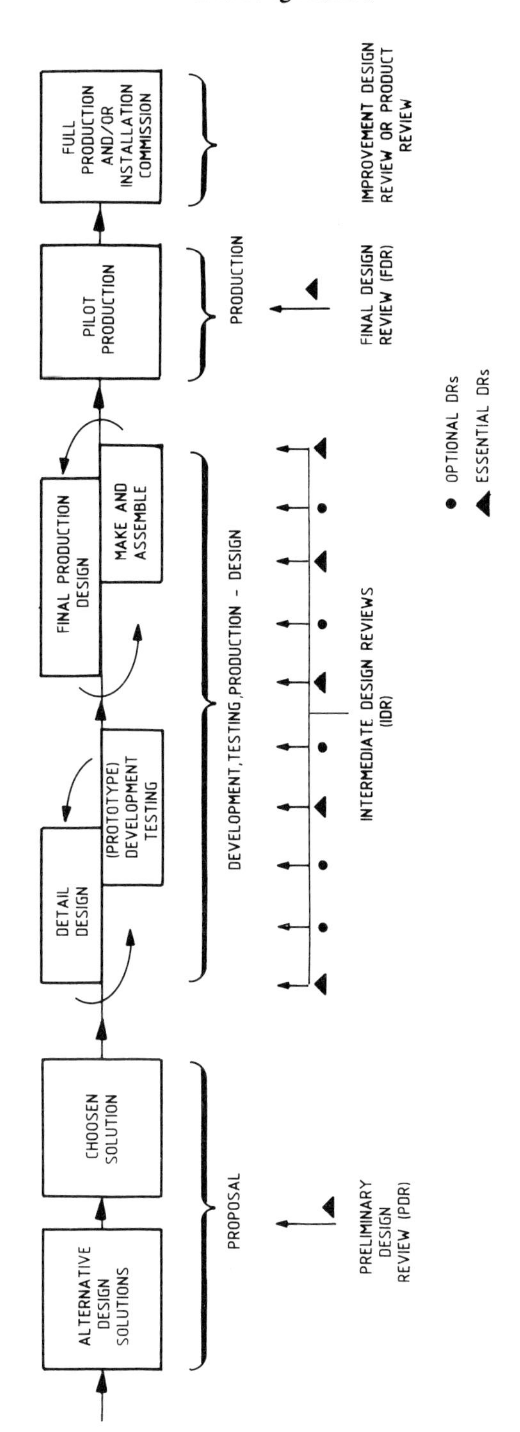

Fig. 12.11 – Product design review schedule.

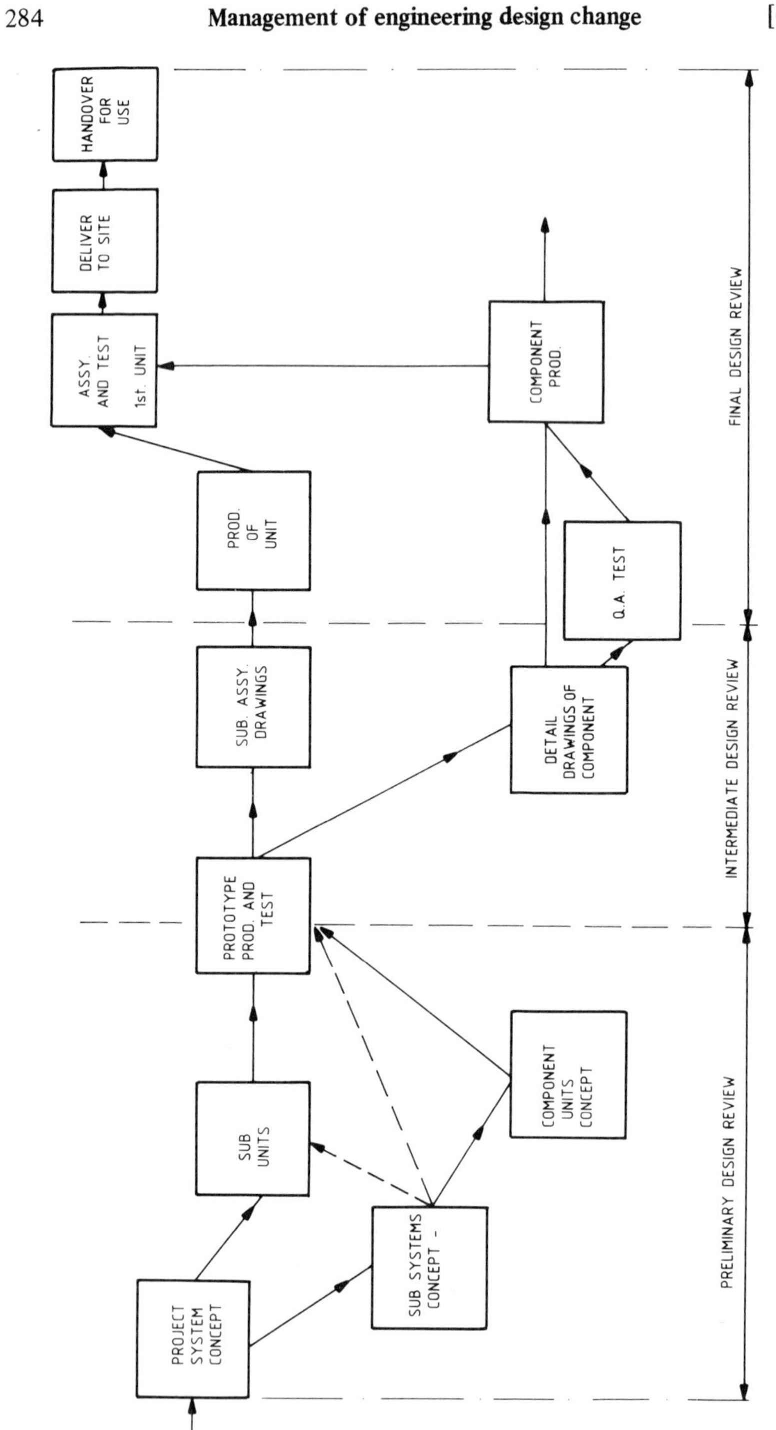

Fig. 12.12 – Project design review schedule.

the review is being taken, but it must define the item being reviewed, set out its requirements, and describe any interfaces with other items. Such data might normally be thought of as layouts, drawings, specifications, anticipated customer needs, cost data, test reports, malfunction reports, relevant QC analysis, and any other pertinent data.

A typical component review for a company made part might include the following:

- Detail drawings – pictorial representation, description of materials, finishes, dimensions, tolerances, fabrication and assembly instructions.
- Installation drawings – General configuration, any attaching hardware, spatial location of mountings, etc.
- Component specification – functional characteristics, test requirements to validate that parameters are met.
- Parts & materials lists – Setting out specifications and class of items.
- Reliability analyses – Failure mode/effect analyses.

The review meeting itself

The organizer of the review meetings should contact each participant concerning the timing and procedure of the meetings. If the process is new to some participants it may help to have a preliminary orientation meeting to explain how the proceedings will be conducted and what will be expected.

The chairman must set the scene and state the specific objective to be achieved and relate this to the overall objectives of the product and project.

Typical items that might be included on the agenda of the various types of review meetings (see Figs. 12.11 and 12.12) are given in Table 12.2.

It is vital to establish the relative importance of each factor in the design early at each review meeting so that appropriate trade-offs may be achieved between reliability, cost, performance, delivery date, appearance, and capability to withstand a particular environment, etc. Each product or item of a project is different, and the combination of crucial factors may well change, depending on the product itself and on the prevailing market conditions.

After the preliminary introduction the design engineers or product manager should describe the product, after which open discussion takes place. It is the responsibility of the chairman to make this flow in a systematic and logical way as laid down in the agenda. Here it is considered useful to have a check list which sets out all matters of product aspects that must be studied at the design review. It has been found helpful if such check lists have a column demanding a YES/NO answer similar to the one shown in Table 12.3. It will be noted that if the answer is No, then the comments column must be filled in, stating who will act and when. It may mean that as a result of design interrogation an investigation has to be undertaken after the review meeting. These are generally the

Table 12.2

Typical items on agenda of various types of design review meetings

PRELIMINARY (PDR)	INTERMEDIATE (IDR)	FINAL (FDR)
1. Marketing describes customer product requirement	1. Review product requirements specification	1. Review product requirement specification
2. Discuss marketing requirements	2. Review factor priority previously established	2. Review factor priority for the product
3. Establish factor priority (ergonomics, performance, reliability, cost, aesthetics, etc.)	3. Discuss mechanical/ electrical design and test results and calculations etc.	3. Discuss validation tests (environmental, etc.)
4. Review design and development plan	4. Discuss civil design layout and detail (if applicable)	4. Discuss commissioning and handover – (special tools and spares)
5. Discuss work load schedule	5. Discuss tooling and packaging	5. Discuss handbooks and manuals
6. Produce summary report	6. Discuss erection and installation	6. Produce summary report and design certification
	7. Discuss handbooks for maintenance and instruction manuals	
	8. Discuss cost estimates and time schedules	
	9. Produce summary reports	

Table 12.3
Design review check list

	YES	NO	(ANSWER 'NO' WITH YOUR COMMENTS)
General requirements: Are all requirements defined sufficiently to permit the design to be completed?			
Are the system specifications up-to-date?			
Are interface specifications up-to-date?			
Are design specifications up-to-date?			
Are repair charts up-to-date?			
Has interchangeability been adequately considered?			
Are test specifications up-to-date?			
Are all specified requirements known to be met because checks and tests have been carried out?			
Are any specified requirements known not to be met?			
Are all the techniques and components used in this design proved through previous use?			
Has the MTBF requirement been specified?			
Function: Have the following effects been evaluated: Changes within specified limits of: Power supply parameter values Environmental conditions Ageing Variation in component tolerances			
Have all circuits in which special attention is required to quality of insulation been shown by theoretical and practical investigation to be satisfactory under all intended conditions of use?			
Have test accepted tolerances been adequately considered?			
Has the provision of special test methods been considered: Factory Field?			

most fruitful results of the early design review meetings, and their results must be carefully evaluated. It may well be that such investigations reveal a new method of tackling a design problem or result in the least liability-prone design.

12.4.6 The benefits from design reviews

Experience has shown that where such design reviews are conducted properly the following benefits have accured:

- Better understanding of the design function and its management.
- Cost improvements – tooling, fabrication methods, materials, finishes, are perfected. Fewer late changes.
- Earlier delivery dates – the design/development cycle is shortened.
- Design capabilities enhanced – designers allowed to respond to constructive criticism.
- Design data base improved – data sheets on manufacturing processes and methods, cost of production, etc.
- Availability of research data – results are put into a form which can be understood by designers.
- Proper product/project logistic support – repair and spares scaling policies have to be clear and specified in detail early in the design process.
- Standardization improved – standard items have to be considered early.
- Quality assurance – good feedback from Q.A. and proper defect analysis ensures better overall designs.

Taken overall, design reviews ensure that proper balanced consideration is given to all factor requirements and that the eventual design will meet the customer's requirements. It should also satisfy senior management that the best information is being used on their latest design work, and that the quality of the design team is continuously being improved.

12.5 CONFIGURATION MANAGEMENT

A further problem associated with the management of change occurs because of equipment complexity which has greatly increased over the years. It is now imperative to define what is made by drawings, specifications, and other documents in more and more detail with the advent of the use of more semi-skilled personnel. Moreover, there may be product variants to meet specific classes of customers. In such cases it is necessary to define the baseline design or lowest common denominator design on to which such variants may be fitted. This is particularly true of work in the automotive and defence fields.

In practice there are so many people whose activities impinge on the design process that their task becomes difficult and changes to the design inevitable, because of the following:

Equipment interface problems
Errors on drawings
Design problems in meeting requirement
Production difficulties
Changing requirements
New markets of operation
Field tests feedback
Environmental problems
User in field problems
New customers

A simple definition of configuration management is:

> To define a base design to meet a customer requirement – control changes and ensure that manufactured equipment is to the correct standard.

Most products or projects today, and, for that matter, any involving advanced technologies, tend to interface with others in a complex way.

People, therefore, often have problems which are not understood by others. Communication is difficult. Configuration management provides a common language which helps to solve some of these problems of communication. It uses factual data, defines what is, and ignores how we got there. Pure logic can be applied without relying on the jargon of each discipline.

12.5.1 Project configuration management

It is essential to consider the role of configuration management at the start of the project, since one of the major problems at this stage is that of defining the requirement. This, at highest level, may be either vague or specific, but must almost inevitably be subdivided into separate requirement specifications which need to be controlled. It is usual to set up a configuration control office or secretariat which has the responsibility for custody of the many documents associated with the design and configuration management task; and to do their job, they must have executive authority delegated to them by the project manager.

Such a configuration control office is responsible for the following kind of functions:

Library of requirement specifications
Release of drawings for manufacture of development models
Release of frozen drawings for production manufacture
Control of changes and re-issue
Collation of design proving evidence in support of design certification.

The manufacturing organization has the responsibility of building equipment to the defined design and of documenting deviations as issued by the configuration control office (CCO).

12.5.2 Software control

An increasingly important aspect of controlling engineering change concerns software work where it is all too easy to make hardware changes without getting the necessary program changes. It is worth noting here the similarity between the different aspects of 'Registration' and 'Design chill' and of 'Bonding' and 'Design freeze'.

SOFTWARE REGISTRATION	HARDWARE 'DESIGN CHILL'
Handover program to project library.	Drawings to CCO for release for manufacture of design proving models.
Program (unit) tests complete.	Early development model tests complete. (A models)
Changes possible but subject to some controls.	Controlled introduction of changes by action of a modification committee.
Some confidence in the program now justified.	Some confidence in the basic design.
'BONDING'	**'DESIGN FREEZE'**
Upgrading status of program previously registered.	Incorporation of all changes found necessary during development.
Extensive integration testing has been done.	Equipment integrated into system and thoroughly tested.
Further changes subject to rigorous assessment and control.	Changes only allowed through formal modification committee.
Only 'Bonded' software can be 'released'.	Accelerated programmes demand production before design of system is fully proven.

SUBJECTS FOR DISCUSSION

(1) Does a formal design review procedure reduce the responsibility of the designer?

(2) What is the difference between a technical audit and a technical review?

(3) Discuss, from the point of view of making a profit, the relative merits of new designs and changes to old designs.

Is innovation useful only with new designs?

(4) How are design changes generated?

SOME USEFUL REFERENCES

[1] *Research report* by Dilip Das, Department of Management Science, University College of Swansea.
[2] *Design aspects of terotechnology – policy & practice.* Department of Industry. HMSO.
[3] *Management aspects of terotechnology – life-cycle costing.* Department of Industry. HMSO.
[4] *Reliability engineering* by D. J. Smith, Pitman 1972.
[5] 'Reliability reviewed' by Prof. A. D. Carter. *Proc. I.Mech.E.,* 1979, **193**(4).
[6] *Maintainability engineering* by D. J. Smith and Alex H. Babb, Pitman, 1973.
[7] *Engineering management and administration* by Val Cronstedt McGraw-Hill Book Co., 1961.
[8] 'Handing over design to the works' by S. C. Dunn and F. J. Adams, *The Radio & Electronic Engineer,* **34**(4) October 1967.
[9] 'The management of cost improvement programmes' in the engineering function by A. E. Swain, *ASM Manual,* January 1975.
[10] *Cost reduction in industry,* O.E.E.C. 1961.
[11] *Value engineering in manufacturing.* American Society of Tool & Manufacturing Engineers, 1967.
[12] 'Designing for improved value' by G. Jacobs *Engineering* February 1980.
[13] 'The specification of quality' by M. E. Bond. *Engineering,* March 1981.
[14] 'The role of the critic in electronics design' by S. C. Dunn, *Electronics & Power,* Journal of the IEE November 1965.
[15] *The design audit – a management tool for 1980s* by B. Jeffries and D. Westgarth, C.M.E. March 1981.
[16] 'Design review as a means of achieving quality in design' by M. A. Napier, *Engineering,* March 1979.
[17] 'Design review a liability preventer' by R. M. Jacobs, *Mechanical Engineering,* August 1975.
[18] *Design review for space systems* NASA. SP-6502. 1967.
[19] Design Audit, B. T. Turner, Engineering Design Education Design Council, Autumn 1984.

CHAPTER 13

The computer aided design and draughting scene

Chapter 13 gives an overview of the uses complexity, and costs of computer aided design systems.

13.1 INTRODUCTION

The correct use and application of CAD is vital to the success and continuance of British industry.

There has been much activity on the CAD/CAM scene over the past few years. Aggressive sales campaigns, conferences, and seminars have all supported the belief that the new technology will enhance organizational performance. Extensive Government support for the introduction of CAD technology in the from of the CADTES schemes† has also been given, but the number of real experts is still low. Hence any prospective customer often does not know what is required in terms of technology and organizational change, and too often the sales force is unaware of potential limitations in the systems on offer.

The current trend amongst those who sell CAD/CAM systems is to tell people what they need and not to determine what the customer really requires. The technical, financial, and managerial implications of the new CAD technology need to be understood by many more design engineers, and senior managers in design and drawing offices must become aware of what CAD can do.

What is meant by 'computer aided design' is not always clear, and this is not surprising because the most economic role of the computer will depend on the product or service that the design office is providing. Sometimes CAD is little more than the extensive use of computers to perform design office calculations, while, at the other extreme, the computer can provide the means of

---

†CADTES scheme provides grants to cover up to one third of the cost of purchase of CAD/CAM by companies. However, although the scheme was launched in 1982 with a working capital of £12M, which has since been doubled, funds are now nearly exhausted.

drawing, the means of standardizing components, the means of modifying drawings, and a direct link with the method of manufacture. Much of this comes about because the information contained in a drawing can be stored by a computer in ways that can be used by manufacturing machinery, and the term 'drawing' is strictly applicable only to the way that the information is presented for the man to study.

When a computer is used to analyse designs its economic advantage should be clear from the beginning. Stressing a compressor disc was a lengthy, iterative procedure when conducted manually, and it should be easy to balance the cost of writing (or buying) a program to do the work and using the program against the man-hours that have previously been necessary. Part load performance calculations for gas turbines, heat exchanger design, or turbine blade designs are further examples of iterative procedures which can usually be done more cheaply and more accurately by a computer program than by a man, once the program has been written. It is probably true, more often than not, that if an algorithm can be written to do a job manually, then the job had better be done by computer.

Sometimes the computer can do a job that we would not have considered doing by hand. An example is the use of finite elements to stress shapes to which we could previously apply only the most approximate of methods. Here, it is less easy to compare the cost of the computer solution with that of the manual solution, as any comparison is between a method that we believe and one which leaves us with a considerable risk.

An extension to analysis is optimization. If we have a program which is capable of determing the values of the parameters of a design scheme so that some performance criterion is met, then an extension to the program may modify those parameters so that some objective is optimized. For example, the environment and required performance of a heat exchanger could be specified and its configuration chosen in general terms; the length, depth, breadth, and number of corrugations could then be determined by a computer program which minimizes weight, subject to meeting the required performance. Flexible pressure connections can be designed by a program which selects the cheapest joint to manufacture while ensuring that fatigue life and operating performance are acceptable.

These computer methods of analysing problems need not be considered as special to design. Procedures often evolve because designers who have computing support see the need for programs to help them stress or analyse the performance of the components that they are designing. Designers who have an interest in computing will sometimes write the programs they need, but as they become more experienced they will know which programs (finite element programs, perhaps) are better bought from specialists.

For much design work, then, it would almost be better not to use the term CAD, because the computing requirements are not particular to design problems. It has been argued, too, that this sort of analysis is not design, although it is computer aided. Design is usually thought to have a creative element, and it is clear that analysis cannot, of itself, be creative. Nevertheless, the creative part of design consists of generating possible solutions to the specified design problem, and analysis plays a major part in rejecting unsuitable designs or showing errors in solutions which may be made suitable by modification. Analysis does not generate ideas but is a catalyst in stimulating the designer to do so.

Computer methods are rather more specific to design when they are directed to the solution of geometrical problems. Such problems may be peculiar to one office, and an example of this could be the determination of the configuration of the Advanced Passenger Train, on a curve, in a tunnel. Some programs are of much more general use, and an example could be the determination of the configuration of pipework in a chemical plant. The great advantages of solving geometrical problems by computer are speed and accuracy. Of course the advantages have to be weighed against the cost of writing or buying the program, but where a company has extensive and repeated geometric layouts to design (as in oil refinery, desalination plant, chemical plant design) it is unlikely that any manual drawing or modelling could compete with the computer. Part of the reason for this is the computer's accuracy which reduces the chance of errors having to be put right when the plant is being built.

Perhaps the application which most people think of as CAD is really computer aided drawing. There are many ways in which the coordinates of a drawing may be picked off and stored in a computer's memory – from light pens to cursors – but, what is more important, these coordinates and the dimensions that they define can be changed easily and quickly. This makes it simple for the draughtsman to generate families of parts which are closely related. It also makes it more likely that a designer asked to create a new detail will base it as closely as possible on an existing successful part. The ability to store drawings leads immediately, then, to a system which encourages standardization. The storage of drawings as coordinates also leads easily to computer aided manufacture, and the stage of manufacture in which specially trained technicians picked dimensions off drawings to make the tapes for NC machines can be eliminated.

A major economic advantage of computerized drawing can be in the management of change. This is particularly the case during the development of high-technology products where, for example, a company has found it economical to use CAD linked with NC machining to produce fabrications which one might expect to be cheaper to produce as castings. The point is that during development many changes are inevitable, and NC machining can permit a change to be made in a couple of days where the lead time for castings would be six months. The costs brought about by delays to a development programme are likely to be many times greater than any saving that the actual process of casting may show over machining.

The computer's ability to store dimensions and to present them as drawings on paper or on a screen can be extended to the production of perspective or isometric views of the part concerned. There are, of course, such technical advantages to this as the ability to check accessibility or the possibility of parts fouling; but, perhaps the greatest value of this property is that it enables the designer to visualize what he has drawn. Reading an engineering drawing is not a normal facility that people possess, and many expensive mistakes (obvious when the operative has made or failed to make the part) could be avoided if the designer had been able to picture what he was drawing.

Conventional engineering drawings are good communications media for describing mechnical parts and are capable of expressing this information in the useful formats for both design and manufacturing processes. When CAD systems were first introduced in the 1970s to improve quality and productivity, they closely emulated the drawing board procedures while providing fast interactive methods for creating and modifying drawing detail. A typical example is shown in Fig. 13.1 which is an extract from a design that has been drawn, dimensioned, and toleranced by computer. However, the very fact that mechanical engineering is concerned with 3D objects of varying size and complexity gives rise to problems in draughting irrespective of whether a drawing board or computer aided approach is taken. In certain cases, the drawing detail could be misrepresented by the drawing office, or even misinterpreted on the shop floor, either of which results in expensive redesign procedures.

Hence the value of 3D modelling. Today, many different types of 3D solutions are available which range from wire frame modelling to solid modelling. Wire frame modelling provides descriptions in which the edges of each object are stored as a collection of 3D lines and circular elements. The resulting stick-like representations are generally the simplest of modelling schemes to create, and in the process, expend very little computer processing time and memory. Whilst this scheme has proved useful for supporting specialist applications such as dynamic visualization, it does exhibit a number of serious deficiencies when used to solve general engineering problems.

Many of the ambiguities of wire frame models are overcome with surface models – the next step forward in 3D modelling. These models are created by connecting various types of surface elements to the edge model, hence their ability to be easily mixed with wire frame representations. The entire surface model may comprise one or more of the surface types. These types of models also present some difficulties in usage, but they have enabled traditional lofting and clay modelling techniques to be replacecd by a mathematical equivalent which allows faired curves and surfaces to be held in a computer in digital form. Some surface modelling applications include turbine blades, ships' hulls, glass bottles, and automobile bodies. Surface modelling has also provided a powerful link with numerical control (NC) in which the part programmer can simulate tool cutter motion over the defined surface.

The solid modelling scheme is the most advanced, in which a complete and unambiguous description is maintained within the computer. For example, solid

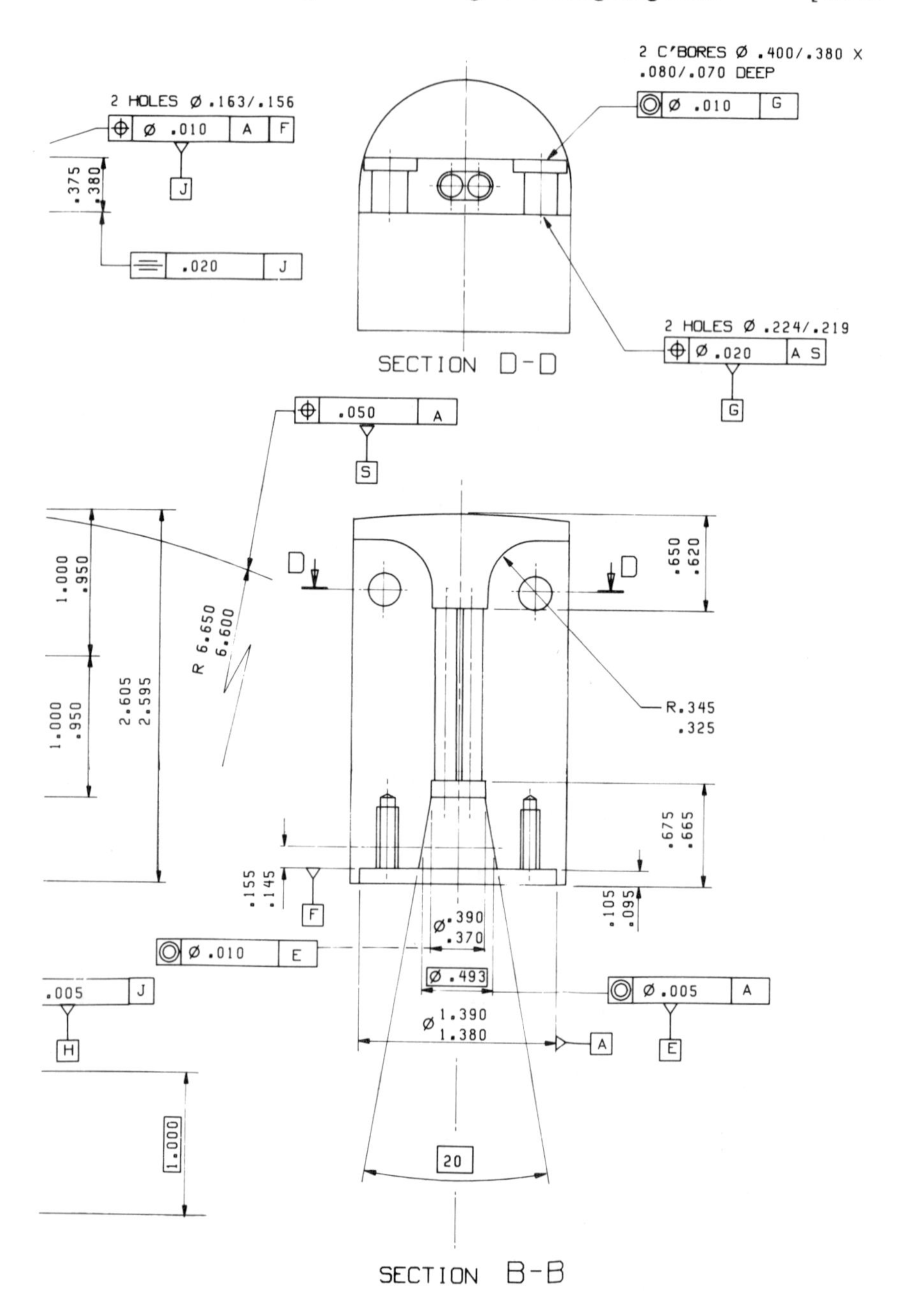

Fig. 13.1 – Extract from a component drawing which has been dimensioned, and toleranced by computer.

models can automatically calculate mechanical properties, perform interference detection, provide sectioned and full views of parts and assemblies with all hidden lines removed, or produce shaded pictures. Fig. 13.2 shows a perspective view of a wing section, with the mesh used for finite element stressing.

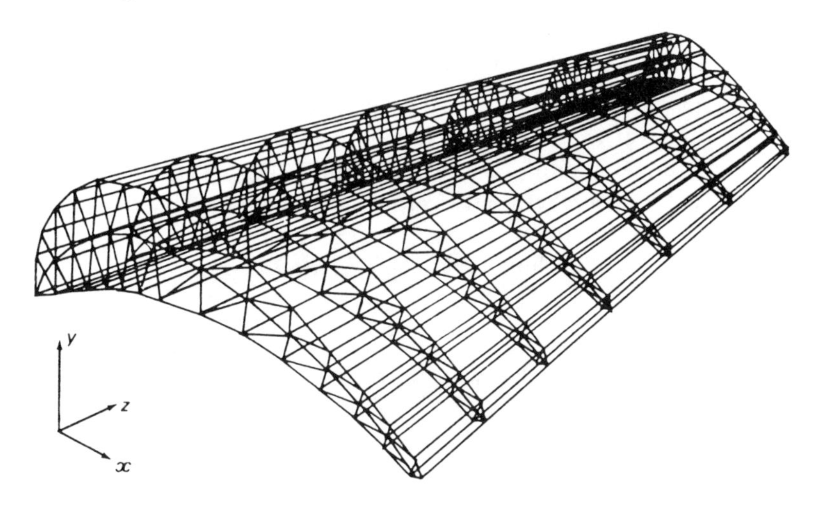

Fig. 13.2 – Meshes generated by the CAD system for an aircraft wing. (reproduced by permission from C. B. Besant: *Computer aided design and manufacture*, Ellis Horwood Ltd, 1980).

Until recently the construction of solid models on large mainframe computers was lengthy and of limited use to industry. But progress has been rapid. The programs are now more robust and support greater functionality, whilst the dramatic drop in the cost of computer power means that the scheme can be run on some of the relatively inexpensive 32-bit minicomputers currently available.

One final point on 3D modelling must be made, and that is the use of colour. The engineering community has been rather slow to see the potential value of full colour graphics – but there is no disputing that a colour display can convey far more information than a monochrome display. It seems natural to design the different layers of a multilayer circuit board using a different colour for each layer; or to design a piece of machinery using different colours for the working parts, the supporting structure, and the housing.

Historically, perhaps, some of the resistance to full colour working is what might be called 'the dye-line syndrome'. Nobody is going to use colour to convey essential information on a drawing, when that colour is liable to be subsequently removed by a monochrome copying machine.

However, colour hard copy is rapidly becoming less expensive. At A4 sizes, multi-pen plotters and (just recently) ink-jet technology have now become available.

The investment necessary for CAD can clearly range from a couple of thousand pounds for a microcomputer to perform common analyses to the hundreds of thousands of pounds required for a system capable of the most sophisticated drawing functions.

13.2 THE CAD/CAM SYSTEM

Senior design staff will at least know what goes on in a well run design office, and they can treat the draughting system as a 'black box' with a functional specification to be scrutinized. But, unfortunately, technicalities will arise, and how many engineers can resist getting interested in the works? It is therefore worth having a look at some of the things that are to be found in the glossy brochures or on the lips of sales representatives.

13.2.1 Software/hardware

The software is what does the work in the CAD system. The hardware is the 'office' in which the software does its work; it provides all the services which the software needs and the communication with that outside world of human beings of which the software is only dimly aware. The hardware of the system will be familiar, consisting of the usual fans, motors, wheels, belts, etc. which wear out, electrical components which burn out, and semiconductors which simply die.

Software, on the other hand, is strange stuff. Labour saving in operation, it is highly labour intensive to produce but ridiculously cheap to reproduce. Software costing £100 000 to develop can be copied for under £10 and distributed on disc and tapes costing under £50. There are no parts to wear out, yet it can break down apparently randomly when a statistically unlikely set of circumstances produces a particular set of data with which it cannot cope properly and the program crashes. For this reason, the longer a particular program has been in use the more reliable it becomes, because the more opportunities there have been for finding out and correcting these exceptional error states. Software needs to mature like wine.

There are two 'departments' to the software: the application software which does the actual graphical manipulations and calculations etc., and the operating system which provides administrative services to the applications software such as file management, accounting, peripheral device handling, etc.

The hardware is composed of the Processor in which the Software sits and various Peripheral Devices such as Disc, Plotter, or Workstation, which basically provide storage or communication with the outside world.

13.2.2 The processor

Memory size is often quoted, but this does not tell you much. Remember, however, that each additional workstation connected to a processor will require additional memory for its own use.

The processing can be either centralized or decentralized. Some vendors supply one powerful processor which is shared by up to six workstations, while others supply a processor with each workstation. The latter configuration has lower cost per workstation for small numbers of workstations and so is easier to get going initially.

However, one cannot avoid a central facility when there are several workstations, since certain functions (such as plotting) need to be shared.

The other variation is the choice between special and general purpose computers. Many vendors employ computers made by computer manufacturers for general use. One notable vendor has designed and manufactured his own computer. The advantage of using a CAD system which employs a general purpose computer is that other types of software can be run as well – but the operating system must also be a generally used one.

13.2.3 The memory

This is where all the drawings are kept while work is being done on them. It is a piece of high-precision machinery needing a nice clean controlled environment. Storage space is always short, so do not stint on capacity.

The traditional face of a computer, distinguished by twin spools of magnetic tape looking like two great eyes, lives on in many cartoons, though some progressive cartoonists now use the keyboard and display screen of a micro as their symbolic computer.

In reality, conventional tape was replaced several years ago by magnetic discs as the primary means of storing computer data; the main reason being that access to specific pieces of information is far faster on a compact, rapidly spinning disc than on a long ribbon of tape.

Among the bewildering variety of sizes and types of disc available, the most important distinction is between the cheap and durable 'floppy'discs at the bottom end of the market used with personal computers, and the far more expensive rigid or hard discs designed for mini and mainframe machines.

The floppy disc has magnetic material deposited on to a flexible plastic disk. Hard disks are made to a far higher degree of precision, using an aluminium platter that is extremely finely ground and polished before the magnetic oxide coating is applied; of course this allows the tiny magnetized regions, which store the bits of computer data, to be packed together much more densely than on the relatively crude floppies.

Your archive of drawings will no longer be microfilm or cabinets of Mylar but magnetic types or discs. Magnetic tapes or discs are also the means by which your vendor sends you versions of his software.

13.2.4 Plotters

All issues of drawings come out on the plotter, which is a relatively slow device, often taking about half an hour to plot a typical drawing. An important question is what kind of pen and medium you are going to use. Liquid ink on Mylar produces the best results but is expensive in running costs and is tricky in operation.

Another important consideration is whether to use sheet or rolls. Remember that a roll-fed plotter can be left to do drawing after drawing without requiring any attention, even overnight. If that is the case then speed is less important. And you pay for speed.

An alternative to the pen plotter is the electrostatic. Do not be taken in by the speed of these devices, which is very high. There is an additional software overhead to convert the data into the raster format which these devices require, so that the total time to convert and print should be investigated. Although printing on opaque paper, satisfactory dye-line prints can be obtained by increasing the exposure, because the line intensity is so high. Some people do not like the slightly jagged affect on circles which electrostatic plotters give.

Your designers will spend a lot of time looking at the screen, so if it is at all shimmery, or difficult to see, any productivity you may have gained will be lost in headaches and eyestrain.

Screens can be either storage or refreshed. Refreshed screens can be either raster or vector and either monochrome or colour. The disadvantage of storage screens is that the effect of an alteration cannot be properly seen until the entire contents of the screen has been repainted, which takes a few seconds. (A few seconds is a long time when you are concentrating.) Their advantage is that they are less expensive than other types. They have been in regular use successfully for many years. Provided that there is the software to exploit them, the refreshed screens will show immediately the effect of any alteration to the drawing. Their main problem at the moment is resolution. Raster screens use the television method of generating the picture with a fixed number, or raster, of horizontal lines which are then intensity modulated. It is therefore not possible to show two points or lines closer together than the distance between two of the horizontal raster lines. The vector screens actually draw the lines on the screen like a pen plotter, but very fast. The more lines they have to draw, the longer it takes, until the picture starts to flicker.

Colour screens are now available, and, provided that the software is available to exploit them, are as good as a large increase in resolution because of the ability of colour to separate one line from another.

Most workstations have a separate alphanumeric screen for messages from the software to the user. Some systems save money by putting the messages onto the graphics screen, and many users find this acceptable. It depends on how it is done.

13.2.5 Gesturing devices

It is impossible to create or alter the drawings without some way of indicating a position in the drawing. The light pen is the best known of these, but is unpopular in CAD on account of its bad ergonomics. It has to be held up in the air against the screen, obscures the item one is trying to indicate, and is inconvenient in the method used to select a new point in a blank area of the screen. The most used device is the tablet and cursor. The tablet is a panel incorporating electronic position sensors placed horizontally on the desk in front of the screen.

A cursor in the form of a cross or similar symbol on the screen is then made to follow the position of an electronic pen as it is moved over the tablet. However, the cheapest device, which many have found quite satisfactory, is the thumb-wheels and cross-hairs. The edges of two small wheels oriented vertically and horizontally project through the surface of the keyboard. Rotating them moves a vertical and a horizontal line passing across the screen, the indicated point being at their intersection.

13.2.6 Software packages available

We are now at the guts of the subject, and the range of facilities available is bewildering. What is genuinely useful and what is just sales gimmick? Firstly, here is a bird's eye view.

1. Geometric constructions
2. Dimensioning
3. Editing and alteration
4. Symbol and standard text libraries
5. Parameterised drawings, families of parts
6. Numerical analysis
7. File management
8. Job accounting
9. 3-Dimensional facilities
10. Parts take-off
11. Numerical control program generation
12. Curve generation
13. Schematics
14. Printed circuit board design

The first eight are well-established facilities which all design offices should find useful. After that we have a grey area of facilities whose value depends very much on the work being performed.

3D facilities are also valuable as they allow designer/draughtsmen to arrange their creations in perspective. Remember, computers work by adding up zeros and ones. Graphics computers work by remembering coordinates either in planes (2D systems) or in mass (3D systems). A typical 2D system would comprise:

a drawing software package;
a desk top microcomputer, with 256K of memory operating a monitor with a 19 in. screen;

a joystick option for graphic screen cursor control;
a digitizing tablet complete with controller;
a desk top plotter, for up to A2 size drawing sheet (USA 'C' size), and
a 'file manager'; this is a floppy disk drive unit, with built-in firmware for file management operating system.

Such a system would provide all the facilities normally associated with large minicomputer based CAD. It would be a cost effect CAD facility capable of producing the full range of 2D engineering drawings. These include assembly drawings, detail or component drawings, wiring circuit schematics, or logic diagrams.

The user can create and store libraries of symbols for use in diagram production.

Standard parts can also be stored as symbols, in symbol libraries, to increase general drawing productivity.

The great value of interpreting the design by means of perspective views for the benefit of non-technical colleagues or clients is not available with a 2D system.

Parts take-off is not as easy as sales presentations would make it seem. The computer simply searches the drawing for instances of various identifiers and tots them up. If the same item is marked twice over, as in two views of an assembly, it will be added in twice.

Dimensioning facilities need scrutiny. What happens if you alter a drawing after the software has automatically put in the dimensions? Does it implement the full British or ISO Standard? Can it do British centre-lines?

If your product line uses parts which are similar in shape but only differ in certain dimensions, then a good graphics programming language will allow you to completely automate the generation of drawings for these parts. A good programming language should include all the commands available to the usual operator plus the ability to take decisions (that is, test a variable and branch accordingly), plus the ability to call up subroutines and pass parameters to them. Parameterised drawing languages vary considerably in the facilities provided.

File management facilities are often not too good for real configuration management compared with the drawing numbering, issue control, and change control systems operated by some offices. They should at least keep track of issue numbers and new drawing numbers.

13.3 WHY USE CAD/CAM

Applications to date have shown how broad and varied the scope is for the use of computers in design and manufacture, and how significant are the benefits.

Specifically, CADCAM is the application of computers in design and manufacture. Many of the techniques required in such systems are unique. Unlike commercial and production control systems a CADCAM system must be able to respond to an instruction and respond to it immediately. For example, a designer is not prepared to wait for several minutes to see his drawing modified, and it is certainly not acceptable for plant and machinery to wait to be told what to do next. The design and development of this type of software has taken time, but the software is now generally available. The application of CADCAM techniques can now progress more generally and bring significant benefits to companies still waiting to make a start or modernize their design and manufacturing practices.

There are many different types of company who can make use of CADCAM. A company may design but not manufacture. At the other extreme, some companies manufacture but have no design requirement. The scope of CADCAM is well documented. However, it is necessary to consider the major areas of application to see how CADCAM can be used in your company. At the highest level CADCAM involves:

- the ability to aid design from initial conception;
- the extensive analysis of that design from performance, appearance, manufacture, and cost;
- the production of drawings for issue to manufacture;
- conversion of geometric data into plant, process, and machine instructions;
- the presentation of manufacturing instructions either directly to plant and machinery or indirectly to operators;
- collection of shop floor information from source either directly or via the operator, setter, foreman, etc. as manufacture proceeds.
- feedback of relevant information to design, planning, and manufacturing engineer;
- production of performance, status, and management reports;
- integration of commercial and production control with CADCAM, through to manufacture and back;
- the application of the computer to aid control of manufacture and subsequently to provide certain direct control.

13.3.1 Modular investment

There are few companies who will apply and require all of the CADCAM modules. There are many who will only need and benefit from a few selected applications. The starting point can be selected as the application which provides the greatest benefit. The CADCAM installation can then be built up as additional needs are identified or applications can be justified.

To achieve the major benefits from this approach it is essential that the investment is modular. The technology is already available to provide such systems. It is the application of the technology which lags behind. There are many constraints which require this modular approach, not least of which are learning, changes in working practices, and cost. It follows that investment in CADCAM will take several years.

The advantages of applying CADCAM far outweigh the disadvantages. The benefits are unlimited although the pay-back will differ from company to company.

Companies can now optimize design, prior to manufacture of an expensive prototype, for appearance, performance, automation, and cost. Designs partly or totally constrained by engineering standards can be drawn automatically or with a minimum of designer input. Lead times in getting a new design into manufacture and on to the market can be substantially reduced.

Combinations of CADCAM with suitable machinery can provide prototypes for engineering and market evaluation in days rather than months. It makes sense to get a new product on to the market early, especially if CADCAM can help make sure it is a high-quality product, it is reliable, and profitable. It is no good releasing a new product if it does not work. CADCAM techniques are an aid to becoming or staying competitive, and allow a company to respond under control to the many changes and variations demanded today. All this can be done quickly and cost-effectively.

Once the design is complete, it has still to be made. Computer-based production control is becoming generally applied in the planning of material requirements and resource scheduling. Process planning systems are available which convert geometry into a series of operations and instructions.

CAM utilizes interactive graphics to assist in the preparation of plant and machine instructions. It is an inefficient use of modern machinery to have it standing idle waiting to be programmed to do its next job. Even then the programming time involved can be significant. Off-line programming can aid in minimizing the time required in machine change over, and is a necessary discipline which ensures the flexibility and utilization of modern machines.

Direct numerical control

The ability to link the CADCAM computer directly with manufacture is available. This technique, generally termed 'direct numerical control' or D.N.C., becomes significant when a two-way data transfer is available. Instructions generated in the CADCAM system can be down-loaded directly to the plant and machinery, minimizing or even eliminating the need for manual input or the use of unreliable shop floor consumables such as paper or magnetic cassettes. Along with machine instructions the details required to set up a job can be passed to the operator or setter, often prior to the work being required, so that he can start to prepare the next job. The necessary material and tools can be selected, pre-setting on to pallets or loading carousels can be carried out, and shortages identified. Early identification of problems will result in increased utilization of equipment.

However, it still has to be recognized that with all the planning and care in the world things still go wrong throughout the work process. Sometimes it is a problem of scheduling a rush job; more often it is a breakage, breakdown, or shortage. The CAM system utilizing shop floor data collection techniques can immediately warn of a machine stoppage. At minimum, this warning can be used to ensure a fast manual response, or if the CAM system is able, a message can be sent to the appropriate department, perhaps works engineers, to ensure remedial work is started immediately.

By directly monitoring the operation of both plant and machinery its utilization is known. The reason for down time can be logged and manufacturing trends established. The ability of the CAM system to record and analyse manufacturing records will eventually lead to the introduction of preventive systems with the associated gains in efficiency and productivity.

The benefits of CADCAM are clear, and the advantages are specific. CADCAM is as meaningful to the small manufacturing company which needs to increase production as it is to the vast multinational company. The application of CADCAM is not easy. However, if it is implemented in the same way as any other capital investment project it can be done successfully and safely. It requires companies with completely different skills to work together to achieve the major benefits. The user must decide what it is he wants from his system and prepare a detailed specification of his requirement. It is not necessary to look at the vast selection of vendors on the market until the specification is ready for issue. Much of the initial development work is now complete, and it is no longer necessary to start from scratch. Nevertheless, suppliers still need companies to work with in order to input the necessary practical experience. Constrained by the applications specification, user and supplier can work together to realize the considerable advantages of CADCAM.

13.4 LOW-COST COMPUTER AIDED DESIGN SYSTEMS

Low-cost computer aided design systems will soon be commonplace in the drawing office. They will be used in much the same way as word processors are today in the administrative offices. In fact, with processing power increasing at at steady rate small systems capable of carrying out many functions may well be regarded as normal 'tools of the trade'.

It is necessary to define what is meant by a low-cost computer aided design system. A number of low-cost systems currently available appear to offer both simple operation and rapid drawing capabilities. However, very often they lack some of the essential features required of a proper engineering system. Generally, drawing accuracy and/or quality are sacrificed to make way for apparent user-friendlinesss. The use of the word 'apparent' in this context is deliberate because the average user would very quickly master the basic capabilities of the system and then not be able to go any further. In other words, he would be frustrated by a lack of capabilities rather than being grateful for the additional user-friendliness. In the computer business it is well known that the inclusion of menu-driving slows the system operation down, and provision should always be made for the more advanced user to dispense with menus if he chooses. Unfortunately, with many low-cost systems such flexibility is not given.

Of course, when evaluating computer aided design for application to the drawing office functions it is necessary to be quite clear about the type of work the system will be expected to do. Most systems will handle very simple schematic layouts, data sheets, etc., etc. On the other hand, for those wanting to produce complex drawings, it is necessary for the CAD system to offer as many functions as the high-cost systems do, even though they cannot work as quickly. For example, the major advantage of a CAD system is the ease with which standard drawing parts or library parts can be manipulated and positioned on the drawing sheet. The better systems will offer complete flexibility in this respect, both in terms of size and complexity of symbols and in terms of the number that can be stored.

Low-cost CAD systems which offer all the functions of the high-cost counterparts are available. However, as with all things that are worthwhile, they take a little bit of getting used to, and some application on the part of the user in setting up his library drawings and symbols during the initial introductory period. Unfortunately, with the over-enthusiastic publicity given to microcomputers recently, many potential users believe that a CAD system will actually carry out the design work for them. Of course a CAD system, regardless of cost, will only be as creative as the designer.

Considering the type of equipment used, instead of looking at the computer first, let us look at the plotter. The output or printer will make or mar whatever system is used. As with all electromechanical devices the price of plotters is likely to stay fairly constant. There are some bargins to be had, but unfortunately these do not lie in the very low-cost end of the plotter range. It is worth paying as much as you can possibly afford when purchasing the pen plotter because the drawing quality is important (see Appendix 13.4).

Next we look at the graphics screen. This is where most of your drawing work is done, and, as with the plotter, it is advisable to buy the best and highest resolution screen you can afford. There is nothing worse than using a CAD system where you need to strain your eyes to try and to figure out the details of any drawing on a screen. The zoom and pan capabilities offered by most systems are of limited help. For example, if you are laying out a large drawing you need to see as much as possible of that drawing. In fact, it is advisable to regularly submit drawings to the pen plotter and get a full-size draft copy for checking. This allows you to gain a perspective of the drawing which is not possible when using the screen only.

The heart of the system is usually considered to be the computer itself. However, even this is misleading because in fact the operating characteristics of a system are determined not by the computer but by the software controlling the computer. It is far more important, therefore, for the potential user to look at software that satisfies his requirements and select a computer that will run the software afterwards. All too often the situation is reversed, and the user finds himself with a machine that has limited software available for it. In general, the range of small computers available today with the performance necessary for use in low-cost CAD systems is narrow, but they do exist.

APPENDIX 13.1
SEVEN BASIC POINTS TO CONSIDER FOR SUCCESSFULLY INSTALLING CAD

- Be sure you know what you want to achieve before choosing hardware and software.
  Decide what the 'D' in CAD is to stand for – 'design' or 'draughting' or both.
- Establish a task force to carry out a feasibility and evaluation study. If a CAD system is to be bought, make the task force responsible for implementing the system.
- Where possible visit and inspect existing CAD users, preferably in a similar business.
- Aim for a single supplier being responsible for the total system.
- Choose an installation contractor with a good track record, preferably in computer-type installations. Specify clearly the responsibility interfaces between the system supplier and the installer.
- Keep all employees in the picture, and pay particular attention to communicating with the unions concerned.
- Find out if the government have any financial assistance schemes.

APPENDIX 13.2
QUESTIONS TO ASK A SOFTWARE SUPPLIER

(a) What does the program do?
(b) How does the program perform its task?
(c) What source data are incorporated, and are these of proven reliability?
(d) Who designed and wrote the program, and why was it written?
(e) What is special about the software?
(f) What savings can be made from the program's use?
(g) What graphics, printing, or plotting features are incorporated?
(h) Is the documentation helpful and easy to understand?
(i) Can the program be 'crashed' easily and, if so, does it recover?
(j) Is instant technical assistance available from the supplier?
(k) How long does it take to obtain new or replacement programs?
(l) Is a program amendment service available?
(m) Can the program be used comprehensively prior to purchase or lease?
(n) How long has the supplier been in business, and what are its plans for the future?

APPENDIX 13.3

GUIDE TO CAD SUPPLIERS

Supplier of CAD system	Name of system	APPLICATIONS General (architectural mapping etc.)	Mech.	PCBs	Typical number of terminals	Approximate price
Applicon (UK)	AGS	*	*	*	4	From £100K
Applied Research of Cambridge	GDS	*	*		5	£70K – 1 terminal £100K – 2 terminals £200K – 5 terminals
British Olivetti	IGS 60	*	*		1	£11K to £15K (excl. plotter)
CalComp Ltd	IGS 400	*	*		1	From 55K
	IGS 500	*	*		4	£90K to £200K
Calma (General Electric Co.)	CARDS			*	12 (Inc. interactive graphic terminals & others)	£75K to £450K
	DDM/CADEC	*	*		12	£75K to 450K
Computervision	Designer systems	*	*	*	Up to 48	£110K – 2 terminals

Davy Computing	Auto-Trol AD/380	*	*	*	12	£120K – 2 terminals £250K – 6 terminals £450K – 12 terminals
Ferranti Cetec Graphics	CAM-X		*		Typical 8 seat system	£150K – £200 – 1 terminal £350K – 8 terminals
M & S Intergraph	IGDS	*	*	*	16	£100K – 1 terminal £400K – 8 terminals
Perkin Elmer Data Systems	CAES	*	*	*	2–16	£125K – 2 terminals £300K – 8 terminals
Quest/GEAC	Q-Plot		*	*	2	£103K
	Q-Draft		*	*	2	£80K
RACAL-Redac	CADET			*	1	From £20K
	MAXI		*	*	8	From £90K
	RIDCS	*	*	*	24	According to configuration
Radan Computational	Radraft		*		1	£23K (excl. plotter) Addtnl. stations from £17K
Software Sciences	SWIFT II		*		8	£80K to £250K
SD Computing Services	IDDS 2000	*	*		4	£180K to £200K

APPENDIX 13.4
PRINTERS

For design and D.O. work it is imperative to have a good quality printer and plotter. What are the options?

1. Dot matrix

The dot matrix printer mimics the way a computer constructs screen output by building up characters or graphics from tiny dots produced by needles hitting a print ribbon. Most machines use typewriter ribbons and ordinary paper. Modern versions are capable of 'letter quality' output – they can produce good financial reports and business correspondence – and very high resolution graphics. A bottom end machine would cost around £150; a computer-aided design printer could easily cost over £1500. Microcomputer printers will produce text at speeds ranging from 100 to about 600 characters per second (cps). Normally they print one character before starting the next. Output is slower in graphics modes because a line of dots (across the page) is printed.

Probably the largest manufacturer of microcomputer dot matrix printers is Epson, whose machines are used by many suppliers. Epsons seem to be consistently reliable and rugged. The LQ-1500, which costs around £1200, produces good quality text (to rival the best electric typewriters) by overlapping dots to produce fully faced characters. The LQ-1500 is also capable of good monochrome grphics. Epson printers were among the first to offer multi-font capability, and graphics for micros, in the same machine. The FX-80 and FX-100 give test and graphics of a lower quality and resolution than the LQ-1500, and are correspondingly cheaper at four or five hundred pounds.

2. Daisywheels

Daisywheel printers are common to word-processing applications. The machine uses a revolving, many-spoked wheel to produce print. Each spoke holds the raised image of character in reverse (rather like a manual typewriter), and when that character is to be printed, the wheel zips into position and a hammer knocks the spoke against the ribbon on to paper.

They are generally not very fast (on average 40 to 60 cps) but produce excellent quality text output and (as a generalization) are solid, reliable machines. Usually, a change of typeface is achieved simply by changing the wheel. By using a two-colour ribbon, some will print in an extra colour. Daisywheels tend to be expensive; a long-lasting model is likely to cost more than £500, and very few worthwhile machines come for less, though a rash of Japanese machines is lowering the bottom end price to £200.

3. Drum printers

Drum printers are the mini/mainframe equivalents of daisywheels – they use a drum instead of a wheel. They are extremely fast but do not suit graphics-oriented applications.

4. Band and chain printers

Band and chain printers use the same basic principle; a revolving band of chain holds the character impressions, and a tiny hammer hits them to print characters. These are essential for the high-speed, high-volume output needed to interface with minicomputers and mainframes.

5. Ink jet machines

Ink jet machines, as the name implies, squirt tiny spots of ink on to paper. With micros they can be very slow, but they are quiet (it is surprising how much noise and therefore disturbance a printer can create). The ink is supplied in small cartridges. Colour ink jet printers give quite reasonable quality: because the dot size is quite fine, they can produce output equal to a good business graphics system. Ink jet printers were previously very expensive – high-speed machines cost over £10 000. Today's machines are based on similar technological principles to their highbrow counterparts, and most come from Japan. Canon is a large supplier of these small units: its A1210 is a neat machine capable of printing in (and mixing to some extent) seven colours at 640 dots/line. Retail price is about £550. It uses one print head which moves back and forth across the paper, printing a few dots at a time. Canon is pioneering an ultra-fast, high-resolution ink jet technology called Bubblejet – a full-line printer employing a line of ink nozzles. The company reckons these machines will be available shortly.

Siemens has also put a lot of effort into ink jet printing. Its printer terminal PT80 will print at 300 cps. Higher up the range the PT88 and 89 offer a choice of needle (dot matrix or ink jet). All three models are happy producing graphics of a good, but not exceptional, quality.

6. Thermal and electrostatic printers

Thermal and electrostatic printers for small micros are generally cheap, low-quality machines. They can also present problems, not least because they use special paper often only available through a few outlets. They work by firing heat or static charges at sensitive paper. In general they do not produce top-quality output (except, again, for very expensive machines), and are often used for rough draughting or program listing. These types are usually produced for the very bottom end of the computer market. However, some better quality models can produce almost exact copies of graphics from a computer's screen. At 15p for each paper copy, quality is excellent and output fast.

7. Laser printers

Laser printers are the latest technology development – the first commercial model was offered in the USA in 1976. Suppliers include Burroughs, Canon, DEC, IBM, and Xerox. They produce sharp, high-contrast output at high speeds (DEC's LN01 can manage 12 pages/min) to extremely high resolution – as much as 300 x 300 dots/in. They are particularly useful for producing large volumes of quality output quickly, and as such could be used as a substitute for typesetting machines. Laser printers are very large, expensive pieces of equipment, but they can handle most text or graphics tasks.

Ink jet and dot matrix machines are best suited for colour output. Dataproduct's P series printer is a modular dot matrix system which can be expanded to produce colour graphics suitable for engineering, design, and business purposes. The colour addition allows printing in the four primary colours (black, cyan, magenta, and yellow) to make up a variety of shades by overprinting. Graphics resolution is usually 84 × 84 dots/in. Lear Siegler makes a model with similar capabilities called VersaPrint. The Diablo Series C printers can produce four-colour output and cost around £850. IBM has realized the possibilities of colour with a 3270-compatible colour matrix printer, the 3268.

Facit, using a technology it calls Flexhammer, produces a colour printer (the 4544) of an excellent quality. Again, it uses the four primary colour-mix principle and prints at 225 cps.

The most noticeable development trend is using dot matrix printers as sophisticated graphics machines, capable of handling complex colour pictures and producing non-standard mathematical and scientific symbols easily. Another development is multi-mode and modular printers. Many more printers of all types are capable of printing in several character sets and/or to different graphics resolutions. And, like the Facit 4544, many can be bought as a basic model and upgraded as needs dictate. More and more microcomputer printers are bi-directional and logic-seeking – they print forwards over one line then backwards over the next. This offers improved speed – an important criterion, particularly with large volumes of output.

Ink jet printers are becoming more sophisticated, and for mini- and micro-computer users viable for the production of the graphs and charts.

Printer buffers are also increasing in size. The buffer is an important component; it stores data from the computer prior to printing. It is necessary because the computer and printer work at different speeds, the computer sending information out faster than the printer can print. A buffer will reduce the amount of wasted computer time according to its size. Siemens' PT80 can take up to a 4K buffer. Epson's LQ-1500 comes with a 2K as standard. Until recently, micro-printers might contain 512 bytes at the most. This might sound like a small difference, but in practice even 1K can be a great time-saver.

It is worth noting that many microprinters can link up to minicomputers. Interfacing has been something of a bugbear – some buyers use parallel interfacing, others use serial. But most now offer one of the 'industry standards' – RS 232 or Centronics-type parallel data interfaces.

Finally, it is worth stressing that when buying a printer it is essential to satisfy yourself that it can produce the kind of quality you are looking for at the volume you want.

The authors have to thank the journal, Eureka Innovative Engineering Design who supplied much of the information on printers.

APPENDIX 13.5
SOFTWARE

The development of super computers, low-cost microcomputers, powerful and inexpensive graphics terminals, and the availability of powerful workstations with large storage capacity, has provided engineers with a wide range of hardware to use. However, because of problems in identifying the software required, and, in some cases, developing it, the growth in computer applications has not proceeded as fast as many pundits predicted a few years ago.

Everyone involved with computers knows how horrendously expensive software development is. As hardware costs fall, programmers' salaries rise, and the systems they produce get more complex. It is estimated that the UK now spends £2000 million a year on writing program codes. But even after arduous coding, testing, and debugging, costs do not end there: maintenance of programs (a euphemism for fixing mistakes that crop up and making necessary alterations) is a continuing heavy burden, and can amount to half the cost of a system.

Experts have long been convinced that the bulk of the serious faults in programs are caused not by errors in coding but by failures to get the specifications right at the beginning. Now, a formal method for making sure that a specification actually does describe what the customer wants is being brought out of the laboratory, in a campaign to get the computer industry to adopt it on a large scale.

The technique is called VDM, for Vienna Development Method, and has grown out of work started at IBM's laboratories in Vienna over ten years ago.

But with all this, remember that the decision to buy a computer is the easiest part of the process – actually implementing it is much more difficult. It takes a lot of planning and coordination and careful checking to make sure that the correct software is available.

Applications software is designed for specific uses by different types of users. When you select applications software make sure it operates correctly with your hardware. Before you choose a program, check the memory, storage (such as the number and type of disk drives and disks), monitor (colour or monochrome), printer, and other requirements to make sure they match your machine. Try to see a demonstration, and read about the software to ensure it does the work you need to do at the moment, and the work you expect to do later on.

CHAPTER 14

Education, training and development of designers

INTRODUCTION

The importance of good engineering design to the economic performance of the nation was emphasized by the DSIR Committee of Mechanical Engineering Design which published its report (the Feilden Report) [1] in 1963. The Moulton Report [2] in 1976 again emphasized the economic importance of the subject, and considered the educational implications in some detail. The importance of good engineering design was stressed yet again in the Corfield Report [3] in 1979 and again in the Finniston Report [4] in 1980. The most recent of these reports, Finniston, gives considerable space to the education of engineers and makes far-reaching proposals for changes in university (and other) curricula. Among these proposals is the setting up of four-year degree schemes, and some universities acted on this.

In spite of decades of official concern, however, the education and training of engineering designers has largely been left to chance. Many say that, like managers, designers are born not made. There is some truth in this, but it does not mean to say that they would not be better managers and better designers if they were given better training.

Unfortunately, the conventional higher education system does not, on the whole, cater for creative engineering designers. There are some exceptional cases where undergraduates are given design projects in their last year. But the greatest progress has been made in the postgraduate courses, and some of these are producing encouraging results; but they are few and far between.

For a long time engineering education has concentrated on analysis: courses lay much emphasis on fundamental laws and derived theory. But designers have to synthesize as well as analyse. By concentrating on theory, many teachers have unconsciously stifled creative ability. By contrast, in architects' courses much more time is devoted to practising design skills. They include abstract exercises, limited-objectives design exercises, and total design projects, as well as live projects. Perhaps some engineers should be trained in this manner.

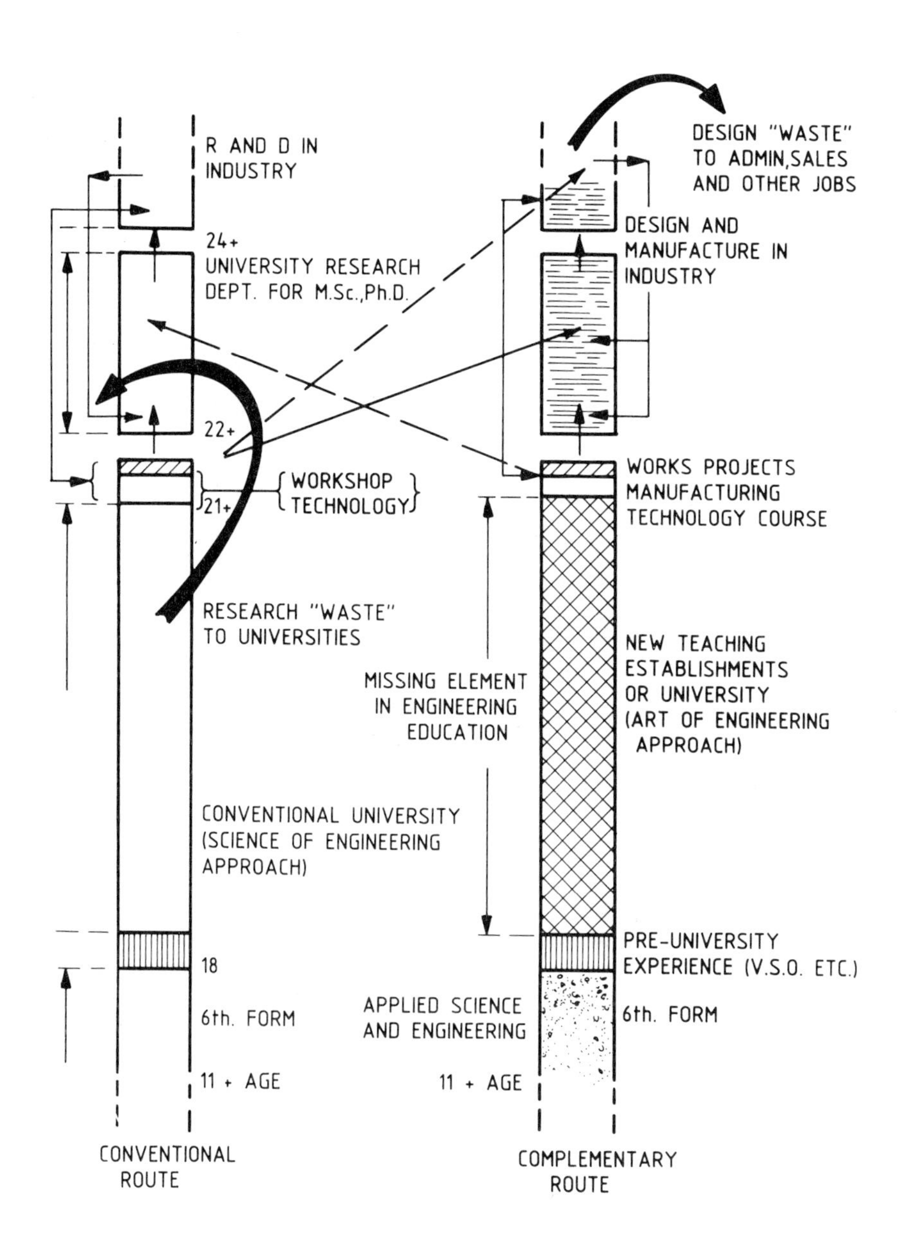

Fig. 14.1 – Routes into design engineering.

The education process is a lengthy one and starts very early in life. The personal attributes required by engineering designers begin to develop in childhood.

The conventional education route to industry is shown in Fig. 14.1. Many efforts have been made to correct its deficiencies from the design point of view at the postgraduate stage. These have varied from the creation of Institutes of Product Technology to special design courses, but too often they have become mere appendages to the conventional route.

What is really needed is a new way of teaching engineering, with the central core being design. This would concentrate more on what might be called the 'art' of engineering. All students would be exposed to the process of design so that they can appreciate its iterative and creative nature. Students must realize the need to balance conflicting requirements of performance, quality, and appearance within the general framework of the time and cost. Certainly post-experience education is also required for practising designers, but this needs improving.

There are, however, considerable problems connected with teaching design. For example: there is the whole question of the acquisition and utilization of information and data, the problem of inaccurate and inadequate terminology, and the large number of variables with which a designer has to deal, to say nothing of an appreciation of creativity and how it can be developed. And, again, most realistic design involves specialized knowledge.

14.1 UNDERSTANDING AND ENCOURAGING CREATIVITY

In a survey carried out on engineering creativity in the field of aero engine design [5] some useful facts emerged. Firstly, of the engineers interviewed, 22 felt that they had little creative ability, 58 fair, and 29 good, creative ability. On examination of this it was found that the respondents' attitudes to their own creative potential was very much conditioned by the work they did. Those engaged on routine calculations and detail drawing did not feel thay had much creative ability. In other words, ability was equated to activity – not necessarily with any justification.

The truth is that creativity, like intelligence, or any other human trait and ability, is normally distributed in the general population and needs to be stimulated and encouraged. Intelligence is not, however, the same as creative ability, nor is it necessary to be a specialist in a particular field in order to contribute useful ideas to it.

It has been suggested that creativity can be measured, and that some simple tests should be devised for new entrants to the profession. But what is essential is that design managers develop the creative abilities that are already available in their teams. One way of doing this is to encourage a systematic approach to design through such methodologies as PABLA [6], morphological analysis [7], brainstorming [8], etc., and the use of the formal specifications and procedures discussed in earlier chapters. Another is to create a climate that will foster creativity. The policies and procedures of the design office are often more important than the differences between the individual designers. Highly creative people appear to have needs additional to those suggested by Maslow (see Chapter 5), such as the love of inventing, work freedom, constructive discontent, and curiosity.

The manager must seek to support activities that will assist his people to satisfy these needs. Possibly the most important thing here is to be a good listener, to be sympathetic and show a sincere interest in understanding a designers' ideas; and to avoid personal antagonism or preferences.

If design is the heart of the engineering business, it follows that the very best creative people should be employed in it. The quality and reliability of a product stem from the design office. Indeed, the fate of an engineering enterprise is largely determined by the competitiveness of its design.

14.2 UNDERSTANDING AND ENCOURAGING AN ECONOMIC APPROACH TO DESIGN

It has been the thesis of earlier chapters that the success or failure of a design is largely determined by economics; whether the product creates a profit for both the manufacturer and the customer. But few engineers are taught to understand the economic consequences of design decisions. It is true that most professional institutions require the student to have read 'engineer in society' subjects before he can be admitted to corporate membership, but these subjects are only broadly defined, and, although economics is required reading, the approach may be very theoretical or may be concerned mainly with macroeconomics.

In the USA an engineer who wishes to become an accredited professional engineer is required to have studied engineering economy, which is directly concerned with the cash flow generated by engineering decisions. In Japan it seems (from Finniston) that economics does not form a significant part of the undergraduates' curriculum, but that developing the engineer, in his employment, is a major objective of management. It is also clear that projects are defined very precisely to a market need and price, so that the young engineer will be exposed to matters of cost and the market. In France and Germany there is no obvious teaching of engineering economy, but courses are longer, broader, and involve considerable practical work. Engineers are also more commonly recruited to general management and therefore expected to have a broader knowledge than the merely technical.

A serious difficulty is that the present three-year degree schemes cannot be expected to broaden their curricula without sacrificing technical content. It is true that four-year courses of study are being introduced (as suggested by Finniston), but, while more realism and industrial experience is to be injected into the curricula, there will be little time for the direct teaching of economics or management topics. Experience of engineering project work [9] suggests that, if the student can be involved in managing the money as well as the technology, the economic lessons will be learned as quickly as the technical. Much of the business of learning the economics of design will thus be left to indirect teaching by project work and by continuing training when the engineer is working for his living.

14.3 TRAINING METHODS

These can be conveniently divided into on-the-job and off-the-job training (see Fig. 14.2).

14.3.1 On-the-job training

Some firms have started to employ design tutors with the responsibility of aiding designers. The tutor can act as an informed critic, thus causing the designer to defend his efforts and possibly re-evaluate them. However, it is the design manager who must be mainly responsible for developing those who work for him. Some of the ways of doing this are as follows.

By special assignments

It may be possible from time to time to give a designer a special assignment in addition to his normal job: say, to investigate ways of improving the productivity of the drawing office; or the information flow into the design office, etc. This causes a designer to move around and talk to people outside his immediate section. It also obliges him to gather information and arrange it in a readable report. If he is made to report back his findings to a panel of engineers, so much the better.

Planned visits

Too many designers are restricted in their vision, and much can be done to help them by arranging a series of visits to the sites where their products are in use, so that they can speak to the users and obtain first-hand information on any difficulties experienced. Visits may also be made to exhibitions and research establishments to keep in touch with competitor's products and gain an insight into future developments.

Secondments

A designer may be sent to another department for a time to see how it works. It is important to ensure, however, that he is given a positive job to do. One example where this proved of great benefit was in connection with purchasing. While helping a subcontractor to produce a satisfactory product, the designer learnt about the problems of purchasing, such as negotiating prices, vendor rating, and the importantance of transport, packaging, etc. Perhaps even more advantageous would be the seconding of a designer to after-sales service so that he can learn at first hand the effects of design errors and the consequent customer irritation.

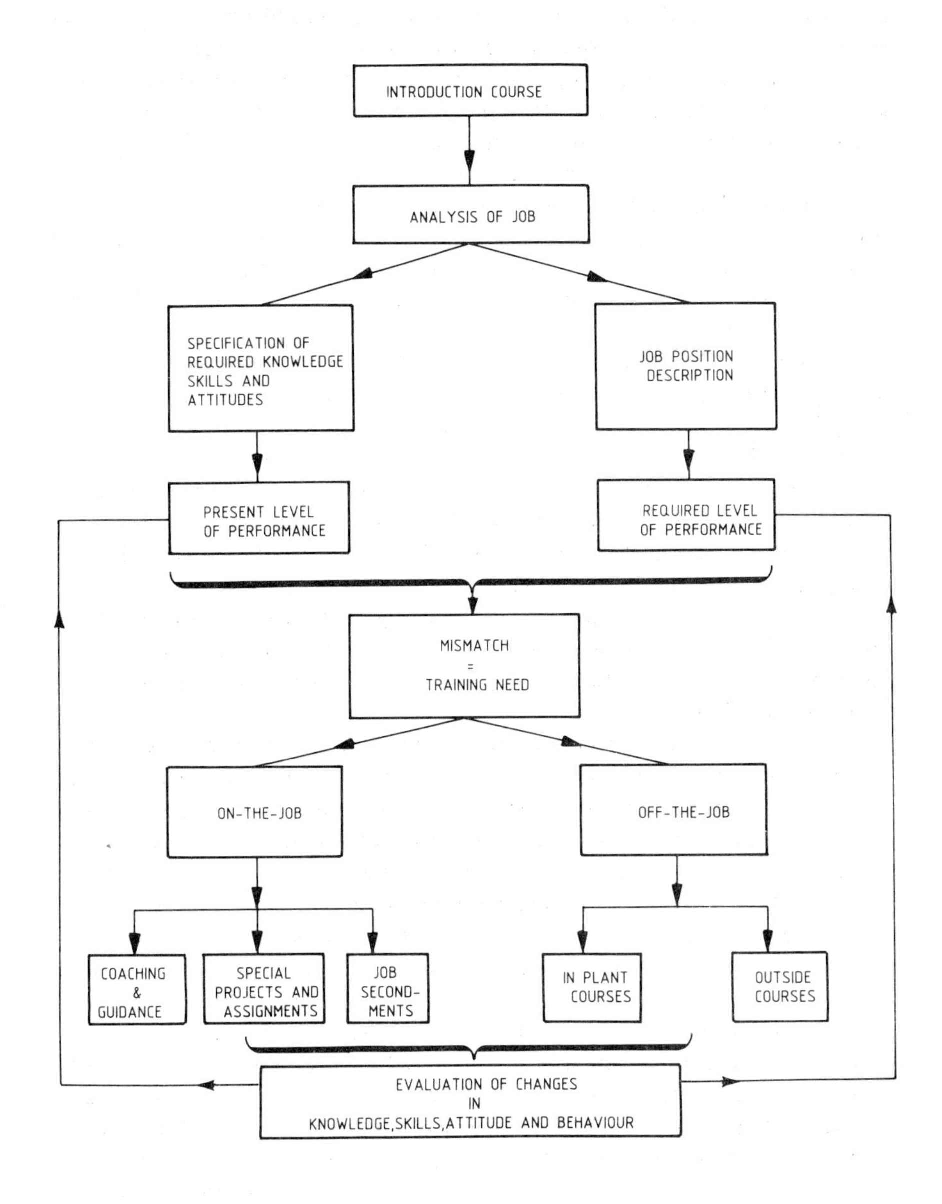

Fig. 14.2 – Design training process.

14.3.2 Off-the-job training

It is likely that designers who move up to become section leaders and then heads of departments will require supervisory and management training. This may be conducted by a company's own training staff, or by some outside agency. For large enterprises both are probably necessary, for knowledge of a particular company – its products, policies, organization, trades union matters, etc. – is best imparted by its own people, whereas a broad design knowledge and new ideas may be better obtained elsewhere. Smaller companies may well have to rely entirely upon outside business schools, where opportunity will be afforded for potential managers to rub shoulders with men from other concerns.

Before sending a candidate on an outside course he must be told when he is going and what it is hoped to achieve by sending him there. After the course is over it is highly desirable for the candidate to be 'de-breifed' so that his performance may be assessed. This will require hard work by both his boss and the course tutor.

It is vitally important to ensure that off-the-job training be wedded to on-the-job training. The latter should identify a designer's weaknesses and strengths, and this can be done by choosing an appropriate outside course for him to attend.

However, it is important to realize that in the last analysis management is learned by managing and being managed. From the way one is managed one can see some of the pitfalls to avoid and the good points to cultivate; how to encourage and bring on people as well as how to correct them.

For training in special techniques and design aids, several methods may be used. The first might be called the 'project method': here a design/make and build exercise can be used to teach, say, the use of value-engineering in achieving an economical design.

The second is by means of design workshops. A number of designers can be brought together to tackle problems on a group basis. They are asked to bring a specific design problem of their own and state its context. Alternative ways of tackling the problem are then propounded by the group. Such work may be interspersed with lectures, introducing new materials and processes, etc.

In both these methods of training it is important to have a real customer for the project if at all possible, so that the project can be properly evaluated on completion. Besides, everybody works better if the result is likely to be of some real use.

Another useful training method is to use design case histories. These are an excellent means for developing confidence and its counterpart, humility. This results from comparison of the students' own efforts with those of a fairly large number of engineers in industry. The comparison is especially valuable when the student also has some experience in a project or in several projects which give him a standard of comparison. Confidence is developed by observing the trials through which mature engineers often must go to arrive at a satisfactory solution.

After studying a few cases, students generally come to belive that engineers in industry are an especially stupid breed of men. The cases report effort in arriving at a solution which often looks quite obvious after it has been worked out and which often can be improved with the wisdom of hindsight. Only after the study of a larger number of cases do students learn that hindsight is always able to improve what others have done. They also learn that as future designers they will be exposed to criticism by the hindsight of their customers and through design reviews and by the production people who must manufacture what the designer has planned. They learn that criticism is no reflection on their ability, but a desirable source of information. They learn this more easily by criticizing the designs of others than by having their own designs reviewed by the teacher.

Design methodologies also need to be presented to young designers.

In teaching design it is necessary to stress that designers have different ways of tackling their work. Engineers tend to do much more calculating and express most of their findings on sketches or orthographic technical drawings. Architects, on the other hand, tend to do more sketching and rendering, and their calculations are of a different kind. Industrial designers often work directly on 3D models. However, despite these apparent differences of approach, investigations have revealed that however systematically or non-systematically a designer might work, the logical structure of design problem solving and the pattern of design procedure do have common features.

There is a case for applying some formal method to design so that its execution beomes manifest and accessible to comment, and to ensure that no important factor has been omitted. However, such formality must never be allowed to stifle the use of the imagination for generating ideas.

No methodology will replace hard concentrated thinking or provide the creative flair. Few good designs derive solely from logic.

The subconscious comes into play in all creative acts. In fact, there are methods used to further creativity which deliberately move information stored in the subconscious mind on to a conscious level. But by applying systematic design methods a certain attitude of mind can be inculcated; further suitable procedures ensure that design is not only undertaken but also talked about.

Some of the features which the various systematic methods have in common are:

1. They attempt to separate the logical and imaginative aspects of design so that each plays its part, with interaction but without interference.
2. A record is obtained of the design progress which can later be checked, not only for calculations but for information inputs.
3. At several stages a large number of people can take part in the design activity.
4. All parameters can be kept continually under review.

5. The design proceeds by predetermined stages, which makes it easier to plan cost and time.

6. The distribution of effort between information retrieval and problem solving is made explicit.

Thus, one overall aim of any design method is to make the process more manageable. Teachers must not claim that any method can improve on what the best designers achieve unaided; but that often such methods do seem to improve the performance of many average designers.

No one system can be relied upon to achieve a perfect design, and, naturally, any system used must provide greater benefits than the cost incurred in adopting it. Certain design methods can be very time-consuming, but short cuts may cause possible valid solutions to be missed.

Some of the possible ways of generating new ideas or searching for alternatives should be given to all those learning about design work. They are particularly useful when a designer has a mental blockage, for they act as goads to the imagination by prompting novel approaches – they also ensure that no feasible possibilities are overlooked.

14.4 PERSONNEL DEVELOPMENT

Education and training must be looked upon as a continuous process in all design offices, part of developing existing potential. It should be conducted at three levels, which are discussed in the following sections.

14.4.1 New recruits and young designers

Young designers with any real drive will wish to extend their studies to expand their technical ability. It may well be possible to help them do this by means of part-time release or even full-time courses. It is important to have agreed training programmes with proper assignments scheduled for each team in technical college or industry.

Assignments in other industries may be desirable. In this connection the E.I.T.Bs industrial liaison officers and Design Council's liaison officers may be able to help in arranging suitable cooperation.

14.4.2 Junior designers

It is often very difficult for a junior designer to keep abreast of the latest developments or even of other work being carried out in his own section or department. To overcome this problem it is useful to conduct in-plant seminars. Junior designers should be asked to present lecturettes on their own work at those gatherings. Time should be given to them and opportunities to prepare visual aids, models, etc. Such seminars can be conducted after working hours.

14.4.3 Senior designers

This is possibly one of the weakest areas in training, and our senior Institutions have done little to help. Time must be given to senior staff to attend lectures, seminars, and conferences to talk with their colleagues. The Institution of Engineering Designers has doen much to make this possible. In some places Designers' Clubs have been formed. Anyone who attends these informal meetings cannot fail to gain an insight into new design work.

All senior designers should be encouraged to prepare papers for publication. The main benefits are not from the kudos obtained, although this may be quite considerable, but from the discipline required in crystallizing their thoughts and presenting them in a logical way. In addition, many new ideas may arise, both in the preparation of the paper itself and the discussion after it has been given. Designers should not be afraid to report their failures as well as their successes; for it is only by overcoming obstacles that real design development takes place.

It is vitally important that every design manager should consider carefully how he can develop his staff, and he must devote time to do this; in particular, to encourage and foster creativity in them.

The goal of any engineer should be to utilize the knowledge and material of the physical world for the benefit of society. To achieve this end, he designs and constructs physical objects, devices, structures, processes, and systems. The problems which engineers seek to solve generally have many possible answers from which an appropriate solution must be chosen. An engineering designer's aim is really to optimize the usefulness of resources for mankind's benefit.

It is a tragedy that so much of our engineering education and training is not directed specifically towards the designer's main aim. Much of each new intake to industry proves unsatisfactory for this reason. The day cannot be far off when a proper science of design will emerge as a teachable discipline, and it is to be hoped that this will percolate into our established teaching units.

There is considerable evidence that engineering management does not recognise the central importance of design. We hear a great deal about efficient manufacture and marketing, delivery dates and aftersales-service, but perhaps not enough about technical excellence and quality. Design is the key to all technical success. In this connection the resent CNAA report on managing design is to be commended [11].

14.5 DESIGN TRAINING IN SCHOOLS

Increasing numbers of school children are being introduced to design. In particular, craft, design, and technology courses are being developed for GCE A level examinations. University admissions tutors are mixed in their views about the suitability of craft, design, and technology as a subject with which to gain admission to a university engineering course, although this seems to be less because of its content than because it might take time that could be devoted to the study of mathematics, physics, and other subjects thought to provide a necessary foundation for engineering science.

Again, if craft, design, and technology courses concentrate on craft, (that is, if they derive mainly from earlier woodwork and metalwork courses), they may be unable to provide time for the technological and economic approach needed to underpin a later study of engineering.

The attack is through teacher training, and there are now courses which aim to develop the teacher of craft, design, and technology, and it is clear that some of these courses are concerned with serious, manufacturing design and design management. The bridge between school and university has still to be crossed, but, at least, sixth form training in CDT may increase the awareness of the significance of engineering design and create a climate in which wider engineering training will be encouraged and rewarded by society.

REFERENCES

[1] Feilden, G. B. R., *Report of a Committee appointed by the Council for Scientific and Industrial Research to consider the present standing of mechanical engineering design,* London, HMSO, 1963.

[2] Moulton, A., *Engineering design education,* The Design Council, London, 1976.

[3] Corfield, K., *Design for manufacture,* N.E.D.O., 1979.

[4] Finniston, M., *Engineering our future,* Cmnd 7794, London, HMSO, 1980.

[5] Turner, B. T., *Senior Clayton Fellowship Report to the I.Mech.E.,* London, 1978.

[6] Latham, R., Taylor, H., & Terry, G., *The problem analysis by logical approach system,* Engineering Materials and Design Conference, London, 1965.

[7] Watts, R. D., *The elements of design, the design method* (ed. Gregory), Butterworths, London, 1966.

[8] Osborn, A., *Applied imagination,* Scribners, N.Y., 1957.

[9] Leech, D. J., 'Undergraduate training in design project management', *Int. Journal of Mech. Engng. Educ.,* I.Mech.E. and UMIST, **3**, (1) 1975.

[10] Turner, B. T., *Creativity in engineering – an overview of some methodoligies of engineering design work,* CEI committee on creativity, paper 4, 1974.

[11] CNAA report, *Managing design – an initiative in management education,* Oct. 1984.

CHAPTER 15

Computer program listings

The computer programs listed in this appendix have been used to provide some of the information in the text of this book.

In every case, the program is written in MBASIC and has been run on a simple microcomputer (generally a DRAGON 32). In some cases this is all that is necessary to obtain any required information, but in other cases the program listed is a simplification of one that has been used on a larger computer, although still generally a microcomputer (such as the IBM personal computer).

Notes against each program listing suggest where the use of a larger computer would desirable in management and teaching.

Program 1. see Fig. 6.1
CPM and Resources
Program name: "CPM . BAS"

```
10 CLS:INPUT"NO.OF ACTIVITIES";NA
20 DIM B(NA),E(NA),T(NA),LS(NA),EF(NA),TS(NA),FS(NA),RE(NA)
30 CLS:INPUT"TO DO RESOURCE ANALYSIS TYPE 'Y'";B$
35 CLS:PRINT"NOTE :":PRINT"FINAL NODE MUST HAVE THE HIGHESTNODE NO.":PRINT
40 IF B$="Y" THEN GOTO 80
50 FOR I=1 TO NA:INPUT"BEGINODE,ENDNODE,ACTIME";B(I),E(I),T(I)
60 NEXT I
70 GOTO 100
80 FOR L=1 TO NA :INPUT"BEGINODE,ENDNODE,ACTIME,RESOURCE";B(L),E(L),T(L),RE(L)
90 NEXT L
92 NN=0
93 FOR I1=1 TO NA
95 IF NN<E(I1) THEN NN=E(I1)
97 NEXT I1
98 DIM ES(NN),LF(NN)
100 FOR J=1 TO NN:ES(J)=0:NEXT J
110 FOR K=1 TO NA:L=B(K):M=E(K):FES=ES(L)+T(K):IF FES>ES(M) THEN 120 ELSE 130
120 ES(M)=FES:GOTO 110
130 NEXT K
140 CLS:PRINT:PRINT"DUE DATE IS =";ES(NN):PRINT:INPUT"TO CHANGE THE DUE DATE TYP
E 'Y'";A$
150 IF A$="Y" THEN INPUT"NEW DUE DATE =";DD ELSE DD=ES(NN)
160 FOR I=1 TO NN:LF(I)=DD:NEXT I
170 FOR G=0 TO NA-1:F=NA-G:L=B(F):M=E(F):FLF=LF(M)-T(F):IF FLF<LF(L) THEN 180 EL
SE 190
180 LF(L)=FLG:GOTO 170
190 NEXT G
200 FOR K=1 TO NA:L=B(K):M=E(K):EF(K)=ES(L)+T(K):LS(K)=LF(M)-T(K):TS(K)=LF(M)-EF
(K):FS(K)=ES(M)-EF(K):NEXT K
210 CLS:PRINT"BN-EN AT  ES  LS  EF  LF  TS  FS"
220 FOR H=1 TO NA:L=B(H):M=E(H):  PRINT USING"## ## ## ### ### ### ### ### ###";
B(H),E(H),T(H),ES(L),LS(H),EF(H),LF(M),TS(H),FS(H);
230 IF INKEY$="" THEN 230
240 NEXT H
250 PRINT:PRINT "TO SEE CRITICAL PATH PRESS A KEY"
260 IF INKEY$="" THEN 260
270 CLS:PRINT"CRITICAL PATH ACTIVITIES":PRINT
280 FOR K=1 TO NA:L=B(K):M=E(K)
290 IF TS(K)=0 THEN PRINT USING"         ### -###";L,M
300 NEXT K
310 IF B$<> "Y" THEN GOTO 490
320 PRINT:PRINT"TO SEE RES. ANALYSIS PRESS A KEY"
330 IF INKEY$="" THEN 330
335 DIM SR(DD)
340 FOR J=1 TO DD
350 SR(J)=0
360 FOR K=1 TO NA:L=B(K)
370 IF J>ES(L) AND J<=EF(K) THEN SR(J)=SR(J)+RE(K)
380 NEXT K
390 NEXT J
400 CLS:M$=" TIME     NO. RES.   ACT. IN HAND":PRINT M$
410 FOR JJ=1 TO DD
420 PRINT USING"####     #####        ";JJ,SR(JJ);
430 FOR K=1 TO NA:L=B(K):M=E(K)
440 IF JJ>ES(L) AND JJ<=EF(K) THEN PRINT USING"#### -####
;L,M;
450 NEXT K
460 PRINT
470 IF INKEY$="" THEN 470
480 NEXT JJ
490 END
```

This program will solve CP networks and will list resource use. It will run on almost any simple computer.

The program can be extended to re-schedule resources by manual intervention.

Program 2. see Fig. 6.4
Work Scheduling
Program name: "JOBKRATIN . BAS"

```
100 CLS:INPUT"TIME LIMIT=";TL
110 INPUT"LAMBDA (N.E.)=";LA
120 INPUT"MUE (N.D.)=";MU
130 INPUT"ZIGMA (N.D.)=";ZI
140 INPUT"SEED 0<SE<1 SE=";SE
142 PRINT#-2,"LAMBDA=";LA
143 PRINT#-2,"MUE    =";MU
144 PRINT#-2,"ZIGMA =";ZI
145 PRINT#-2,"TIME L=";TL
146 PRINT#-2,"SEED   =";SE
147 PRINT#-2:PRINT#-2
148 PRINT#-2,"J.A. TAFS TAS1 STA1 FTA1  FRT TAS2 STA2 FTA2  FRT TAS3 STA3 FTA3
FRT     K"
149 PRINT#-2
150 W1E=SE*(11/17)
152 X2E=SE*(13/19)
157 C3E=SE*(17/19)
158 AJ=0.0:PI=1:GOTO 200
159 IF PI = TL GOTO 1500
160 RQ=(3.141593+SE)^2.4
170 RQ=RQ-INT(RQ)
180 SE=RQ
190 AJ=(1.0/LA)*LOG(1.0/(1.0-RQ))
195 AJ=INT(AJ):IF AJ<1 THEN GOTO   160
197 PI=PI+1
200 V=W1E:GOSUB 2000:W1E=V:T1=TJ:H1=H1+T1
210 V=X2E:GOSUB 2000:X2E=V:T2=TJ:H2=H2+T2
220 V=C3E:GOSUB 2000:C3E=V:T3=TJ:H3=H3+T3
225 Y3=T1+T2+T3
230 B1F=0.0:D1S=0.0
235 TAJ=TAJ+AJ
240 IF TAJ<= F1 THEN D1S=F1-TAJ:S1=F1
250 IF TAJ > F1 THEN B1F=TAJ-F1:S1=F1+B1F
270 F1=F1+T1+B1F
280 DS=DS+D1S
290 BF=BF+B1F
300 B2F=0.0:D2S=0.0
310 IF F1 <= F2 THEN D2S=F2-F1:S2=F2
320 IF F1 >  F2 THEN B2F=F1-F2:S2=F2+B2F
330 F2=F2+T2+B2F
340 BG=BG+B2F
350 DT=DT+D2S
360 B3F=0.0:D3S=0.0
370 IF F2 <= F3 THEN D3S=F3-F2:S3=F3
380 IF F2 >  F3 THEN B3F=F2-F3:S3=F3+B3F
390 F3=F3+T3+B3F
400 BH=BH+B3F
410 DU=DU+D3S
420 U1=F3-TAJ:K7=U1/Y3
1000 PRINT#-2,USING"#### #### #### #### #### #### #### #### #### #### #### ####
#### #### ###.###";AJ,TAJ,T1,S1,F1,B1F,T2,S2,F2,B2F,T3,S3,F3,B3F,K7
1499 GOTO 159
1500 PRINT#-2
1510 PRINT#-2,"                         --------         --------         --
-------"
1520 PRINT#-2
1535 PRINT#-2, USING"TOTAL FREE TIME          ####                ####
      ####";BF,BG,BH
1537 PRINT#-2
1540 U1M=((F1-BF)/F1)*100.0
1550 U2M=((F2-BG)/F2)*100.0
1560 U3M=((F3-BH)/F3)*100.0
1570 PRINT#-2, USING"USE OF MAN               ###.##%             ###.##%
      ###.##%";U1M,U2M,U3M
1580 G1=F1/H1
1590 G2=F2/H2
1600 G3=F3/H3
```

```
1610 PRINT#-2
1620 PRINT#-2,"AVE. TURN AROUND TIME"
1630 PRINT#-2, USING"-----------------------        ##.###              ##.###
     ##.###";G1,G2,G3
1640 PRINT#-2,"AVE. WORK CONTENT"
1999 END
2000 REM SUB LOGNORMAL GENERATOR
2005 Q=0.0
2010 FOR K9=1 TO 12
2020 Q1=(3.141593+V)^2.4
2030 Q1=Q1-INT(Q1)
2040 V=Q1
2050 Q=Q+Q1
2060 NEXT K9
2070 TJ=((Q-6.0)*ZI)+MU
2080 TJ=EXP(TJ)
2085 TJ=INT(TJ):IF TJ<1 THEN GOTO   2005
2090 RETURN
```

This program is a simple simulation, specific to the example of Fig. 6.4. It will run a almost any simple computer.

The program is not of general application but may be extended to cope with the conditions of any particular design situation.

Program 3. see Fig. 8.5
Net Present Value
Program name: "NPV . BAS"

```
10 REM    "DCF"
20 CLS:INPUT "N ";N
30 PRINT "N=";N:PRINT
40 X=0
50 DIM C(25),X(25)
60 FOR J= 1 TO N+1
70 K=J-1
80 PRINT "C(";K;")?"
90 INPUT C(J)
100 PRINT"C(";K;")=";C(J)
110 NEXT J
120 INPUT"I ";I
130 PRINT"                          I =";I
140 X=0.0
150 FOR L=1 TO N+1
160 K=L-1
170 X(L)=C(L)/((1+I)^K)
180 X=X+X(L)
190 NEXT L
200 PRINT "                         PV=";X
210 INPUT"NEW RATE Y/N";Q$
220 IF Q$="Y" THEN GOTO 120
```

This program will calculate the net present value of a simple cash flow stream. It may be extended to calculate the rate of return of a simple cash flow stream. The program will work on most simple computers.

Extensions which give graphical outputs, equivalent annual cash flows and other information can be supplied to run on most microcomputers.

Program 4. see Figs. 10.A2 and 10.A7
Plotting WEIBULL lives
Program name: "DRALIF . BAS"

```
100 DIM WL(500),Y(500),HY(20),A(32,5),US(10)
105 CLS:GOSUB 3100
110 CLS0:INPUT"SAMPLE SIZE  N=";N
120 FOR J2=1 TO N
130 PRINT"FAILURE NO. ( ";J2;" )"
140 INPUT"                          =";WL(J2)
145 WL(J2)=LOG(WL(J2))
150 NEXT J2
160 SGL=WL(1):GGL=WL(N)
220 GOSUB 900
230 FOR II=1 TO N
240 TPA=(II-0.5)/N
250 Y(II)=(AY+EY)-(((LOG(LOG(1/(1-TPA))))*SY)+(ABS(KB)*SY))
260 NEXT II
300 GOSUB 3000
315 IF INKEY$="" THEN 315
320 INPUT"RUN AGAIN    Y/N ";R$
330 IF R$="Y" THEN GOTO 110
340 STOP
900 REM SUB DRAW AXIS
910 JS=INT(SGL):JL=INT(GGL)
920 IF(GGL-JL)>0.0 THEN JL=JL+1
930 IF JS<=0 AND JL>=0 THEN IX=JL+ABS(JS)
940 IF JS<0 AND JL<=0 THEN IX=ABS(JS)-ABS(JL)
950 IF JS>=0 AND JL>0 THEN IX=JL-JS
960 KT=2:TP=(1-0.5)/N
970 V9=LOG(LOG(1/(1-TP))):KB=INT(V9)
975 IF (V9-KB)<0.1 THEN KB=KB-1
977 IY=KT+ABS(KB)
980 AX=200:AY=150
990 EX=28:EY=21
1000 SX=AX/IX:SY=AY/IY
1010 PMODE4
1015 PCLS
1020 SCREEN 1,0
1030 LINE(EX,EY)-(EX,AY+EY),PSET
1035 LINE(EX,AY+EY)-(AX+EX,AY+EY),PSET
1040 LINE(AX+EX,AY+EY)-(AX+EX,EY),PSET
1045 LINE(AX+EX,EY)-(EX,EY),PSET
1100 REM R. H. SCALE
1110 HX=AX+EX:OHX=HX+4
1120 COZ=SY*ABS(KB)
1130 FOR I=1 TO 9
1140 HY(I)=AY+EY-(((LOG(LOG(100/(100-I))))*SY)+COZ)
1150 J=I+9
1160 HY(J)=AY+EY-(((LOG(LOG(100/(100-(I*10)))))*SY)+COZ)
1170 NEXT I
1180 HY(19)=AY+EY-(((LOG(LOG(100/(100-95))))*SY)+COZ)
1190 HY(20)=AY+EY-(((LOG(LOG(100/(100-99))))*SY)+COZ)
1200 FOR K=1 TO 20
1210 IF HY(K)>AY+EY THEN 1230
1220 LINE (HX,HY(K))-(OHX,HY(K)),PSET
1230 NEXT K
1280 OH=OHX+3:OF=OH+7
1290 D$="9":X7=OH:Y7=HY(20)-3:GOSUB 3330:X7=OF:GOSUB 3330
1300 D$="9":X7=OH:Y7=HY(18)-3:GOSUB 3330:D$="0":X7=OF:GOSUB 3330
1310 D$="6":X7=OH:Y7=HY(15)-3:GOSUB 3330:D$="0":X7=OF:GOSUB 3330
1320 D$="4":X7=OH:Y7=HY(13)-3:GOSUB 3330:D$="0":X7=OF:GOSUB 3330
1330 D$="2":X7=OH:Y7=HY(11)-3:IF Y7>AY+EY THEN 1500:ELSE GOSUB 3330:D$="0":X7=OF
:GOSUB 3330
1340 D$="1":X7=OH:Y7=HY(10)-3:IF Y7>AY+EY THEN 1500:ELSE GOSUB 3330:D$="0":X7=OF
:GOSUB 3330
1350 D$="5":X7=OH:Y7=HY(5)-3:IF Y7>AY+EY THEN 1500:ELSE GOSUB 3330
1360 D$="1":X7=OH:Y7=HY(1)-3:IF Y7>AY+EY THEN 1500:ELSE GOSUB 3330
```

```
1500 GOSUB 4000
1600 RETURN
3000 REM SUB DRAW SAMPES
3010 DS=ABS(JS):IF JS>0 THEN DS=(-1)*JS
3020 FOR M=1 TO N
3030 WL(M)=(WL(M)*SX)+(DS*SX)+EX
3040 NEXT M
3050 FOR MM=2 TO N
3060 L=MM-1
3070 LINE (WL(L),Y(L))-(WL(MM),Y(MM)),PSET
3080 NEXT MM
3090 RETURN
3100 REM SUB STRUCTURE
3110 FOR I=1 TO 16
3120 FOR J=1 TO 5
3130 READ A(I,J)
3140 A(16+I,J)=A(I,J)+1
3150 IF A(16+I,J)=2 THEN A(16+I,J)=0
3160 NEXT J:NEXT I
3170 DATA 1,1,1,1,1
3180 DATA 1,1,1,1,0
3190 DATA 1,1,1,0,1
3200 DATA 1,1,0,1,1
3210 DATA 1,0,1,1,1
3220 DATA 0,1,1,1,1
3230 DATA 1,1,1,0,0
3240 DATA 1,1,0,0,1
3250 DATA 1,0,0,1,1
3260 DATA 0,0,1,1,1
3270 DATA 1,1,0,1,0
3280 DATA 1,0,1,0,1
3290 DATA 0,1,0,1,1
3300 DATA 1,0,1,1,0
3310 DATA 0,1,1,0,1
3320 DATA 0,1,1,1,0
3325 RETURN
3330 REM SUB DRAW LETTER
3350 IF D$=CHR$(46) THEN 3510
3360 FOR I=1 TO 10
3370 IF D$=CHR$(47+I) THEN 3390
3380 NEXT I
3390 PRESET (X7,Y7)
3400 ON I GOTO 3410,3420,3430,3440,3450,3460,3470,3480,3490,3500
3410 L(1)=16:L(2)=32:L(3)=32:L(4)=32:L(5)=32:L(6)=16:GOTO 3520
3420 L(1)=20:L(2)=25:L(3)=20:L(4)=20:L(5)=20:L(6)=16:GOTO 3520
3430 L(1)=16:L(2)=32:L(3)=19:L(4)=20:L(5)=21:L(6)=1:GOTO 3520
3440 L(1)=16:L(2)=32:L(3)=24:L(4)=19:L(5)=32:L(6)=16:GOTO 3520
3450 L(1)=22:L(2)=22:L(3)=22:L(4)=31:L(5)=1:L(6)=19:GOTO 3520
3460 L(1)=1:L(2)=22:L(3)=2:L(4)=18:L(5)=18:L(6)=2:GOTO 3520
3470 L(1)=16:L(2)=22:L(3)=2:L(4)=32:L(5)=32:L(6)=16:GOTO 3520
3480 L(1)=1:L(2)=18:L(3)=19:L(4)=20:L(5)=21:L(6)=21:GOTO 3520
3490 L(1)=16:L(2)=32:L(3)=16:L(4)=32:L(5)=32:L(6)=16:GOTO 3520
3500 L(1)=16:L(2)=32:L(3)=32:L(4)=6:L(5)=18:L(6)=16:GOTO 3520
3510 L(1)=17:L(2)=17:L(3)=17:L(4)=17:L(5)=16:L(6)=16
3520 M=1
3530 CJ=(M-1)
3540 FOR I=1 TO 5
3550 FI=(I-1)
3580 IF A(L(M),I)=0 THEN 3620
3590 PSET(X7+FI,Y7+CJ,1)
3620 NEXT I
3625 M=M+1
3630 IF M>6 THEN 3640:ELSE GOTO    3530
3640 RETURN
4000 REM SUB LOG SCALE
4010 YL=AY+EY:DY=YL+4:TY=DY+4
4020 US(1)=0:SUM=US(1)
4030 FOR L1=2 TO 10
4040 US(L1)=(LOG(L1)-SUM):SUM=SUM+US(L1)
4045 US(L1)=US(L1)*SX
```

```
4050 NEXT L1
4060 XCOZ=SX*(-1)*JS+EX
4070 IF XCOZ>EX THEN 4140
4080 IF XCOZ=EX THEN 4210
4085 PP=XCOZ
4090 TTY=OY
4095 J=2
4100 PP=PP+US(J)
4110 IF PP>=EX THEN 4240
4120 J=J+1
4130 IF J=10 THEN TTY=TY:GOTO 4100
4135 IF J>10 THEN 4090:ELSE GOTO 4100
4140 PP=XCOZ-US(10)
4150 J=9
4160 TTY=OY
4165 IF PP<EX THEN 4210
4167 IF PP>AX+EX THEN 4180
4170 LINE (PP,YL)-(PP,TTY),PSET
4180 PP=PP-US(J)
4190 J=J-1
4200 IF J=1 THEN TTY=TY:J=10:GOTO 4165:ELSE GOTO 4160
4210 PP=XCOZ
4215 IF PP>AX+EX THEN 4300
4220 J=1
4230 TTY=TY
4240 LINE (PP,YL)-(PP,TTY),PSET
4245 J=J+1
4250 PP=PP+US(J)
4260 IF PP>AX+EX THEN 4300
4275 TTY=OY
4280 IF J=10 THEN 4220:ELSE GOTO 4240
4300 Y7=TY+2:Z5=XCOZ:EC=EX
4310 IF Z5>AX+EC THEN 4500
4320 IF Z5<EC THEN 4340
4330 D$="1":X7=Z5:GOSUB 3330
4340 Z5=(LOG(10)*SX)+XCOZ
4350 IF Z5<EC THEN 4380
4360 IF Z5>AX+EC THEN 4500
4370 D$="1":X7=Z5:GOSUB 3330:D$="0":X7=X7+6:GOSUB 3330
4380 Z5=(LOG(100)*SX)+XCOZ
4390 IF Z5<EC THEN 4420
4400 IF Z5>AX+EC THEN 4500
4410 D$="1":X7=Z5::GOSUB 3330:D$="0":X7=Z5+6:GOSUB 3330:X7=X7+6:GOSUB 3330
4420 Z5=(LOG(1000)*SX)+XCOZ
4430 IF Z5<EC THEN 4460
4440 IF Z5>AX+EC THEN 4500
4450 D$="1":X7=Z5:GOSUB 3330:D$="0":X7=X7+6:GOSUB 3330:X7=X7+6:GOSUB 3330:X7=X7+
6:GOSUB 3330
4460 Z5=(LOG(10000)*SX)+XCOZ
4470 IF Z5<EC THEN 4500
4480 IF Z5>AX+EC THEN 4500
4490 D$="1":X7=Z5:GOSUB 3330:D$="0":X7=X7+6:GOSUB 3330:X7=X7+6:GOSUB 3330:X7=X7+
6:GOSUB 3330:X7=X7+6:GOSUB 3330
4500 Z5=XCOZ:IF Z5<=EC THEN 5000
4530 Z5=(LOG(0.1)*SX)+XCOZ
4540 IF Z5>AX+EC THEN 4570
4550 IF Z5<EC THEN 5000
4560 D$="0":X7=Z5:GOSUB 3330:D$=".":X7=X7+5:GOSUB 3330:D$="1":X7=X7+5:GOSUB 3330
4570 Z5=(LOG(0.01)*SX)+XCOZ
4580 IF Z5>AX+EC THEN 4610
4590 IF Z5<EC THEN 5000
4600 D$="0":X7=Z5:GOSUB 3330:D$=".":X7=X7+5:GOSUB 3330:D$="0":X7=X7+5:GOSUB 3330
:D$="1":X7=X7+6:GOSUB 3330
4610 Z5=(LOG(0.001)*SX)+XCOZ
4620 IF Z5>AX+EC THEN 4650
4630 IF Z5<EC THEN 5000
4640 D$="0":X7=Z5:GOSUB 3330:D$=".":X7=X7+5:GOSUB 3330:D$="0":X7=X7+5:GOSUB 3330
:X7=X7+6:GOSUB 3330:D$="1":X7=X7+6:GOSUB 3330
4650 Z5=(LOG(0.0001)*SX)+XCOZ
4660 IF Z5>AX+EC THEN 5000
```

```
4670 IF Z5<EC THEN 5000
4680 D$="0":X7=Z5:GOSUB 3330:D$=".":X7=X7+5:GOSUB 3330:D$="0":X7=X7+5:GOSUB 3330
:X7=X7+6:GOSUB 3330:X7=X7+6:GOSUB 3330:D$="1":X7=X7+6:GOSUB 3330
5000 RETURN
```

This program will choose appropriate axes and scales and plot a given sample of lives on what is, effectively, Weibull graph paper.

The program is suitable for use on a Dragon 32 computer.

A version of the program that is suitable for use with a BBC microcomputer; an IBM, PC; a Rainbow; a Sinclair QL and such other, larger computers as the Prime 750, can be supplied.

Program 5. see Fig. 10.A4
Poisson Distribution
Program name: "POISSON . BAS"

```
10 CLS:INPUT"LAMBDA=";LAM
20 INPUT"N=";N
30 INPUT"T=";T
40 Y=1:S=0.0
50 FOR K=0 TO N
60 Y=Y*K:IF Y=0 THEN Y=1
70 X=(((LAM*T)^K)*(EXP(-LAM*T)))/Y
80 S=S+X
90 NEXT K
100 F=1-S
110 PRINT:PRINT"F=";F
115 PRINT:INPUT"AGAIN  Y/N  ";A$
117 IF A$="Y" THEN GOTO 10
120 END
```

Although used for calculating the probability of running out of spares, the program is trivial and could be run on the simplest of computers.

Program 6. see Fig. 10.A8
The Accumulated Life of a number of spares (with WEIBULL lives)
Program name: "SUMLIV . BAS"

```
10 DIM X(1000)
20 CLS:INPUT"BETA=";B
30 INPUT"ETA =";E
40 INPUT"NO.OF LIVES IP MAX=1000    ";IP
50 INPUT"SEED 0>SE>1    SE=";SE
55 Z=SE
60 INPUT"SUM OF HOW MANY    NN=";NN
70 INPUT"RESULTS EVERY KSTH,   KS=";KS
80 P2=B*LOG(E)
90 FOR L=1 TO IP
100 UM=0.0
110 FOR J=1 TO NN
120 RQ=(3.141593+SE)^2.4
130 RQ=RQ-INT(RQ)
140 SE=RQ
150 P1=LOG(LOG(1.0/(1.0-RQ)))
160 TX=EXP((P1+P2)/B)
170 UM=UM+TX
180 NEXT J
190 X(L)=UM
200 NEXT L
210 MO1=IP-1
220 FOR K3=1 TO MO1
230 MIN=K3
240 J3=K3+1
250 FOR L3=J3 TO IP
260 IF X(L3)<X(MIN) THEN MIN=L3
270 NEXT L3
280 WEE=X(K3)
290 X(K3)=X(MIN)
300 X(MIN)=WEE
310 NEXT K3
320 PRINT
330 PRINT"BETA=";B;"  ETA=";E
332 PRINT"NO. OF LIVES=";IP
335 PRINT"SUM OF ";NN;" LIVES."
336 PRINT"RESULTS EVERY ";KS;"TH"
337 PRINT"SEED=";Z
338 PRINT
340 FOR LJ=KS TO IP STEP KS
350 PRINT"X(";LJ;") =";X(LJ)
360 NEXT LJ
370 INPUT"RESULTS ON PAPER  Y/N ";A$
380 IF A$="N" THEN GOTO 999
390 PRINT#-2,"BETA=";B;"  ETA=";E;"  NO. OF LIVES=";IP
400 PRINT#-2,"SUM OF ";NN;" LIVES."
410 PRINT#-2,"RESULTS EVERY ";KS;" TH"
420 PRINT#-2,"SEED=";Z
430 PRINT#-2
440 FOR LJ=KS TO IP STEP KS
450 PRINT#-2,"X(";LJ;")=";X(LJ)
460 NEXT LJ
999 END
```

This program is a simulation which will run on the simplest of computers if only about 1000 simulations are required. A version will provide 4000 simulations with the BBC micro computer. Ten thousand simulations are desirable if industrial use is to be made of the program. Versions of the program can be supplied for use with the Prime 750; the CIFER; the VAX series and, with some limitations for the IBM, PC; the Sinclair QL; the Rainbow; the BBC micro.

Program 7. see Fig. 10.A9
Optimum Replacement Life
Program name: "OPTLIF . BAS"

```
10 CLS:INPUT"N,BETA,ETA";N,B,E
20 INPUT"COSTS,COSTSCH";C,CS
30 PRINT#-2,"N=";N;"  BETA=";B;"  ETA=";E;
40 PRINT#-2,"  COSTS=";C;"  COSTSCH=";CS:PRINT#-2
50 INPUT"SEED  0>SE>1 ";SE
70 INPUT"T";T
75 ACL=0.0:CO=0.0
80 P2=B*LOG(E)
90 FOR I=1 TO N
100 RQ=(3.141593+SE)^2.4
110 RQ=RQ-INT(RQ)
120 SE=RQ
130 P1=LOG(LOG(1/(1-RQ)))
140 WL=EXP((P1+P2)/B)
150 IF WL<T THEN ACL=ACL+WL:CO=CO+C ELSE ACL=ACL+T:CO=CO+CS
160 NEXT I
170 AVC=CO/ACL
180 PRINT USING" T= ####      AVC= ###.#####";T,AVC
190 PRINT#-2,USING" T= ####      AVC= ###.#####";T,AVC
200 INPUT"NEXT T   Y/N";A$
210 IF A$="Y" THEN GOTO 70
```

This program will run on the simplest of computers if only a small number (N) of simulations is required.

For commercial use, the number of simulations needs to be several thousand and more sophisticated computers are required. Versions suitable for the BBC micro; the Sinclair QL; the IBM, PC; the Rainbow; the CIFER; the VAX series and the Prime 750; may be made available.

Program 8. see Fig. 10.A10
Plotting Samples of Lives
Program name: "SC GR . BAS"

```
100 DIM WL(500),Y(500),HY(20),A(32,5),US(10)
105 GOSUB 3100
110 CLS0:INPUT"SAMPLE SIZE  N=";N
120 INPUT"NO. OF SAMPLES  NS=";NS
130 INPUT"BETA=";B
140 INPUT"THETA=";E
150 INPUT"SEED RND GEN. O>SE>1  SE=";SE
160 TSE=SE:GOSUB 500:SGL=WL(1):GGL=WL(N)
165 LS=NS-1
170 FOR J3=1 TO LS
180 GOSUB 500
190 IF WL(1)<SGL THEN SGL=WL(1)
200 IF WL(N)>GGL THEN GGL=WL(N)
210 NEXT J3
220 GOSUB 900
230 FOR II=1 TO N
240 TPA=(II-0.5)/N
250 Y(II)=(AY+EY)-(((LOG(LOG(1/(1-TPA))))*SY)+(ABS(KB)*SY))
260 NEXT II
270 TSE=SE
280 FOR JJ=1 TO NS
290 GOSUB 500
300 GOSUB 3000
310 NEXT JJ
315 IF INKEY$="" THEN 315
320 INPUT"RUN AGAIN    Y/N ";R$
330 IF R$="Y" THEN GOTO 110
340 STOP
500 REM SUB GEN WLS
510 P2=B*LOG(E)
520 FOR L2=1 TO N
530 RQ=(3.141593+TSE)^2.4
540 RQ=RQ-INT(RQ)
550 TSE=RQ
560 P1=LOG(LOG(1/(1-RQ)))
570 WL(L2)=EXP((P1+P2)/B):WL(L2)=LOG(WL(L2))
580 NEXT L2
590 GOSUB 700
600 RETURN
700 REM SUB SORTING
710 N1=N-1
720 FOR I=1 TO N1
730 MIN=I:J=I+1
740 FOR K=J TO N
750 IF WL(K)<WL(MIN) THEN MIN=K
760 NEXT K
770 BB=WL(I)
780 WL(I)=WL(MIN)
790 WL(MIN)=BB
800 NEXT I
810 RETURN
900 REM SUB DRAW AXIS
910 JS=INT(SGL):JL=INT(GGL)
920 IF(GGL-JL)>0.0 THEN JL=JL+1
930 IF JS<=0 AND JL>=0 THEN IX=JL+ABS(JS)
940 IF JS<0 AND JL<=0 THEN IX=ABS(JS)-ABS(JL)
950 IF JS>=0 AND JL>0 THEN IX=JL-JS
960 KT=2:TP=(1-0.5)/N
970 V9=LOG(LOG(1/(1-TP))):KB=INT(V9)
975 IF (V9-KB)<0.1 THEN KB=KB-1
977 IY=KT+ABS(KB)
980 AX=200:AY=150
990 EX=28:EY=21
1000 SX=AX/IX:SY=AY/IY
1010 PMODE4
```

```
1015 PCLS
1020 SCREEN 1,0
1030 LINE(EX,EY)-(EX,AY+EY),PSET
1035 LINE(EX,AY+EY)-(AX+EX,AY+EY),PSET
1040 LINE(AX+EX,AY+EY)-(AX+EX,EY),PSET
1045 LINE(AX+EX,EY)-(EX,EY),PSET
1100 REM R. H. SCALE
1110 HX=AX+EX:OHX=HX+4
1120 COZ=SY*ABS(KB)
1130 FOR I=1 TO 9
1140 HY(I)=AY+EY-(((LOG(LOG(100/(100-I))))*SY)+COZ)
1150 J=I+9
1160 HY(J)=AY+EY-(((LOG(LOG(100/(100-(I*10)))))*SY)+COZ)
1170 NEXT I
1180 HY(19)=AY+EY-(((LOG(LOG(100/(100-95))))*SY)+COZ)
1190 HY(20)=AY+EY-(((LOG(LOG(100/(100-99))))*SY)+COZ)
1200 FOR K-1 TO 20
1210 IF HY(K)>AY+EY THEN 1230
1220 LINE (HX,HY(K))-(OHX,HY(K)),PSET
1230 NEXT K
1280 OH=OHX+3:OF=OH+7
1290 D$="9":X7=OH:Y7=HY(20)-3:GOSUB 3330:X7=OF:GOSUB 3330
1300 D$="9":X7=OH:Y7=HY(18)-3:GOSUB 3330:D$="0":X7=OF:GOSUB 3330
1310 D$="6":X7=OH:Y7=HY(15)-3:GOSUB 3330:D$="0":X7=OF:GOSUB 3330
1320 D$="4":X7=OH:Y7=HY(13)-3:GOSUB 3330:D$="0":X7=OF:GOSUB 3330
1330 D$="2":X7=OH:Y7=HY(11)-3:IF Y7>AY+EY THEN 1500;ELSE GOSUB 3330:D$="0":X7=OF
:GOSUB 3330
1340 D$="1":X7=OH:Y7=HY(10)-3:IF Y7>AY+EY THEN 1500:ELSE GOSUB 3330:D$="0":X7=OF
:GOSUB 3330
1340 D$="1":X7=OH:Y7=HY(10)-3:IF Y7>AY+EY THEN 1500:ELSE GOSUB 3330:D$="0":X7=OF
:GOSUB 3330
1350 D$="5":X7=OH:Y7=HY(5)-3:IF Y7>AY+EY THEN 1500:ELSE GOSUB 3330
1360 D$="1":X7=OH:Y7=HY(1)-3:IF Y7>AY+EY THEN 1500:ELSE GOSUB 3330
1500 GOSUB 4000
1600 RETURN
3000 REM SUB DRAW SAMPES
3010 DS=ABS(JS):IF JS>0 THEN DS=(-1)*JS
3020 FOR M=1 TO N
3030 WL(M)=(WL(M)*SX)+(DS*SX)+EX
3040 NEXT M
3050 FOR MM=2 TO N
3060 L=MM-1
3070 LINE (WL(L),Y(L))-(WL(MM),Y(MM)),PSET
3080 NEXT MM
3090 RETURN
3100 REM SUB STRUCTURE
3110 FOR I=1 TO 16
3120 FOR J=1 TO 5
3130 READ A(I,J)
3140 A(16+I,J)=A(I,J)+1
3150 IF A(16+I,J)=2 THEN A(16+I,J)=0
3160 NEXT J:NEXT I
3170 DATA 1,1,1,1,1
3180 DATA 1,1,1,1,0
3190 DATA 1,1,1,0,1
3200 DATA 1,1,0,1,1
3210 DATA 1,0,1,1,1
3220 DATA 0,1,1,1,1
3230 DATA 1,1,1,0,0
3240 DATA 1,1,0,0,1
3250 DATA 1,0,0,1,1
3260 DATA 0,0,1,1,1
3270 DATA 1,1,0,1,0
3280 DATA 1,0,1,0,1
3290 DATA 0,1,0,1,1
3300 DATA 1,0,1,1,0
3310 DATA 0,1,1,0,1
3320 DATA 0,1,1,1,0
3325 RETURN
3330 REM SUB DRAW LETTER
3350 IF D$=CHR$(46) THEN 3510
```

```
3360 FOR I=1 TO 10
3370 IF D$=CHR$(47+I) THEN 3390
3380 NEXT I
3390 PRESET (X7,Y7)
3400 ON I GOTO 3410,3420,3430,3440,3450,3460,3470,3480,3490,3500
3410 L(1)=16:L(2)=32:L(3)=32:L(4)=32:L(5)=32:L(6)=16:GOTO 3520
3420 L(1)=20:L(2)=25:L(3)=20:L(4)=20:L(5)=20:L(6)=16:GOTO 3520
3430 L(1)=16:L(2)=32:L(3)=19:L(4)=20:L(5)=21:L(6)=1:GOTO 3520
3440 L(1)=16:L(2)=32:L(3)=24:L(4)=19:L(5)=32:L(6)=16:GOTO 3520
3450 L(1)=22:L(2)=22:L(3)=22:L(4)=31:L(5)=1:L(6)=19:GOTO 3520
3460 L(1)=1:L(2)=22:L(3)=2:L(4)=18:L(5)=18:L(6)=2:GOTO 3520
3470 L(1)=16:L(2)=22:L(3)=2:L(4)=32:L(5)=32:L(6)=16:GOTO 3520
3480 L(1)=1:L(2)=18:L(3)=19:L(4)=20:L(5)=21:L(6)=21:GOTO 3520
3490 L(1)=16:L(2)=32:L(3)=16:L(4)=32:L(5)=32:L(6)=16:GOTO 3520
3500 L(1)=16:L(2)=32:L(3)=32:L(4)=6:L(5)=18:L(6)=16:GOTO 3520
3510 L(1)=17:L(2)=17:L(3)=17:L(4)=17:L(5)=16:L(6)=16
3520 M=1
3530 CJ=(M-1)
3540 FOR I=1 TO 5
3550 FI=(I-1)
3580 IF A(L(M),I)=0 THEN 3620
3590 PSET(X7+FI,Y7+CJ,1)
3620 NEXT I
3625 M=M+1
3630 IF M>6 THEN 3640:ELSE GOTO     3530
3640 RETURN
4000 REM SUB LOG SCALE
4010 YL=AY+EY:OY=YL+4:TY=OY+4
4020 US(1)=0:SUM=US(1)
4030 FOR L1=2 TO 10
4040 US(L1)=(LOG(L1)-SUM):SUM=SUM+US(L1)
4045 US(L1)=US(L1)*SX
4050 NEXT L1
4060 XCOZ=SX*(-1)*JS+EX
4070 IF XCOZ>EX THEN 4140
4080 IF XCOZ=EX THEN 4210
4085 PP=XCOZ
4090 TTY=OY
4095 J=2
4100 PP=PP+US(J)
4110 IF PP>=EX THEN 4240
4120 J=J+1
4130 IF J=10 THEN TTY=TY:GOTO 4100
4135 IF J>10 THEN 4090:ELSE GOTO 4100
4140 PP=XCOZ-US(10)
4150 J=9
4160 TTY=OY
4165 IF PP<EX THEN 4210
4167 IF PP>AX+EX THEN 4180
4170 LINE (PP,YL)-(PP,TTY),PSET
4180 PP=PP-US(J)
4190 J=J-1
4200 IF J=1 THEN TTY=TY:J=10:GOTO 4165:ELSE GOTO 4160
4210 PP=XCOZ
4215 IF PP>AX+EX THEN 4300
4220 J=1
4230 TTY=TY
4240 LINE (PP,YL)-(PP,TTY),PSET
4245 J=J+1
4250 PP=PP+US(J)
4260 IF PP>AX+EX THEN 4300
4275 TTY=OY
4280 IF J=10 THEN 4220:ELSE GOTO 4240
4300 Y7=TY+2:Z5=XCOZ:EC=EX
4310 IF Z5>AX+EC THEN 4500
4320 IF Z5<EC THEN 4340
4330 D$="1":X7=Z5:GOSUB 3330
4340 Z5=(LOG(10)*SX)+XCOZ
4350 IF Z5<EC THEN 4380
4360 IF Z5>AX+EC THEN 4500
4370 D$="1":X7=Z5:GOSUB 3330:D$="0":X7=X7+6:GOSUB 3330
```

```
4380 Z5=(LOG(100)*SX)+XCOZ
4390 IF Z5<EC THEN 4420
4400 IF Z5>AX+EC THEN 4500
4410 D$="1":X7=Z5::GOSUB 3330:D$="0":X7=Z5+6:GOSUB 3330:X7=X7+6:GOSUB 3330
4420 Z5=(LOG(1000)*SX)+XCOZ
4430 IF Z5<EC THEN 4460
4440 IF Z5>AX+EC THEN 4500
4450 D$="1":X7=Z5:GOSUB 3330:D$="0":X7=X7+6:GOSUB 3330:X7=X7+6:GOSUB 3330:X7=X7+
6:GOSUB 3330
4460 Z5=(LOG(10000)*SX)+XCOZ
4470 IF Z5<EC THEN 4500
4480 IF Z5>AX+EC THEN 4500
4490 D$="1":X7=Z5:GOSUB 3330:D$="0":X7=X7+6:GOSUB 3330:X7=X7+6:GOSUB 3330:X7=X7+
6:GOSUB 3330:X7=X7+6:GOSUB 3330
4500 Z5=XCOZ:IF Z5<=EC THEN 5000
4530 Z5=(LOG(0.1)*SX)+XCOZ
4540 IF Z5>AX+EC THEN 4570
4550 IF Z5<EC THEN 5000
4560 D$="0":X7=Z5:GOSUB 3330:D$=".":X7=X7+5:GOSUB 3330:D$="1":X7=X7+5:GOSUB 3330
4570 Z5=(LOG(0.01)*SX)+XCOZ
4580 IF Z5>AX+EC THEN 4610
4590 IF Z5<EC THEN 5000
4600 D$="0":X7=Z5:GOSUB 3330:D$=".":X7=X7+5:GOSUB 3330:D$="0":X7=X7+5:GOSUB 3330
:D$="1":X7=X7+6:GOSUB 3330
4610 Z5=(LOG(0.001)*SX)+XCOZ
4620 IF Z5>AX+EC THEN 4650
4630 IF Z5<EC THEN 5000
4640 D$="0":X7=Z5:GOSUB 3330:D$=".":X7=X7+5:GOSUB 3330:D$="0":X7=X7+5:GOSUB 3330
:X7=X7+6:GOSUB 3330:D$="1":X7=X7+6:GOSUB 3330
4650 Z5=(LOG(0.0001)*SX)+XCOZ
4660 IF Z5>AX+EC THEN 5000
4670 IF Z5<EC THEN 5000
4680 D$="0":X7=Z5:GOSUB 3330:D$=".":X7=X7+5:GOSUB 3330:D$="0":X7=X7+5:GOSUB 3330
:X7=X7+6:GOSUB 3330:X7=X7+6:GOSUB 3330:D$="1":X7=X7+6:GOSUB 3330
5000 RETURN
```

This program generates samples of lives and plots them on what is, effectively, Weibull graph paper. The program is written to run with Dragon 32 graphics but is slow. Versions for use with the BBC micro computer; the IBM, PC; the Rainbow; and larger computers can be supplied.

Index